T0226246

Lecture Notes in Artificial Intelligence 13200

Subseries of Lecture Notes in Computer Science

More information about this subseries at https://link.springer.com/bookseries/1244

Andreas Holzinger · Randy Goebel · Ruth Fong ·
Taesup Moon · Klaus-Robert Müller ·
Wojciech Samek (Eds.)

xxAI - Beyond Explainable AI

International Workshop
Held in Conjunction with ICML 2020
July 18, 2020, Vienna, Austria
Revised and Extended Papers

 Springer

Editors
Andreas Holzinger
University of Natural Resources
and Life Sciences Vienna
Vienna, Austria

Ruth Fong
Princeton University
Princeton, NJ, USA

Klaus-Robert Müller
Technische Universität Berlin
Berlin, Germany

Randy Goebel
University of Alberta
Edmonton, AB, Canada

Taesup Moon
Seoul National University
Seoul, Korea (Republic of)

Wojciech Samek
Fraunhofer Heinrich Hertz Institute
Berlin, Germany

ISSN 0302-9743   ISSN 1611-3349 (electronic)
Lecture Notes in Artificial Intelligence
ISBN 978-3-031-04082-5   ISBN 978-3-031-04083-2 (eBook)
https://doi.org/10.1007/978-3-031-04083-2

LNCS Sublibrary: SL7 – Artificial Intelligence

Preface

In recent years, statistical machine learning (ML) has become very successful, it has triggered a renaissance of artificial intelligence (AI) and has improved enormously in predictivity. Sophisticated models have steadily increased in complexity, which has often happened at the expense of human interpretability (correlation vs. causality). Consequently, an active field of research called explainable AI (xAI) has emerged with the goal of creating tools and models that are both predictive and interpretable and understandable for humans. The growing xAI community has already achieved important advances, such as robust heatmap-based explanations of DNN classifiers. From applications in digital transformation (e.g., agriculture, climate, forest operations, medical applications, cyber-physical systems, automation tools and robotics, sustainable living, sustainable cities, etc.), there is now a need to massively engage in new scenarios, such as explaining unsupervised and intensified learning and creating explanations that are optimally structured for human decision makers. While explainable AI fundamentally deals with the implementation of transparency and traceability of statistical black-box ML methods, there is an urgent need to go beyond explainable AI, e.g., to extend explainable AI with causability, to measure the quality of explanations, and to find solutions for building efficient human-AI interfaces for these novel interactions between artificial intelligence and human intelligence. For certain tasks, interactive machine learning with the human-in-the-loop can be advantageous because a human domain expert can sometimes complement the AI with implicit knowledge. Such a human-in-the-loop can sometimes – not always of course – contribute to an artificial intelligence with experience, conceptual understanding, context awareness and causal reasoning. Formalized, this human knowledge can be used to create structural causal models of human decision making, and features can be traced back to train AI – and thus contribute to making AI even more successful – beyond the current state-of-the-art. The field of explainable AI has received exponential interest in the international machine learning and AI research community. Awareness of the need to explain ML models has grown in similar proportions in industry, academia and government. With the substantial explainable AI research community that has been formed, there is now a great opportunity to make this push towards successful explainable AI applications.

With this volume of Springer Lecture Notes in Artificial Intelligence (LNAI), we will help the international research community to accelerate this process, promote a more systematic use of explainable AI to improve models in diverse applications, and ultimately help to better understand how current explainable AI methods need to be improved and what kind of theory of explainable AI is needed.

The contributions in this volume were very carefully selected by the editors together with help from the Scientific Committee and each paper was reviewed by three international experts in the field.

March 2022

Andreas Holzinger
Randy Goebel
Ruth Fong
Taesup Moon
Klaus-Robert Müller
Wojciech Samek

Organization

Scientific Committee

Osbert Bastani	Trustworthy Machine Learning Group, University of Pennsylvania, Philadelphia, USA
Tarek R. Besold	Neurocat.ai, Artificial Intelligence Safety, Security and Privacy, Berlin, Germany
Przemyslaw Biecek	Faculty of Mathematics and Information Science, Warsaw University of Technology, Poland
Alexander Binder	Information Systems Technology and Design, Singapore University of Technology and Design, Singapore
John M. Carroll	Penn State's University Center for Human-Computer Interaction, University Park, USA
Sanjoy Dasgupta	Artificial Intelligence Group, School of Engineering, University of California, San Diego, USA
Amit Dhurandhar	Machine Learning and Data Mining Group, Thomas J. Watson Research Center, Yorktown Heights, USA
David Evans	Department of Computer Science, University of Virginia, Charlottesville, USA
Alexander Felfernig	Applied Artificial Intelligence (AIG) Research Group, Graz University of Technology, Austria
Aldo Faisal	Brain and Behaviour Lab, Imperial College London, UK
Hani Hagras	School of Computer Science and Electronic Engineering, University of Essex, UK
Sepp Hochreiter	Institute for Machine Learning, Johannes Kepler University Linz, Austria
Xiaowei Huang	Department of Computer Science, University of Liverpool, UK
Krishnaram Kenthapadi	Amazon AWS Artificial Intelligence, Fairness Transparency and Explainability Group, Sunnyvale, USA
Gitta Kutyniok	Mathematical Data Science and Artificial Intelligence, Ludwig-Maximilians-Universität München, Germany
Himabindu Lakkaraju	AI4LIFE Group and TrustML, Department of Computer Science, Harvard University, USA

Gregoire Montavon

Machine Learning & Intelligent Data Analysis Group, Faculy of Electrical Engineering & Computer Science, TU Berlin, Germany

Sang Min Park

Data Science Lab, Department of Biomedical Science, Seoul National University, Seoul, Korea

Natalia Diaz-Rodriguez

Autonomous Systems and Robotics Lab, École Nationale Supérieure de Techniques Avancées, Paris, France

Lior Rokach

Dep. of Software & Information Systems Engineering, Faculty of Engineering Sciences, Ben-Gurion University of the Negev, Israel

Ribana Roscher

Institute for Geodesy and Geoinformation, University of Bonn, Germany

Kate Saenko

Computer Vision and Learning Group, Boston University, MA, USA

Sameer Singh

Department of Computer Science, University of California, Irvine, USA

Ankur Taly

Google Research, Mountain View, CA, USA

Andrea Vedaldi

Visual Geometry Group, Engineering Science Department, University of Oxford, UK

Ramakrishna Vedantam

Facebook AI Research (FAIR), New York, USA

Bolei Zhou

Department of Information Engineering, The Chinese University of Hong Kong, China

Jianlong Zhou

Faculty of Engineering and Information Technology, University of Technology, Sydney, Australia

Sponsor

Fraunhofer
Heinrich Hertz Institute

Contents

Editorial

xxAI - Beyond Explainable Artificial Intelligence

Andreas Holzinger[1,2,3(✉)] (iD), Randy Goebel[3], Ruth Fong[4], Taesup Moon[5],
Klaus-Robert Müller[6,7,8,10] (iD), and Wojciech Samek[9,10] (iD)

[1] Human-Centered AI Lab, University of Natural Resources and Life Sciences,
Vienna, Austria
andreas.holzinger@human-centered.ai
[2] Medical University Graz, Graz, Austria
[3] xAI Lab, Alberta Machine Intelligence Institute, Edmonton, Canada
[4] Princeton University, Princeton, USA
[5] Seoul National University, Seoul, Korea
[6] Department of Artificial Intelligence, Korea University, Seoul, Korea
[7] Max Planck Institute for Informatics, Saarbrücken, Germany
[8] Machine Learning Group, Technical University of Berlin, Berlin, Germany
[9] Department of Artificial Intelligence, Fraunhofer Heinrich Hertz Institute,
Berlin, Germany
[10] BIFOLD – Berlin Institute for the Foundations of Data and Learning,
Berlin, Germany

Abstract. The success of statistical machine learning from big data,
especially of deep learning, has made artificial intelligence (AI) very
popular. Unfortunately, especially with the most successful methods, the
results are very difficult to comprehend by human experts. The application of AI in areas that impact human life (e.g., agriculture, climate,
forestry, health, etc.) has therefore led to an demand for trust, which
can be fostered if the methods can be interpreted and thus explained
to humans. The research field of explainable artificial intelligence (XAI)
provides the necessary foundations and methods. Historically, XAI has
focused on the development of methods to explain the decisions and
internal mechanisms of complex AI systems, with much initial research
concentrating on explaining how convolutional neural networks produce
image classification predictions by producing visualizations which highlight what input patterns are most influential in activating hidden units,
or are most responsible for a model's decision. In this volume, we summarize research that outlines and takes next steps towards a broader
vision for explainable AI in moving beyond explaining classifiers via such
methods, to include explaining other kinds of models (e.g., unsupervised
and reinforcement learning models) via a diverse array of XAI techniques
(e.g., question-and-answering systems, structured explanations). In addition, we also intend to move beyond simply providing model explanations
to directly improving the transparency, efficiency and generalization ability of models. We hope this volume presents not only exciting research
developments in explainable AI but also a guide for what next areas to
focus on within this fascinating and highly relevant research field as we

A. Holzinger et al. (Eds.): xxAI 2020, LNAI 13200, pp. 3–10, 2022.
https://doi.org/10.1007/978-3-031-04083-2_1

enter the second decade of the deep learning revolution. This volume is an outcome of the ICML 2020 workshop on "XXAI: Extending Explainable AI Beyond Deep Models and Classifiers."

Keywords: Artificial intelligence · Explainable AI · Machine learning · Explainability

1 Introduction and Motivation for Explainable AI

In the past decade, deep learning has re-invigorated the machine learning research by demonstrating its power in learning from vast amounts of data in order to solve complex tasks - making AI extremely popular [5], often even beyond human level performance [24]. However, its power is also its peril: deep learning models are composed of millions of parameters; their high complexity [17] makes such "black-box" models challenging for humans to understand [20]. As such "black-box" approaches are increasingly applied to high-impact, high-risk domains, such as medical AI or autonomous driving, the impact of its failures also increases (e.g., medical misdiagnoses, vehicle crashes, etc.).

Consequently, there is an increasing demand for a diverse toolbox of methods that help AI researchers and practitioners design and understand complex AI models. Such tools could provide explanations for model decisions, suggest corrections for failures, and ensure that protected features, such as race and gender, are not misinforming or biasing model decisions. The field of explainable AI (XAI) [32] focuses on the development of such tools and is crucial to the safe, responsible, ethical and accountable deployment of AI technology in our wider world. Based on the increased application of AI in practically all domains which affects human life (e.g., agriculture, climate, forestry, health, sustainable living, etc.), there is also a need to address new scenarios in the future, e.g., explaining unsupervised and intensified learning and creating explanations that are optimally structured for human decision makers with respect to their individual previous knowledge. While explainable AI is essentially concerned with implementing transparency and tractability of black-box statistical ML methods, there is an urgent need in the future to go beyond explainable AI, e.g., to extend explainable AI to include causality and to measure the quality of explanations [12]. A good example is the medical domain where there is a need to ask "what-if" questions (counterfactuals) to gain insight into the underlying independent explanatory factors of a result [14]. In such domains, and for certain tasks, a human-in-the-loop can be beneficial, because such a human expert can sometimes augment the AI with tacit knowledge, i.e. contribute to an AI with human experience, conceptual understanding, context awareness, and causal reasoning. Humans are very good at multi-modal thinking and can integrate new insights into their conceptual knowledge space shaped by experience. Humans are also robust, can generalize from a few examples, and are able to understand context from even a small amount of data. Formalized, this human knowledge can be used to build structural causal models of human decision making, and the features can be traced back to train AI - helping to make current AI even more successful beyond the current state of the art.

In such sensitive and safety-critical application domains, there will be an increasing need for trustworthy AI solutions in the future [13]. Trusted AI requires both robustness and *explainability* and should be balanced with human values, ethical principles [25], and legal requirements [36], to ensure privacy, security, and safety for each individual person. The international XAI community is making great contributions to this end.

2 Explainable AI: Past and Present

In tandem with impressive advances in AI research, there have been numerous methods introduced in the past decade that aim to explain the decisions and inner workings of deep neural networks. Many such methods can be described along the following two axes: (1) whether an XAI method produces *local* or *global* explanations, that is, whether its explanations explain individual model decisions or instead characterize whole components of a model (e.g., a neuron, layer, entire network); and (2) whether an XAI method is *post-hoc* or *ante-hoc*, that is, whether it explains a deep neural network after it has been trained using standard training procedures or it introduces a novel network architecture that produces an explanation as part of its decision. For a brief overview on XAI methods please refer to [15]. Of the research that focuses on explaining specific predictions, the most active area of research has been on the problem of feature attribution [31], which aims to identify what parts of an input are responsible for a model's output decision. For computer vision models such as object classification networks, such work typically produce *heatmaps* that highlight which regions of an input image most influence a model's prediction [3, 8, 28, 33–35, 38, 41].

Similarly, *feature visualization* methods have been the most popular research stream within explainable techniques that provide global explanations. Such techniques typically explain hidden units or activation tensors by showing either real or generated images that most activate the given unit [4, 27, 35, 38, 40] or set of units [10, 18, 42] or are most similar to the given tensor [21].

In the past decade, most explainable AI research has focused on the development of post-hoc explanatory methods like feature attribution and visualization.

That said, more recently, there have been several methods that introduce novel, *interpretable-by-design* models that were intentionally designed to produce an explanation, for example as a decision tree [26], via graph neural networks [29], by comparing to prototypical examples [7], by constraining neurons to correspond to interpretable attributes [19, 22], or by summing up evidence from multiple image patches [6].

As researchers have continued to develop explainable AI methods, some work has also focused on the development of disciplined evaluation benchmarks for explainable AI and have highlighted some shortcomings of popular methods and the need for such metrics [1–3, 9, 11, 16, 23, 28, 30, 37, 39].

In tandem with the increased research in explainable AI, there have been a number of research outputs [32] and gatherings (e.g., tutorials, workshops, and conferences) that have focused on this research area, which have included some of the following:

- NeurIPS workshop on "Interpreting, Explaining and Visualizing Deep Learning – Now what?" (2017)
- ICLR workshop on "Debugging Machine Learning Models" (2019)
- ICCV workshop on "Workshop on Interpretating and Explaining Visual AI Models" (2019)
- CVPR tutorial on "Interpretable Machine Learning for Computer Vision" (2018–ongoing)
- ACM Conference on Fairness, Accountability, and Transparency (FAccT) (2018–ongoing)
- CD-MAKE conference with Workshop on xAI (2017–ongoing)

Through these community discussions, some have recognized that there were still many under-explored yet important areas within explainable AI.

Beyond Explainability. To that end, we organized the ICML 2020 workshop "XXAI: Extending Explainable AI Beyond Deep Models and Classifiers," which focused on the following topics:

1. *Explaining beyond neural network classifiers* and explaining other kinds of models such as random forests and models trained via unsupervised or reinforcement learning.
2. *Explaining beyond heatmaps* and using other forms of explanation such as structured explanations, question-and-answer and/or dialog systems, and human-in-the-loop paradigms.
3. *Explaining beyond explaining* and developing other research to improve the transparency of AI models, such as model development and model verification techniques.

This workshop fostered many productive discussions, and this book is a follow-up to our gathering and contains some of the work presented at the workshop along with a few other relevant chapters.

3   Book Structure

We organized this book into three parts:

1. Part 1: Current Methods and Challenges
2. Part 2: New Developments in Explainable AI
3. Part 3: An Interdisciplinary Approach to Explainable AI

Part 1 gives an overview of the current state-of-the-art of XAI methods as well as their pitfalls and challenges. In Chapter 1, Holzinger, Samek and colleagues give a general overview on popular XAI methods. In Chapter 2, Bhatt et al. point out that current explanation techniques are mainly used by the internal stakeholders who develop the learning models, not by the external end-users who actually get the service. They give nice take away messages learned from an interview study on how to deploy XAI in practice. In Chapter 3, Molnar et al. describe the general pitfalls a practitioner can encounter when employing model agnostic interpretation methods. They point out that the pitfalls exist when there are issues with model generalization, interactions between features etc., and called for a more cautious application of explanation methods. In Chapter 4, Salewski et al. introduce a new dataset that can be used for generating natural language explanations for visual reasoning tasks.

In Part 2, several novel XAI approaches are given. In Chapter 5, Kolek et al. propose a novel rate-distortion framework that combines mathematical rigor with maximal flexibility when explaining decisions of black-box models. In Chapter 6, Montavon et al. present an interesting approach, dubbed as neuralization-propagation (NEON), to explain unsupervised learning models, for which directly applying the supervised explanation techniques is not straightforward. In Chapter 7, Karimi et al. consider a causal effect in the algorithmic recourse problem and presents a framework of using structural causal models and a novel optimization formulation. The next three chapters in Part 2 mainly focus on XAI methods for problems beyond simple classification. In Chapter 8, Zhou gives a brief summary on recent work on interpreting deep generative models, like Generative Adversarial Networks (GANs), and show how human-understandable concepts can be identified and utilized for interactive image generation. In Chapter 9, Dinu et al. apply explanation methods to reinforcement learning and use the recently developed RUDDER framework in order to extract meaningful strategies that an agent has learned via reward redistribution. In Chapter 10, Bastani et al. also focus on interpretable reinforcement learning and describe recent progress on the programmatic policies that are easily verifiable and robust. The next three chapters focus on using XAI beyond simple explanation of a model's decision, e.g., pruning or improving models with the aid of explanation techniques. In Chapter 11, Singh et al. present the PDR framework that considers three aspects: devising a new XAI method, improving a given model with the XAI methods, and verifying the developed methods with real-world problems. In Chapter 12, Bargal et al. describe the recent approaches that utilize spatial and spatiotemporal visual explainability to train models that generalize better and possess more desirable characteristics. In Chapter 13, Becking et al. show how explanation techniques like Layer-wise Relevance Propagation [3] can be leveraged with information theory concepts and can lead to a better network quantization strategy. The next two chapters then exemplify how XAI methods can be applied to various kinds of science problems and extract new findings. In Chapter 14, Marcos et al. apply explanation methods to marine science and show how a landmark-based approach can generate heatmaps to monitor migration of whales in the ocean. In Chapter 15, Mamalakis et al. survey interesting recent results that applied explanation techniques to meteorology and climate science, e.g., weather prediction.

Part 3 presents more interdisciplinary application of XAI methods beyond technical domains. In Chapter 16, Hacker and Passoth provide an overview of legal obligations to explain AI and evaluate current policy proposals.

In Chapter 17, Zhou et al. provide a state-of-the-art overview on the relations between explanation and AI fairness and especially the roles of explanation on human's fairness judgement. Finally, in Chapter 18, Tsai and Carroll review logical approaches to explainable AI (XAI) and problems/challenges raised for explaining AI using genetic algorithms. They argue that XAI is more than a matter of accurate and complete explanation, and that it requires pragmatics of explanation to address the issues it seeks to address.

Most of the chapters fall under Part 2, and we are excited by the variety of XAI research presented in this volume. While by no means an exhaustive collection, we hope this book presents both quality research and vision for the current challenges, next steps, and future promise of explainable AI research.

Acknowledgements. The authors declare that there are no conflict of interests. This work does not raise any ethical issues. Parts of this work have been funded by the Austrian Science Fund (FWF), Project: P-32554, explainable AI, by the German Ministry for Education and Research (BMBF) through BIFOLD (refs. 01IS18025A and 01IS18037A), and by the European Union's Horizon 2020 programme (grant no. 965221), Project iToBoS.

References

1. Adebayo, J., Gilmer, J., Muelly, M., Goodfellow, I., Hardt, M., Kim, B.: Sanity checks for saliency maps. In: NeurIPS (2018)
2. Adebayo, J., Muelly, M., Liccardi, I., Kim, B.: Debugging tests for model explanations. In: NeurIPS (2020)
3. Bach, S., Binder, A., Montavon, G., Klauschen, F., Müller, K.R., Samek, W.: On pixel-wise explanations for non-linear classifier decisions by layer-wise relevance propagation. PLoS ONE **10**(7), e0130140 (2015)
4. Bau, D., Zhou, B., Khosla, A., Oliva, A., Torralba, A.: Network dissection: quantifying interpretability of deep visual representations. In: CVPR (2017)
5. Bengio, Y., Lecun, Y., Hinton, G.: Deep learning for AI. Commun. ACM **64**(7), 58–65 (2021). https://doi.org/10.1145/3448250
6. Brendel, W., Bethge, M.: Approximating CNNs with bag-of-local-features models works surprisingly well on ImageNet. In: ICLR (2019)
7. Chen, C., Li, O., Tao, D., Barnett, A., Rudin, C., Su, J.K.: This looks like that: deep learning for interpretable image recognition. In: NeurIPS (2019)
8. Fong, R., Patrick, M., Vedaldi, A.: Understanding deep networks via extremal perturbations and smooth masks. In: ICCV (2019)
9. Fong, R., Vedaldi, A.: Interpretable explanations of black boxes by meaningful perturbation. In: ICCV (2017)
10. Fong, R., Vedaldi, A.: Net2Vec: quantifying and explaining how concepts are encoded by filters in deep neural networks. In: Proceedings of the CVPR (2018)
11. Hoffmann, A., Fanconi, C., Rade, R., Kohler, J.: This looks like that... does it? Shortcomings of latent space prototype interpretability in deep networks. In: ICML Workshop on Theoretic Foundation, Criticism, and Application Trend of Explainable AI (2021)
12. Holzinger, A., Carrington, A., Müller, H.: Measuring the quality of explanations: the System Causability Scale (SCS). KI - Künstliche Intelligenz **34**(2), 193–198 (2020). https://doi.org/10.1007/s13218-020-00636-z

13. Holzinger, A., et al.: Information fusion as an integrative cross-cutting enabler to achieve robust, explainable, and trustworthy medical artificial intelligence. Inf. Fusion **79**(3), 263–278 (2022). https://doi.org/10.1016/j.inffus.2021.10.007

14. Holzinger, A., Malle, B., Saranti, A., Pfeifer, B.: Towards multi-modal causability with graph neural networks enabling information fusion for explainable AI. Inf. Fusion **71**(7), 28–37 (2021). https://doi.org/10.1016/j.inffus.2021.01.008

15. Holzinger, A., Saranti, A., Molnar, C., Biececk, P., Samek, W.: Explainable AI methods - a brief overview. In: Holzinger, A., et al. (eds.) xxAI 2020. LNAI, vol. 13200, pp. 13–38. Springer, Cham (2022)

16. Hooker, S., Erhan, D., Kindermans, P.J., Kim, B.: A benchmark for interpretability methods in deep neural networks. In: NeurIPS (2019)

17. Hu, X., Chu, L., Pei, J., Liu, W., Bian, J.: Model complexity of deep learning: a survey. Knowl. Inf. Syst. **63**(10), 2585–2619 (2021). https://doi.org/10.1007/s10115-021-01605-0

18. Kim, B., et al.: Interpretability beyond feature attribution: quantitative testing with concept activation vectors (TCAV). In: Proceedings of the ICML (2018)

19. Koh, P.W., et al.: Concept bottleneck models. In: ICML (2020)

20. Lakkaraju, H., Arsov, N., Bastani, O.: Robust and stable black box explanations. In: Daumé, H., Singh, A. (eds.) International Conference on Machine Learning (ICML 2020), pp. 5628–5638. PMLR (2020)

21. Mahendran, A., Vedaldi, A.: Visualizing deep convolutional neural networks using natural pre-images. Int. J. Comput. Vis. **120**(3), 233–255 (2016)

22. Marcos, D., Fong, R., Lobry, S., Flamary, R., Courty, N., Tuia, D.: Contextual semantic interpretability. In: Ishikawa, H., Liu, C.-L., Pajdla, T., Shi, J. (eds.) ACCV 2020. LNCS, vol. 12625, pp. 351–368. Springer, Cham (2021). https://doi.org/10.1007/978-3-030-69538-5_22

23. Margeloiu, A., Ashman, M., Bhatt, U., Chen, Y., Jamnik, M., Weller, A.: Do concept bottleneck models learn as intended? In: ICLR Workshop on Responsible AI (2021)

24. Mnih, V., et al.: Human-level control through deep reinforcement learning. Nature **518**(7540), 529–533 (2015)

25. Mueller, H., Mayrhofer, M.T., Veen, E.B.V., Holzinger, A.: The ten commandments of ethical medical AI. IEEE Comput. **54**(7), 119–123 (2021). https://doi.org/10.1109/MC.2021.3074263

26. Nauta, M., van Bree, R., Seifert, C.: Neural prototype trees for interpretable fine-grained image recognition. In: Proceedings of the IEEE/CVF Conference on Computer Vision and Pattern Recognition, pp. 14933–14943 (2021)

27. Olah, C., Mordvintsev, A., Schubert, L.: Feature visualization. Distill **2**(11), e7 (2017)

28. Petsiuk, V., Das, A., Saenko, K.: Rise: randomized input sampling for explanation of black-box models. In: Proceedings of the BMVC (2018)

29. Pfeifer, B., Secic, A., Saranti, A., Holzinger, A.: GNN-subnet: disease subnetwork detection with explainable graph neural networks. bioRxiv, pp. 1–8 (2022). https://doi.org/10.1101/2022.01.12.475995

30. Poppi, S., Cornia, M., Baraldi, L., Cucchiara, R.: Revisiting the evaluation of class activation mapping for explainability: a novel metric and experimental analysis. In: CVPR Workshop on Responsible Computer Vision (2021)

31. Samek, W., Montavon, G., Lapuschkin, S., Anders, C.J., Müller, K.R.: Explaining deep neural networks and beyond: a review of methods and applications. Proc. IEEE **109**(3), 247–278 (2021)

32. Samek, W., Montavon, G., Vedaldi, A., Hansen, L.K., Müller, K.-R. (eds.): Explainable AI: Interpreting, Explaining and Visualizing Deep Learning. LNCS (LNAI), vol. 11700. Springer, Cham (2019). https://doi.org/10.1007/978-3-030-28954-6

33. Selvaraju, R.R., Cogswell, M., Das, A., Vedantam, R., Parikh, D., Batra, D.: Grad-CAM: visual explanations from deep networks via gradient-based localization. In: ICCV (2017)

34. Shitole, V., Li, F., Kahng, M., Tadepalli, P., Fern, A.: One explanation is not enough: structured attention graphs for image classification. In: NeurIPS (2021)

35. Simonyan, K., Vedaldi, A., Zisserman, A.: Deep inside convolutional networks: visualising image classification models and saliency maps. In: ICLR Workshop (2014)

36. Stoeger, K., Schneeberger, D., Holzinger, A.: Medical artificial intelligence: the European legal perspective. Commun. ACM **64**(11), 34–36 (2021). https://doi.org/10.1145/3458652

37. Yang, M., Kim, B.: Benchmarking attribution methods with relative feature importance (2019)

38. Zeiler, M.D., Fergus, R.: Visualizing and understanding convolutional networks. In: Fleet, D., Pajdla, T., Schiele, B., Tuytelaars, T. (eds.) ECCV 2014. LNCS, vol. 8689, pp. 818–833. Springer, Cham (2014). https://doi.org/10.1007/978-3-319-10590-1_53

39. Zhang, J., Lin, Z., Brandt, J., Shen, X., Sclaroff, S.: Top-down neural attention by excitation backprop. In: Leibe, B., Matas, J., Sebe, N., Welling, M. (eds.) ECCV 2016. LNCS, vol. 9908, pp. 543–559. Springer, Cham (2016). https://doi.org/10.1007/978-3-319-46493-0_33

40. Zhou, B., Khosla, A., Lapedriza, A., Oliva, A., Torralba, A.: Object detectors emerge in deep scene CNNs. In: Proceedings of the ICLR (2015)

41. Zhou, B., Khosla, A., Lapedriza, A., Oliva, A., Torralba, A.: Learning deep features for discriminative localization. In: CVPR (2016)

42. Zhou, B., Sun, Y., Bau, D., Torralba, A.: Interpretable basis decomposition for visual explanation. In: Ferrari, V., Hebert, M., Sminchisescu, C., Weiss, Y. (eds.) ECCV 2018. LNCS, vol. 11212, pp. 122–138. Springer, Cham (2018). https://doi.org/10.1007/978-3-030-01237-3_8

Current Methods and Challenges

Explainable AI Methods - A Brief Overview

Andreas Holzinger[1,2,3(✉)] (iD), Anna Saranti[1,2] (iD), Christoph Molnar[4] (iD),
Przemyslaw Biecek[5,6] (iD), and Wojciech Samek[7,8] (iD)

[1] Human-Centered AI Lab, University of Natural Resources and Life Sciences,
Vienna, Austria
`andreas.holzinger@human-centered.ai`
[2] Medical University Graz, Graz, Austria
[3] xAI Lab, Alberta Machine Intelligence Institute, Edmonton, Canada
[4] Leibniz Institute for Prevention Research and Epidemiology - BIPS GmbH,
Bremen, Germany
[5] Warsaw University of Technology, Warsaw, Poland
[6] University of Warsaw, Warsaw, Poland
[7] Department of Artificial Intelligence, Fraunhofer Heinrich Hertz Institute,
Berlin, Germany
[8] BIFOLD – Berlin Institute for the Foundations of Data and Learning,
Berlin, Germany

Abstract. Explainable Artificial Intelligence (xAI) is an established
field with a vibrant community that has developed a variety of very
successful approaches to explain and interpret predictions of complex
machine learning models such as deep neural networks. In this article, we
briefly introduce a few selected methods and discuss them in a short, clear
and concise way. The goal of this article is to give beginners, especially
application engineers and data scientists, a quick overview of the state
of the art in this current topic. The following 17 methods are covered
in this chapter: LIME, Anchors, GraphLIME, LRP, DTD, PDA, TCAV,
XGNN, SHAP, ASV, Break-Down, Shapley Flow, Textual Explanations
of Visual Models, Integrated Gradients, Causal Models, Meaningful Perturbations, and X-NeSyL.

Keywords: Explainable AI · Methods · Evaluation

1 Introduction

Artificial intelligence (AI) has a long tradition in computer science. Machine
learning (ML) and particularly the success of "deep learning" in the last decade
made AI extremely popular again [15,25,90].

The great success came with additional costs and responsibilities: the most
successful methods are so complex that it is difficult for a human to re-trace,
to understand, and to interpret how a certain result was achieved. Consequently, explainability/interpretability/understandability is motivated by the

A. Holzinger et al. (Eds.): xxAI 2020, LNAI 13200, pp. 13–38, 2022.
https://doi.org/10.1007/978-3-031-04083-2_2

lack of transparency of these black-box approaches, which do not foster trust and acceptance of AI in general and ML in particular. Increasing legal and data protection aspects, e.g., due to the new European General Data Protection Regulation (GDPR, in force since May 2018), complicate the use of black-box approaches, particularly in domains that affect human life, such as the medical field [56,63,73,76].

The term explainable AI (xAI) was coined by DARPA [28] and gained meanwhile a lot of popularity. However, xAI is not a new buzzword. It can be seen as a new name for a very old quest in science to help to provide answers to questions of why [66]. The goal is to enable human experts to understand the underlying explanatory factors of why an AI decision has been made [64]. This is highly relevant for causal understanding and thus enabling ethical responsible AI and transparent verifiable machine learning in decision support [74].

The international community has developed a very broad range of different methods and approaches and here we provide a short concise overview to help engineers but also students to select the best possible method. Figure 1 shows the most popular XAI toolboxes.

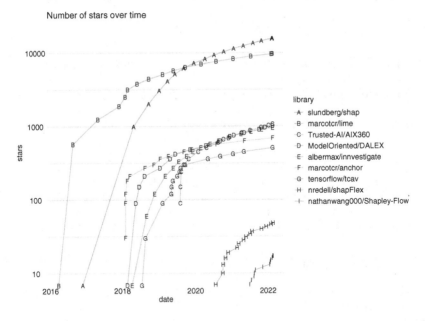

Fig. 1. Number of stars on GitHub for the most popular repositories presented in this paper. While these repositories focus on the explanation task, the new Quantus toolbox [30] offers a collection of methods for evaluating and comparing explanations.

In the following we provide a short overview of some of the most popular methods for explaining complex models. We hope that this list will help both practitioners in choosing the right method for model explanation and

XAI method developers in noting the shortcomings of currently available methods. Figure 2 gives an overview of the chronology of development of successive explanatory methods. Methods such as LRP and LIME were among the first[1] generic techniques to explain decisions of complex ML models. In addition to the overview of explanation techniques, we would also like to hint the interested reader at work that developed methods and offered datasets to objectively evaluate and systematically compare explanations. To mention here is Quantus[2] [30], a new toolbox offering an exhaustive collection of evaluation methods and metrics for explanations, and CLEVR-XAI[3] [8], a benchmark dataset for the ground truth evaluation of neural network explanations.

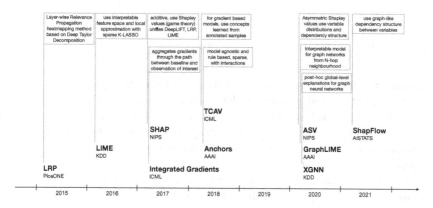

Fig. 2. Chronology of the development of successive explanatory methods described in this paper. Initially, the methods were focused on model analysis based on the model itself or on sample data. Subsequent methods used more and more information about the structure and relationships between the analysed variables.

2   Explainable AI Methods - Overview

2.1   LIME (Local Interpretable Model Agnostic Explanations)

Idea: By treating the machine learning models as black-box functions, model agnostic explanation methods typically only have access to the model's output. The fact that these methods do not require any information about the model's internals, e.g., in the case of neural networks the topology, learned parameters (weights, biases) and activation values, makes them widely applicable and very flexible.

---

[1] We are aware that gradient-based sensitivity analysis and occlusion-based techniques have been proposed even earlier [11,62,75,89]. However, theses techniques have various disadvantages (see [61,70]) and are therefore not considered in this paper.

[2] https://github.com/understandable-machine-intelligence-lab/quantus.

[3] https://github.com/ahmedmagdiosman/clevr-xai.

One prominent representative of this class of explanation techniques is the Local Interpretable Model-agnostic Explanations (LIME) method [67]. The main idea of LIME is to explain a prediction of a complex model f_M, e.g., a deep neural network, by fitting a local surrogate model f_S, whose predictions are easy to explain. Therefore, LIME is also often referred to as surrogate-based explanation technique [70]. Technically, LIME generates samples in the neighborhood $\mathcal{N}_{\mathbf{x}_i}$ of the input of interest $\mathbf{x}_i$, evaluates them using the target model, and subsequently approximates the target model in this local vicinity by a simple linear function, i.e., a surrogate model which is easy to interpret. Thus, LIME does not directly explain the prediction of the target model $f_M(\mathbf{x}_i)$, but rather the predictions of a surrogate model $f_S(\mathbf{x}_i)$, which locally approximates the target model (i.e., $f_M(\mathbf{x}) \approx f_S(\mathbf{x})$ for $\mathbf{x} \in \mathcal{N}_{\mathbf{x}_i}$).

GitHub Repo: https://github.com/marcotcr

Discussion: There are meanwhile many successful applications of LIME in different application domains which demonstrates the popularity of this model agnostic method. As a limitation can be seen that LIME only indirectly solves the explanation problem by relying on a surrogate model. Thus, the quality of the explanation largely depends on the quality of the surrogate fit, which itself may require dense sampling and thus may result in large computational costs. Furthermore, sampling always introduces uncertainty, which can lead to non-deterministic behaviours and result in variable explanations for the same input sample.

2.2 Anchors

Idea: The basic idea is that individual predictions of any black-box classification model are explained by finding a decision rule that sufficiently "anchors" the prediction - hence the name "anchors" [68]. The resulting explanations are decision rules in the form of IF-THEN statements, which define regions in the feature space. In these regions, the predictions are fixed (or "anchored") to the class of the data point to be explained. Consequently, the classification remains the same no matter how much the other feature values of the data point that are not part of the anchor are changed.

Good anchors should have high precision and high coverage. *Precision* is the proportion of data points in the region defined by the anchor that have the same class as the data point being explained. *Coverage* describes how many data points an anchor's decision rule applies to. The more data points an anchor covers, the better, because the anchor then covers a larger area of the feature space and thus represents a more general rule. Anchors is a model-agnostic explanation method, i.e., it can be applied to any prediction model without requiring knowledge about the internals. Search and construction of decision rules is done by reinforcement learning (RL) [32,81] in combination with a modified beam search, a heuristic search algorithm that extends the most promising nodes in a graph. The anchors algorithm cycles through different steps: produce candidate anchors, select the best candidates, then use beam search to extend the anchor rules. To select the best candidate, it is necessary to call the model many times, which can be seen as an exploration or multi-armed bandit problem.

GitHub Repo: https://github.com/marcotcr/anchor

Discussion: The anchors are model-independent and can be applied to different domains such as tabular data, images and text, depending on the perturbation strategy. However, in the current Python implementation, anchors only supports tabular and text data. Compared to LIME, the scope of interpretation is clearer as the anchors specify the boundaries within which they should be interpreted. The coverage of an anchor decision rule can be used as a measure of the model fidelity of the anchor. Furthermore, the decision rules are easy to understand, but there are many hyper-parameters in the calculation of anchors, such as the width of the beam and the precision threshold, which need to be tuned individually. The perturbation strategies also need to be carefully selected depending on the application and model. The calculation of anchors requires many calls to the prediction function, which makes the anchors computationally intensive. Data instances that are close to the decision boundary of the model may require more complex rules with more features and less coverage. Unbalanced classification problems can produce trivial decision rules, such as classifying each data point as the majority class. A possible remedy is to adapt the perturbation strategy to a more balanced distribution.

2.3   GraphLIME

Idea: GraphLIME [38] is a method that takes the basic idea of LIME (see Sect. 2.1) but is not linear. It is applied to a special type of neural network architecture, namely graph neural networks (GNN). These models can process non-Euclidean data as they are organised in a graph structure [9]. The main tasks that GNNs perform are node classification, link prediction and graph classification. Like LIME, this method tries to find an interpretable model, which in this case is the Hilbert-Schmidt Independence Criterion (HSIC) Lasso model, for explaining a particular node in the input graph. It takes into account the fact that during the training of the GNN, several nonlinear aggregation and combination methods use the features of neighbouring nodes to determine the representative embedding of each node. This embedding is used to distinguish nodes into different classes in the case of node classification and to collectively distinguish graphs in graph classification tasks.

Since for this type of model a linear explanation as LIME would return unfaithful results, the main idea of GraphLIME is to sample from the N-hop neighbourhood of the node and collect features w.r.t. to the node prediction. Those are used to train the HSIC Lasso model, which is a kernel method - thereby interpretable - that can compute on which node features the output prediction depends on. This is similar to the perturbation method that LIME uses, while the comparison, in this case, is based on HSIC estimation between the random variables representing the features and the prediction distributions. This method learns correlations between the features of the neighbours which also underline its explanation capabilities.

GitHub Repo: https://github.com/WilliamCCHuang/GraphLIME

Discussion: The developers compared GraphLIME with one of the first xAI methods for GNNs at the time, namely GNNExplainer, w.r.t. three criteria: (1) ability to detect useless features, (2) ability to decide whether the prediction is trustworthy, and (3) ability to identify the better model among two GNN classifiers. They show that for synthetic data and human-labelled annotations, GraphLIME exceeds the GNNExplainer by far in the last two criteria. They arrive at a very interesting insight, namely that models that have fewer untrustworthy features in their explanation have better classification performance. Furthermore, GraphLIME is shown to be computationally much more efficient than GNNExplainer. It would be beneficial - and is considered future work - if GraphLIME was also trying to find important graph substructures instead of just features, if it was compared with other methods like PGExplainer [52], PGMExplainer [83], GNN-LRP [72], and if it were extended to multiple instance explanations. Finally, it is important to note that GraphLIME is successfully used for the investigation of backdoor attacks on GNNs by uncovering the relevant features of the graph's nodes [86].

2.4 Method: LRP (Layer-wise Relevance Propagation)

Idea: Layer-wise Relevance Propagation (LRP) [10] is a propagation-based explanation method, i.e., it requires access to the model's internals (topology, weights, activations etc.). This additional information about the model, however, allows LRP to simplify and thus more efficiently solve the explanation problem. More precisely, LRP does not explain the prediction of a deep neural network in one step (as model agnostic methods would do), but exploits the network structure and redistributes the explanatory factors (called relevance R) layer by layer, starting from the model's output, onto the input variables (e.g., pixels). Each redistribution can be seen as the solution of a simple (because only between two adjacent layers) explanation problem (see interpretation of LRP as Deep Taylor Decomposition in Sect. 2.5).

Thus, the main idea of LRP is to explain by decomposition, i.e., to iteratively redistribute the total evidence of the prediction $f(\mathbf{x})$, e.g., indicating that there is a cat in the image, in a conservative manner from the upper to the next lower layer, i.e.,

$$\sum_i R_i^{(0)} = \ldots = \sum_j R_j^{(l)} = \sum_k R_k^{(l+1)} = \ldots = f(\mathbf{x}). \tag{1}$$

Note that $R_i^{(0)}$ denotes the relevance assigned to the ith input element (e.g., pixel), while $R_j^{(l)}$ stands for relevance assigned to the jth neuron at the lth layer. This conservative redistribution not only ensures that no relevance is added or lost on the way (analogous to energy conservation principle or Kirchhoff's law in physics), but also allows for signed explanations, where positive relevance values hint at relevant information supporting the prediction and negative relevance values indicate evidence speaking against it. Different redistribution rules, adapted to the specific properties of particular neural network layers, have been proposed for LRP [42,58]. In contrast to other XAI techniques which are purely based on heuristics, the LRP rules have a clear theoretical foundation, namely they result from the Deep Taylor Decomposition (DTD) [60] of the relevance function with a particular choice of root point (see Sect. 2.5).

While LRP has been originally developed for convolutional neural networks and bag-of-words type of models, various extensions have been proposed, making it a widely applicable XAI techniqe today. For instance, Arras et al. [6,7] developed meaningful LRP redistribution rules for LSTM models. Also LRP variants for GNN and Transformer models have been recently proposed [3,72]. Finally, through the "neuralization trick", i.e., by converting a non-neural network model into a neural network, various other classical ML algorithms have been made explainable with LRP, including k-means clustering [39], one-class SVM [40] as well as kernel density estimation [59]. Furthermore, meta analysis methods such as spectral relevance analysis (SpRAy) [47] have been proposed to cluster and systematically analyze sets of explanations computed with LRP (SpRAY is not restricted to LRP explanations though). These analyses have been shown useful to detect artifacts in the dataset and uncover so-called "Clever Hans" behaviours of the model [47].

The recently published Zennit toolbox [4] implements LRP (and other methods) in Python, while the CoRelAy[4] toolboxi offers a collection of meta analysis methods. Furthermore, the GitHub library iNNvestigate provides a common interface and out-of-the-box implementation for many analysis methods, including LRP [2].

GitHub Repo: https://github.com/chr5tphr/zennit
https://github.com/albermax/innvestigate

Discussion: LRP is a very popular explanation method, which has been applied in a broad range of domains, e.g., computer vision [46], natural language processing [7], EEG analysis [78], meteorology [54], among others.

The main advantages of LRP are its high computational efficiency (in the order of one backward pass), its theoretical underpinning making it a trustworthy and robust explanation method (see systematic comparison of different XAI methods [8]), and its long tradition and high popularity (it is one of the first XAI techniques, different highly efficient implementations are available, and it

---

[4] https://github.com/virelay/corelay.

has been successfully applied to various problems and domains). The price to pay for the advantages is a restricted flexibility, i.e., a careful adaptation of the used redistribution rules may be required for novel model architectures. For many popular layers types recommended redistribution rules are described in [42,58].

Finally, various works showed that LRP explanations can be used beyond sheer visualization purposes. For instance, [47] used them to semi-automatically discover artefacts in large image corpora, while [5,79] went one step further and demonstrated that they can be directly (by augmenting the loss) or indirectly (by adapting training data) used to improve the model. Another line of work [14,87] exploits the fact that LRP computes relevance values not only for the input variables, but for all elements of the neural network, including weights, biases and individual neurons, to optimally prune and quantize the neural model. The idea is simple, since LRP explanations tell us which parts of the neural network are relevant, we can simply remove the irrelevant elements and thus improve the coding efficiency and speed up the computation.

2.5   Deep Taylor Decomposition (DTD)

Idea: The Deep Taylor Decomposition (DTD) method [60] is a propagation-based explanation technique, which explains decisions of a neural network by decomposition. It redistributes the function value (i.e., the output of the neural network) to the input variables in a layer-by-layer fashion, while utilizing the mathematical tool of (first-order) Taylor expansion to determine the proportion or relevance assigned to the lower layer elements in the redistribution process (i.e., their respective contributions). This approach is closely connected to the LRP method (see Sect. 2.4). Since most LRP rules can be interpreted as a Taylor decomposition of the relevance function with a specific choice of root point, DTD can be seen as the mathematical framework of LRP.

DTD models the relevance of a neuron k at layer l as a simple relevance function of the lower-layer activations, i.e.,

$$R_k(\mathbf{a}) = \max(0, \sum_i a_i w_{ik}) c_k, \qquad (2)$$

where $\mathbf{a} = [a_1 \dots a_d]$ are the activations at layer $l-1$, w_{ik} are the weights connecting neurons i (at layer $l-1$) and k (at layer l), and c_k is a constant. This model is certainly valid at the output layer (as R_k is initialized with the network output $f(\mathbf{x})$). Through an inductive argument the authors of [60] proved that this model also (approximatively) holds at intermediate layers. By representing this simple function as Taylor expansion around a root point $\tilde{\mathbf{a}}$, i.e.,

$$R_k(\mathbf{a}) = \underbrace{R_k(\tilde{\mathbf{a}})}_{0} + \underbrace{\sum_i (a_i - \tilde{a}_i) \cdot \nabla[R_k(\mathbf{a})]_i}_{\text{redistributed relevance}} + \underbrace{\epsilon}_{0}, \qquad (3)$$

DTD tells us how to meaningfully redistribute relevance from layer l to layer $l-1$. This redistribution process is iterated until the input layer. Different choices of root point are recommended for different types of layers (conv layer, fully connected layer, input layer) and lead to different LRP redistribution rules [58].

GitHub Repo: https://github.com/chr5tphr/zennit
https://github.com/albermax/innvestigate

Discussion: DTD is a theoretically motivated explanation framework, which redistributes relevance from layer to layer in a meaningful manner by utilizing the concept of Taylor expansion. The method is highly efficient in terms of computation and can be adapted to the specific properties of a model and its layers (e.g., by the choice of root point). As for LRP, it is usually not straight forward to adapt DTD to novel model architectures (see e.g. local renormalization layers [18]).

2.6   Prediction Difference Analysis (PDA)

Idea: At the 2017 ICLR conference, Zintgraf et al. [91] presented the Prediction Difference Analysis (PDA) method. The method is based on the previous idea presented by [69] where, for a given prediction, each input feature is assigned a relevance value with respect to a class c. The idea of PDA is that the relevance of a feature x_i can be estimated by simply measuring how the prediction changes when the feature is unknown, i.e., the difference between $p(c|\mathbf{x})$ and $p(c|\mathbf{x}_{\backslash i})$, where $\mathbf{x}_{\backslash i}$ denotes the set of all input features except x_i. Now to evaluate the prediction, specifically to find $p(c|\mathbf{x}_{\backslash i})$ there are three possibilities: (1) label the feature as unknown, (2) re-train the classifier omitting the feature, or (3) simulate the absence of a feature by marginalizing the feature. With that a relevance vector $(\mathrm{WE}_i)_{i=1...m}$ (whereby m represent the number of features) is generated, that is of the same size as the input and thus reflects the relative importance of all features. A large prediction difference indicates that the feature contributed significantly to the classification, while a small difference indicates that the feature was not as important to the decision. So specifically, a positive value WE_i means that the feature contributed to the evidence for the class of interest and much more so that removing the feature would reduce the classifier's confidence in the given class. A negative value, on the other hand, means that the feature provides evidence against the class: Removing the feature also removes potentially contradictory or disturbing information, and makes the classifier more confident in the class under study.

GitHub Repo: https://github.com/lmzintgraf/DeepVis-PredDiff

Discussion: Making neural network decisions interpretable through visualization is important both to improve models and to accelerate the adoption of black-box classifiers in application areas such as medicine. In the original paper the authors illustrate the method in experiments on natural images (ImageNet data), as well as medical images (MRI brain scans). A good discussion can be found in: https://openreview.net/forum?id=BJ5UeU9xx

2.7   TCAV (Testing with Concept Activation Vectors)

Idea: TCAV [41] is a concept-based neural network approach that aims to quantify how strongly a concept, such as colour, influences classification. TCAV is based on the idea of concept activation vectors (CAV), which describe how neural activations influence the presence or absence of a user-specific concept. To calculate such a CAV, two data sets must first be collected and combined: One dataset containing images representing the concept and one dataset consisting of images in which this concept is not present. Then a logistic regression model is trained on the combined dataset to classify whether the concept is present in an image. The activations of the user-defined layer of the neural network serve as features for the classification model. The coefficients of the logistic regression model are then the CAVs. For example, to investigate how much the concept "stripped" contributes to the classification of an image as "zebra" by a convolutional neural network, a dataset representing the concept "stripped" and a random dataset in which the concept "stripped" is not present must be assembled. From the CAVs, the conceptual sensitivity can be calculated, which is the product of the CAV and the derivative of the classification (of the original network) with respect to the specified neural network layer and class. Conceptual sensitivity thus indicates how strongly the presence of a concept contributes to the desired class.

While the CAV is a local explanation as it relates to a single classification, the TCAV combines the CAVs across the data into a global explanation method and thus answers the question of how much a concept contributed overall to a given classification. First, the CAVs are calculated for the entire dataset for the selected class, concept and level. Then TCAV calculates the ratio of images with positive conceptual sensitivity, which indicates for how many images the concept contributed to the class. This ratio is calculated multiple times, each time using a different "negative" sample where the concept is not present, and a two-tailed Student t-test [77] is applied to test whether the conceptual sensitivity is significantly different from zero (the test part is where the "T" in TCAV comes from).

GitHub Repo: https://github.com/tensorflow/tcav

Discussion: TCAV can be applied to detect concept sensitivity for image classifiers that are gradient-based, such as deep neural networks. TCAV can also be used to analyze fairness aspects, e.g. whether gender or attributes of protected groups are used for classification. Very positive is that TCAV can be used by users without machine learning expertise, as the most important part is collecting the concept images, where domain expertise is important. TCAV allows to test a classification model for arbitrary concepts, even if the model was not explicitly trained on them. The technique can be used to study whether a network learned "flawed" concepts, such as spurious correlations. Detecting flawed concepts can help to improve the model. For example, it could be studied how important the presence of snow was for classifying wolves on images, and if it turns out to be important, adding images with wolves without snow might improve the robustness of the model. One drawback can be seen in the effort for labeling and collecting new data. Some concepts might also be too abstract to test, as the collection of a concept dataset might be difficult. How would one, for example, collect a dataset of images representing the concept "happiness"? Furthermore, TCAV may not work well with shallower neural networks, as only deeper networks learn more abstract concepts. Also, the technique is also not applicable to text and tabular data, but mainly to image data[5] (last accessed: 21-Feb-2022). A practical example from the medical domain can be found in [19].

2.8 XGNN (Explainable Graph Neural Networks)

Idea: The XGNN method [88] is a post-hoc method that operates on the model level, meaning that it does not strive to provide individual example-level explanations. RL drives a search to find an adequate graph starting by a randomly chosen node or a relatively small graph, as defined by prior knowledge. The RL algorithm follows two rewards at the same time: first, it tries to increase the performance of the GNN, but secondly to keep generating valid graphs, depending on the domain requirements. The action space contains only edge addition for edges in the existing graph or an enhancement with a new node. In the case where the action has a non-desirable contribution, a negative reward is provided.

GitHub Repo: https://github.com/divelab/DIG/tree/dig/benchmarks/xgraph/supp/XGNN and pseudocode in the paper.

Discussion: This explanation method is invented particularly for the task of graph classifications. The returned graphs are the ones that were the most representative for the GNN decision and usually have a particular property that is ingrained to make the validation possible. It is worth to mention that this is the only method that provides mode-level explanations for GNN architectures. The use of RL is justified by the fact that the search for the explanation graph is non-differentiable, since it is not only driven by the performance but also by the plausibility and validity of the generated graph. Because the training of GNNs involves aggregations and combinations, this is an efficient way to overcome the obstacle of non-differentiation. The provided explanation is considered to be more effective for big datasets, where humans don't have the time to check each example's explanation individually. A disadvantage can be seen by the fact that the research idea is based on the assumption that network motifs that are the result of this explanation method are the ones on which the GNN is most

---

[5] For discussion see: https://openreview.net/forum?id=S1viikbCW.

"responsive"; nevertheless, this is not entirely true, since one does not know if other graph information was also important for the decision of the network. The results of the explanations are also non-concrete since in many cases ground truth is missing. That leads to a rather weak validation that bases on abstract concepts and properties of the discovered graphs, such as if they contain cycles or not.

2.9   SHAP (Shapley Values)

Note that the concepts described in this section also apply to the methods presented in Sects. 2.9–2.12.

Methods in this family are concerned with explanations for the model f at some individual point x^*. They are based on a value function e_S where S is a subset of variable indexes $S \subseteq \{1, ..., p\}$. Typically, this function is defined as the expected value for a conditional distribution in which conditioning applies to all variables in a subset of S

$$e_S = E[f(x)|x_S = x_S^*]. \tag{4}$$

Expected value is typically used for tabular data. In contrast, for other data modalities, this function is also often defined as the model prediction at x^* after zeroing out the values of variables with indices outside S. Whichever definition is used, the value of e_S can be thought of as the model's response once the variables in the subset S are specified.

The purpose of attribution is to decompose the difference $f(x*) - e_\varnothing$ into parts that can be attributed to individual variables (see Fig. 3A).

Idea: Assessing the importance of variable i is based on analysing how adding variable i to the set S will affect the value of the function e_S. The contribution of a variable i is denoted by $\phi(i)$ and calculated as weighted average over all possible subsets S

$$\phi(i) = \sum_{S \subseteq \{1,...,p\}/\{i\}} \frac{|S|!(p - 1 - |S|)!}{p!} \left(e_{S \cup \{i\}} - e_S\right). \tag{5}$$

This formula is equivalent to

$$\phi(i) = \frac{1}{|\Pi|} \sum_{\pi \in \Pi} e_{\text{before}(\pi,i) \cup \{i\}} - e_{\text{before}(\pi,i)}, \tag{6}$$

where Π is a set of all orderings of p variables and $\text{before}(\pi, i)$ stands for subset of variables that are before variable i in the ordering π. Each ordering corresponds to set of values e_S that shift from $e_\varnothing$ to $f(x^*)$ (see Fig. 3B).

In summary, the analysis of a single ordering shows how adding consecutive variables changes the value of the e_S function as presented in Fig. 3B. SHAP [51] arises as an averaging of these contributions over all possible orderings. This algorithm is an adaptation of Shapley values to explain individual predictions of machine learning models. Shapley values were initially proposed to distribute payouts fairly in cooperative games and are the only solution based on axioms of efficiency, symmetry, dummy, and additivity.

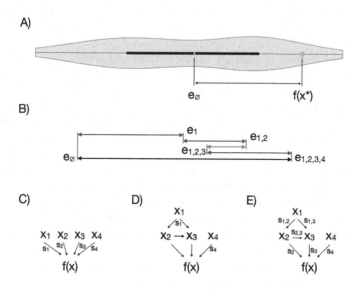

Fig. 3. Panel A. The methods presented in Sects. 2.9–2.12 explain the difference in prediction between a particular observation (x) and a baseline value. Often for the baseline value is taken the expected value from the model's prediction distribution. The methods described here distribute this difference $e_\varnothing - f(x^*)$ among the variables in the model. Panel B. Attributions are based on the changes in the expected value of the model prediction due to successive conditioning. For a given sequence of variable order (here, 1, 2, 3, 4) one can calculate how adding another variable will change the expected prediction of the model. Panel C. For the SHAP method, the variables have no structure, so any sequence of variables is treated as equally likely. Panel D. The ASV method takes into account a causal graph for variables. Only variable orderings that are consistent with this dependency graph are considered in the calculation of attributions. Causal graph controls where to assign attributions in the case of dependent variables. Panel E. The Shapley Flow method also considers a causal graph. It allocates attributions to the edges in this graph, showing how these attributions propagate through the graph.

GitHub Repo: https://github.com/slundberg/shap

Discussion: SHAP values sum up to the model prediction, i.e.

$$f(x^*) = e_\varnothing + \sum_i \phi(i). \tag{7}$$

In some situations, this is a very desirable property, e.g. if a pricing model predicts the value of a certain product, it is desirable to decompose this prediction additively into components attributable to individual variables. SHAP draws from a rich theoretical underpinning in game theory and fulfils desirable axioms, for example, that features that did not contribute to the prediction get an attribution of zero. Shapley values can be further combined to global interpretations of the model, such as feature dependence plots, feature importance and interaction analysis.

One large drawback of Shapley values is their immense computational complexity. For modern models such as deep neural networks and high dimensional inputs, the exact computation of Shapley values is intractable. However, model-specific implementations exist for tree-based methods (random forest, xgboost etc.) or additive models [49]. With care, one should use certain estimation versions of SHAP, such as KernelSHAP, because those are slow to compute. Furthermore, when features are dependent, Shapley values will cause extrapolation to areas with low data density. Conditional versions exist [50] (for tree-based models only), but the interpretation changes which is a common pitfall [57]. SHAP explanations are not sparse since to each feature that changes the prediction, a Shapley value different from zero is attributed, no matter how small the influence. If sparse explanations are required, counterfactual explanations might be preferable.

2.10 Asymmetric Shapley Values (ASV)

Idea: SHAP values are symmetrical. This means that if two variables have the same effect on the model's behaviour, e.g. because they take identical values, they will receive equal attributions. However, this is not always a desirable property. For example, if we knew that one of the variables has a causal effect on the other, then it would make more sense to assign the entire attribution to the source variable.

Asymmetric Shapley values (ASV) [22,23] allow the use of additional knowledge about the causal relations between variables in the model explanation process. A cause-effect relationship described in the form of causal graph allows the attribution of variables to be redistributed in such a way that the source variables have a greater attribution, providing effect on both the other dependent variables and the model predictions (see Fig. 3D). SHAP values are a special case of ASV values, where the casual graph is reduced to a set of unrelated vertices (see Fig. 3C).

The ASV values for variable i are also calculated as the average effect of adding a variable to a coalition of other variables, in the same way as expressed in Eq. (6). The main difference is that not all possible orders of variables are considered, but only the orders are consistent with the casual graph. Thus, a larger effect will be attributed to the source variables.

GitHub Repo: https://github.com/nredell/shapFlex

Discussion: In order to use the ASV, a causal graph for the variables is needed. Such a graph is usually created based on domain knowledge. Examples include applications in bioinformatics with signalling pathways data for which the underlying causal structure is experimentally verified or application in social sciences, where sociodemographic data in which the direction of the relationship can be determined based on expert knowledge (e.g., age affects income rather than income affects age).

A particular application of the ASV value is the model fairness analysis. If a protected attribute, such as age or sex, does not directly affect the model's score, its SHAP attribute will be zero. But if the protected attribute is the cause for other proxy variables, then the ASV values will capture this indirect effect on the model.

2.11 Break-Down

Idea: Variable contribution analysis is based on examining the change in e_S values along with a growing set of variables described by a specific order (see Fig. 3B). If the model f has interactions, different orderings of the variables may lead to different contributions. The SHAP values average over all possible orderings (see Eq. 6), thus leads to additive contributions and neglecting the interactions.

An alternative is to analyze different orderings to detect when one variable has different contributions depending on what other variables precede it. This is a sign of interaction. The Break-Down method (see [16, 17]) analyzes the various orders to identify and visualize interactions in the model. The final attributions are determined based on a single ordering which is chosen based on greedy heuristics.

GitHub Repo: https://github.com/ModelOriented/DALEX

Discussion: Techniques such as SHAP generate explanations in the form of additive contributions. However, these techniques are often used in the analysis of complex models, which are often not additive. [26] shows that for many tabular datasets, an additive explanation may be an oversimplification, and it may lead to a false belief that the model behaves in an additive way.

2.12 Shapley Flow

Idea: Like for Asymmetric Shapley Values (ASV), Shapley Flow [84] also allows the use of the dependency structure between variables in the explanation process. As in ASV, the relationship is described by a causal graph. However, unlike ASV and other methods, attribution is assigned not to the nodes (variables) but the edges (relationships between variables). An edge in a graph is significant if its removal would change the predictions of the model (see Fig. 3E). The edge attribution has the additional property that for each explanation, boundaries hold the classical Shapley values. The most extreme explanation boundary corresponds to the ASV method. The Shapley Flow method determines the attributions for each edge in the causal graph.

GitHub Repo: https://github.com/nathanwang000/Shapley-Flow

Discussion: Shapley Flow attribution analysis carries a lot of information about both the structure of the relationship between variables and its effect of particular groups of variables (explanation boundaries) on the predictions. On the rather disadvantageous side is that it requires knowledge of the dependency structure in the form of a directed causal graph, which limits the number of problems in which it can be applied. For readability reasons, it is limited to small numbers of variables. Also it requires definition of a background case, i.e. reference observation. Potential explanations may vary depending on the reference observation chosen.

2.13   Textual Explanations of Visual Models

Idea: The generation of textual descriptions of images is addressed by several machine learning models that contain both a part that processes the input images - typically a convolutional neural network (CNN) - and one that learns an adequate text sequence, usually a recurrent neural network (RNN). Those two parts cooperate for the production of image descriptive sentences that presupposes that a classification task is successfully accomplished. One of the first benchmark datasets that contained image descriptions was already invented in 2014, the Microsoft COCO (MS-COCO) [48]. The models that achieve a good performance classification, first detect components and concepts of the image and then construct sentences where objects, subjects as well as their characteristics are connected by verbs. The problem of semantic enrichment of images for language-related tasks is addressed in a number of ways (see, for example, the Visual Genome project [45]); however, in most cases, such descriptions are not directly tied to visual recognition tasks. Nevertheless, an advantage is that textual descriptions are easier to analyze and validate than attribution maps.

It is important to note that the mere description of the image's content is not equivalent to an explanation of the decision-making process of the neural network model. Unless the produced sentences contain the unique attributes that help differentiate between the images of each class, the content of the words should not be considered class-relevant content. A solution to this problem is proposed in [31]. This method's main goal is to do exactly that; to find those characteristics that are discriminative, since they were used by the neural network models to accomplish the task - those exactly need to be present in the generated text. To achieve this, the training does not just use the relevance loss, which generates descriptions relevant to the predicted class based on context borrowed from a fine-grained image recognition model through conditional probabilities. A discriminative loss is invented to generate sentences rich in class-discriminative features. The introduced weight update procedure consists of two components, one based on the gradient of relevance loss and the second based on the gradient of discriminative loss, so that descriptions that are both relevant to the predicted class and contain words with high discriminative capacity are rewarded. The reinforcement learning method REINFORCE [85] is used for the backpropagation of the error through sampling during the training process.

GitHub Repo: https://github.com/LisaAnne/ECCV2016

Discussion: High METEOR [13] and CIDEr [82] scores for relevant explanations were measured for the generated sentences. It is necessary to compare the resulting explanations with experts since they only know the difference between sentences that correctly describe the visual content and ones that concentrate on what occurs only in the class the images belong. This is positive and negative at the same time; unfortunately, there is no way to check how much of the generated explanation is consistent without domain knowledge. Furthermore, data artefacts can also influence both the performance and explanation quality negatively. Overall though, even ablation studies where parts of the model were tested separately, showed that the components individually had a higher performance than when trained alone. That indicates that the common training of visual processing and textual explanation generation is beneficial for each part individually.

2.14   Integrated Gradients

Idea: The Integrated Gradients method [80] is based on two fundamental axioms, sensitivity and implementation invariance. Sensitivity means that non-zero attributions are given to every input and baseline that differ in one feature but have different predictions. Implementation invariance means that if two models behave identical/are functionally equivalent, then attributions must be identical. Although these two axioms sound very natural, it turns out that many attribution methods do not have these properties. In particular, when a model has flattened predictions for a specific point of interest, the gradient in the point of interest zeroes out and does not carry information useful for the explanation.

The approach proposed by the Integrated Gradients method for model f aggregates the gradients $\frac{\partial f(x)}{\partial x_i}$ computed along the path connecting the point of interest x to the highlighted observation - the baseline x^* (for computer vision this could be a black image and for text an empty sentence).

More formally, for ith feature, Integrated Gradients are defined as

$$IntegratedGrads_i(x) = (x_i - x_i^*) \int_{\alpha=0}^{1} \frac{\partial f(x^* + \alpha(x - x^*))}{\partial x_i} d\alpha.$$

The integral can be replaced by a sum over a set of alpha values in the interval [0,1].

GitHub Repo: https://github.com/ankurtaly/Integrated-Gradients

Discussion: Integrated Gradients is a widespread technique for explaining deep neural networks or other differentiable models. It is a theoretically sound approach based on two desirable properties: sensitivity and implementation invariance. In addition, it is computationally efficient and uses gradient information at a few selected points α. The three main drawbacks are: (1) need for the baseline observation, selection of which significantly influence the attributions, (2) works only for differentiable models, suitable for neural networks but not, e.g., for decision trees, (3) by default, gradients are integrated along the shortest path between the baseline and the point of interest. Depending on the topology of the data, this path does not always make sense and cover the data. Furthermore, deep models usually suffer from the gradient shattering problem [12], which may negative affect the explanation (see discussion in [70]). Extensions to this method are proposed to overcome the above drawbacks.

2.15   Causal Models

Description: In the work of Madumal et al. [53] a structural causal model [29] is learned, which can be considered an extension of Bayesian Models [44,71] of the RL environment with the use of counterfactuals. It takes into account events that would happen or environment states that would be reached under different actions taken by the RL agent. Ultimately, the goal of any RL agent is to maximize a long-term reward; the explanation provides causal chains until the reward receiving state is reached. The researchers pay attention to keep the explanations minimally complete, by removing some of the intermediate nodes in the causal chains, to conform to the explanation satisfaction conditions according to the Likert scale [33]. The counterfactual explanation is computed by comparing causal chain paths of actions not chosen by the agent (according to the trained policy). To keep the explanation as simple as possible, only the differences between the causal chains comprise the returned counterfactual explanation.

GitHub Repo: No Github Repo

Discussion: Model-free reinforcement learning with a relatively small state and action space has the advantage that we can explain how the RL agent takes its decisions in a causal way; since neural networks base their decisions on correlations, this is one of the first works towards causal explanations. The user has the ability to get answers to the questions "why" and "why not" an action was chosen by the agent. The provided explanations are appropriate according to satisfiability, ethics requirements and are personalized to the human mental model by the use of a dedicated user interface. On the rather negative side is that this explanation method is evaluated on problems with very small state and action space (9 and 4 correspondingly). Current RL problems have much larger state and action spaces and the solution can be found with the use of Deep Reinforcement Learning [27,32]. The reason is, that the structural model of the environment dynamics are not known a priori and must be discovered and approximated during exploration. Furthermore, this work applies only to the finite domain, although the authors note that it will be part of their research work to extend it to continuous spaces.

2.16   Meaningful Perturbations

Idea: This approach was proposed by Fong and Vedaldi [21] and can be regarded as model-agnostic, perturbation-based explanation method. Thus, the explanation is computed solely based on the reaction of the model to a perturbed (or occluded) input sample. For a given sample $\mathbf{x}$, the method aims to synthesize a sparse occlusion map (i.e., the explanation) that leads to the maximum drop of the model's prediction $f(\mathbf{x})$, relative do the original prediction with the unperturbed $\mathbf{x}$. Thus, compared to simple occlusion-based techniques which naively perturb a given sample by sequentially occluding parts of it, the Meaningful Perturbation algorithm aims to directly learn the explanation by formulating the explanation problem as a meta-prediction task and using tools from optimization to solve it. Sparsity constraints ensure that the search focuses on finding the smallest possible perturbation mask that has the larger effect on the certainty of the classification performance.

GitHub Repo: https://github.com/ruthcfong/perturb_explanations

Discussion: As other model agnostic approaches, Meaningful Perturbations is a very flexible method, which can be directly applied to any machine learning model. The approach can be also interpreted from a rate-distortion perspective [43]. Since the Meaningful Perturbations method involves optimization, it is computationally much more demanding than propagation-based techniques such as LRP. Also it is well-known that the perturbation process (occlusion or deletion can be seen as a particular type of perturbation), moves the sample out of the manifold of natural images and thus can introduce artifacts. The use of generative models have been suggested to overcome this out-of-manifold problem [1].

2.17   EXplainable Neural-Symbolic Learning (X-NeSyL)

Idea: Symbolic AI is an emerging field that has been shown to contribute immensely to Explainable AI. Neuro-Symbolic methods [24] incorporate prior human knowledge for various tasks such as concept learning and at the same time they produce output that is more interpretable, such as mathematical equations or Domain-Specific Languages (DSL) [55].

A research work that is dedicated to using symbolic knowledge of the domain experts, expressed as a knowledge graph (KG), to align it with the explanations of a neural network is the EXplainable Neural-Symbolic Learning (X-NeSyL) [20]. The researchers start with the goal to encourage the neural network that performs classification to assign feature importances to the object's parts in a way that corresponds to the compositional way humans classify. After using state-of-the-art CNN architectures and applying methods such as SHAP (see Sect. 2.9) to quantify the positive and negative influence of each detected feature, a graph is built that encompasses constraints and relations elicited from the computed importances. This graph is compared to the KG provided by human experts. A designated loss that punishes non-overlap between these two has been shown to boost explainability and in some cases performance.

GitHub Repo: https://github.com/JulesSanchez/X-NeSyL,
https://github.com/JulesSanchez/MonuMAI-AutomaticStyleClassification

Discussion: This method can be seen as an explainability-by-design approach. That means that at each step of the training process, it is made sure that the end result will be interpretable. This is not an ad-hoc method; the training contains a loss to guide the neural network towards explanations that have a human-expert like structure. Furthermore, SHAP values provide intermediate feature relevance results that are straightforward to understand. Disadvantageous is that the same thing that fosters explainability, contributes to the negatives of this approach, namely that it needs domain-specific knowledge. This is not always easy to gather, it may be contradicting if several experts are involved and in that way constraints the network to compute in a specific way. The researchers comment on that particular issue and exercise their method with many datasets, test several CNN architectures and provide performance results with established as well as newly invented methods to see where and how the human-in-the-loop [37] works in practice.

3   Conclusion and Future Outlook

In the future, we expect that the newly invented xAI methods will capture causal dependencies. Therefore, it will be important to measure the quality of explanations so that an xAI method achieves a certain level of causal understanding [65] for a user with effectiveness, efficiency and satisfaction in a given context of use [34]. Successful xAI models in the future will also require new human-AI interfaces [36] that enable contextual understanding and allow a domain expert to ask questions and counterfactuals [35] ("what-if" questions). This is where a human-in-the-loop can (sometimes - not always, of course) bring human experience and conceptual knowledge to AI processes [37]. Such conceptual understanding is something that the best AI algorithms in the world (still) lack, and this is where the international xAI community will make many valuable contributions in the future.

Acknowledgements. The authors declare no conflict of interests. This work does not raise any ethical issues. The content is partly based on the XAI seminar LV 706.315 provided by Andreas Holzinger since 2015 at Graz University of Technology. Parts of this work have been funded by the Austrian Science Fund (FWF), Project: P-32554, explainable AI, by the German Ministry for Education and Research (BMBF) through BIFOLD (refs. 01IS18025A and 01IS18037A), by the European Union's Horizon 2020 programme (grant no. 965221), Project iToBoS, and by the German Research Foundation (DFG), Emmy Noether Grant 437611051.

References

1. Agarwal, C., Nguyen, A.: Explaining image classifiers by removing input features using generative models. In: Ishikawa, H., Liu, C.-L., Pajdla, T., Shi, J. (eds.) ACCV 2020. LNCS, vol. 12627, pp. 101–118. Springer, Cham (2021). https://doi.org/10.1007/978-3-030-69544-6_7
2. Alber, M., et al.: iNNvestigate neural networks! J. Mach. Learn. Res. (JMLR) **20**(93), 1–8 (2019)
3. Ali, A., Schnake, T., Eberle, O., Montavon, G., Müller, K.R., Wolf, L.: XAI for transformers: better explanations through conservative propagation. arXiv preprint arXiv:2202.07304 (2022)
4. Anders, C.J., Neumann, D., Samek, W., Müller, K.R., Lapuschkin, S.: Software for dataset-wide XAI: from local explanations to global insights with Zennit, CoRelAy, and ViRelAy. arXiv preprint arXiv:2106.13200 (2021)
5. Anders, C.J., Weber, L., Neumann, D., Samek, W., Müller, K.R., Lapuschkin, S.: Finding and removing clever HANs: using explanation methods to debug and improve deep models. Inf. Fusion **77**, 261–295 (2022)
6. Arras, L., et al.: Explaining and interpreting LSTMs. In: Samek, W., Montavon, G., Vedaldi, A., Hansen, L.K., Müller, K.-R. (eds.) Explainable AI: Interpreting, Explaining and Visualizing Deep Learning. LNCS (LNAI), vol. 11700, pp. 211–238. Springer, Cham (2019). https://doi.org/10.1007/978-3-030-28954-6_11
7. Arras, L., Montavon, G., Müller, K.R., Samek, W.: Explaining recurrent neural network predictions in sentiment analysis. In: Proceedings of the EMNLP 2017 Workshop on Computational Approaches to Subjectivity, Sentiment & Social Media Analysis (WASSA), pp. 159–168. Association for Computational Linguistics (2017)
8. Arras, L., Osman, A., Samek, W.: CLEVR-XAI: a benchmark dataset for the ground truth evaluation of neural network explanations. Inf. Fusion **81**, 14–40 (2022)
9. Asif, N.A., et al.: Graph neural network: a comprehensive review on Non-Euclidean space. IEEE Access **9**, 60588–60606 (2021)
10. Bach, S., Binder, A., Montavon, G., Klauschen, F., Müller, K.R., Samek, W.: On pixel-wise explanations for non-linear classifier decisions by layer-wise relevance propagation. PLoS ONE **10**(7), e0130140 (2015)
11. Baehrens, D., Schroeter, T., Harmeling, S., Kawanabe, M., Hansen, K., Müller, K.R.: How to explain individual classification decisions. J. Mach. Learn. Res. **11**, 1803–1831 (2010)

12. Balduzzi, D., Frean, M., Leary, L., Lewis, J., Ma, K.W.D., McWilliams, B.: The shattered gradients problem: if ResNets are the answer, then what is the question? In: International Conference on Machine Learning, pp. 342–350. PMLR (2017)
13. Banerjee, S., Lavie, A.: Meteor: an automatic metric for MT evaluation with improved correlation with human judgments. In: Proceedings of the ACL Workshop on Intrinsic and Extrinsic Evaluation Measures for Machine Translation and/or Summarization, pp. 65–72 (2005)
14. Becking, D., Dreyer, M., Samek, W., Müller, K., Lapuschkin, S.: Ecq^x: explainability-driven quantization for low-bit and sparse DNNs. In: Holzinger, A., et al. (eds.) xxAI 2020. LNAI, vol. 13200, pp. 271–296. Springer, Cham (2022)
15. Bengio, Y., Lecun, Y., Hinton, G.: Deep learning for AI. Commun. ACM **64**(7), 58–65 (2021)
16. Biecek, P.: DALEX: explainers for complex predictive models in R. J. Mach. Learn. Res. **19**(84), 1–5 (2018). http://jmlr.org/papers/v19/18-416.html
17. Biecek, P., Burzykowski, T.: Explanatory Model Analysis. Chapman and Hall/CRC, New York (2021). https://pbiecek.github.io/ema/
18. Binder, A., Montavon, G., Lapuschkin, S., Müller, K.-R., Samek, W.: Layer-wise relevance propagation for neural networks with local renormalization layers. In: Villa, A.E.P., Masulli, P., Pons Rivero, A.J. (eds.) ICANN 2016. LNCS, vol. 9887, pp. 63–71. Springer, Cham (2016). https://doi.org/10.1007/978-3-319-44781-0_8
19. Clough, J.R., Oksuz, I., Puyol-Antón, E., Ruijsink, B., King, A.P., Schnabel, J.A.: Global and local interpretability for cardiac MRI classification. In: Shen, D., et al. (eds.) MICCAI 2019. LNCS, vol. 11767, pp. 656–664. Springer, Cham (2019). https://doi.org/10.1007/978-3-030-32251-9_72
20. Díaz-Rodríguez, N., et al.: Explainable neural-symbolic learning (X-NeSyL) methodology to fuse deep learning representations with expert knowledge graphs: the MonuMAI cultural heritage use case. arXiv preprint arXiv:2104.11914 (2021)
21. Fong, R.C., Vedaldi, A.: Interpretable explanations of black boxes by meaningful perturbation. In: Proceedings of the IEEE International Conference on Computer Vision, pp. 3429–3437 (2017)
22. Frye, C., de Mijolla, D., Cowton, L., Stanley, M., Feige, I.: Shapley-based explainability on the data manifold. arXiv preprint arXiv:2006.01272 (2020)
23. Frye, C., Rowat, C., Feige, I.: Asymmetric shapley values: incorporating causal knowledge into model-agnostic explainability. In: Larochelle, H., Ranzato, M., Hadsell, R., Balcan, M.F., Lin, H. (eds.) Advances in Neural Information Processing Systems, vol. 33, pp. 1229–1239 (2020)
24. d'Avila Garcez, A.S., Broda, K.B., Gabbay, D.M.: Neural-Symbolic Learning Systems: Foundations and Applications. Springer, Heidelberg (2012). https://doi.org/10.1007/978-1-4471-0211-3
25. Goodfellow, I., Bengio, Y., Courville, A.: Deep Learning. MIT Press, Cambridge (2016)
26. Gosiewska, A., Biecek, P.: iBreakDown: Uncertainty of Model Explanations for Non-additive Predictive Models. arXiv preprint arXiv:1903.11420 (2019)
27. Graesser, L., Keng, W.L.: Foundations of Deep Reinforcement Learning: Theory and Practice in Python. Addison-Wesley Professional (2019)
28. Gunning, D., Aha, D.W.: Darpa's explainable artificial intelligence program. AI Mag. **40**(2), 44–58 (2019)
29. Halpern, J.Y., Pearl, J.: Causes and explanations: a structural-model approach. Part II: Explanations. Br. J. Philos. Sci. **56**(4), 889–911 (2005)
30. Hedström, A., et al.: Quantus: an explainable AI toolkit for responsible evaluation of neural network explanations. arXiv preprint arXiv:2202.06861 (2022)

31. Hendricks, L.A., Akata, Z., Rohrbach, M., Donahue, J., Schiele, B., Darrell, T.: Generating visual explanations. In: Leibe, B., Matas, J., Sebe, N., Welling, M. (eds.) ECCV 2016. LNCS, vol. 9908, pp. 3–19. Springer, Cham (2016). https://doi.org/10.1007/978-3-319-46493-0_1

32. Hernandez-Leal, P., Kartal, B., Taylor, M.E.: A survey and critique of multiagent deep reinforcement learning. Auton. Agent. Multi-Agent Syst. **33**(6), 750–797 (2019). https://doi.org/10.1007/s10458-019-09421-1

33. Hoffman, R.R., Mueller, S.T., Klein, G., Litman, J.: Metrics for explainable AI: challenges and prospects. arXiv preprint arXiv:1812.04608 (2018)

34. Holzinger, A., Carrington, A., Mueller, H.: Measuring the quality of explanations: the system causability scale (SCS). Comparing human and machine explanations. KI - Künstliche Intelligenz (German Journal of Artificial intelligence), Special Issue on Interactive Machine Learning, Edited by Kristian Kersting, TU Darmstadt **34**(2), 193–198 (2020)

35. Holzinger, A., Malle, B., Saranti, A., Pfeifer, B.: Towards multi-modal causability with graph neural networks enabling information fusion for explainable AI. Inf. Fusion **71**(7), 28–37 (2021)

36. Holzinger, A., Mueller, H.: Toward human-AI interfaces to support explainability and causability in medical AI. IEEE Comput. **54**(10), 78–86 (2021)

37. Holzinger, A., et al.: Interactive machine learning: experimental evidence for the human in the algorithmic loop. Appl. Intell. **49**(7), 2401–2414 (2018). https://doi.org/10.1007/s10489-018-1361-5

38. Huang, Q., Yamada, M., Tian, Y., Singh, D., Yin, D., Chang, Y.: GraphLIME: local interpretable model explanations for graph neural networks. arXiv preprint arXiv:2001.06216v1 (2020)

39. Kauffmann, J., Esders, M., Montavon, G., Samek, W., Müller, K.R.: From clustering to cluster explanations via neural networks. arXiv preprint arXiv:1906.07633 (2019)

40. Kauffmann, J., Müller, K.R., Montavon, G.: Towards explaining anomalies: a deep Taylor decomposition of one-class models. Pattern Recogn. **101**, 107198 (2020)

41. Kim, B., et al.: Interpretability beyond feature attribution: quantitative testing with concept activation vectors (TCAV). In: International Conference on Machine Learning, pp. 2668–2677. PMLR (2018)

42. Kohlbrenner, M., Bauer, A., Nakajima, S., Binder, A., Samek, W., Lapuschkin, S.: Towards best practice in explaining neural network decisions with LRP. In: 2020 International Joint Conference on Neural Networks (IJCNN), pp. 1–7. IEEE (2020)

43. Kole, S., Bruna, J., Kutyniok, G., Levie, R., Nguyen, D.A.: A rate-distortion framework for explaining neural network decisions. In: Holzinger, A., et al. (eds.) xxAI 2020. LNAI, vol. 13200, pp. 91–115. Springer, Cham (2022)

44. Koller, D., Friedman, N.: Probabilistic Graphical Models: Principles and Techniques. MIT Press (2009)

45. Krishna, R., et al.: Visual genome: connecting language and vision using crowdsourced dense image annotations. Int. J. Comput. Vis. **123**(1), 32–73 (2017)

46. Lapuschkin, S., Binder, A., Müller, K.R., Samek, W.: Understanding and comparing deep neural networks for age and gender classification. In: Proceedings of the IEEE International Conference on Computer Vision Workshops (ICCVW), pp. 1629–1638 (2017)

47. Lapuschkin, S., Wäldchen, S., Binder, A., Montavon, G., Samek, W., Müller, K.R.: Unmasking clever HANs predictors and assessing what machines really learn. Nat. Commun. **10**, 1096 (2019)

48. Lin, T.-Y., et al.: Microsoft COCO: common objects in context. In: Fleet, D., Pajdla, T., Schiele, B., Tuytelaars, T. (eds.) ECCV 2014. LNCS, vol. 8693, pp. 740–755. Springer, Cham (2014). https://doi.org/10.1007/978-3-319-10602-1_48

49. Lundberg, S.M., et al.: From local explanations to global understanding with explainable AI for trees. Nat. Mach. Intell. **2**(1), 56–67 (2020)

50. Lundberg, S.M., Erion, G.G., Lee, S.I.: Consistent individualized feature attribution for tree ensembles. arXiv preprint arXiv:1802.03888 (2018)

51. Lundberg, S.M., Lee, S.I.: A unified approach to interpreting model predictions. In: Advances in Neural Information Processing Systems, vol. 30, pp. 4765–4774 (2017)

52. Luo, D., et al.: Parameterized explainer for graph neural network. In: Advances in Neural Information Processing Systems, vol. 33, pp. 19620–19631 (2020)

53. Madumal, P., Miller, T., Sonenberg, L., Vetere, F.: Explainable reinforcement learning through a causal lens. In: Proceedings of the AAAI Conference on Artificial Intelligence, vol. 34, pp. 2493–2500 (2020)

54. Mamalakis, A., Ebert-Uphoff, I., Barnes, E.: Explainable artificial intelligence in meteorology and climate science: Model fine-tuning, calibrating trust and learning new science. In: Holzinger, A., et al. (eds.) xxAI 2020. LNAI, vol. 13200, pp. 315–339. Springer, Cham (2022)

55. Mao, J., Gan, C., Kohli, P., Tenenbaum, J.B., Wu, J.: The neuro-symbolic concept learner: interpreting scenes, words, and sentences from natural supervision. arXiv preprint arXiv:1904.12584 (2019)

56. Mittelstadt, B.: Principles alone cannot guarantee ethical AI. Nat. Mach. Intell. **1**, 1–7 (2019)

57. Molnar, C., et al.: Pitfalls to avoid when interpreting machine learning models. arXiv preprint arXiv:2007.04131 (2020)

58. Montavon, G., Binder, A., Lapuschkin, S., Samek, W., Müller, K.-R.: Layer-wise relevance propagation: an overview. In: Samek, W., Montavon, G., Vedaldi, A., Hansen, L.K., Müller, K.-R. (eds.) Explainable AI: Interpreting, Explaining and Visualizing Deep Learning. LNCS (LNAI), vol. 11700, pp. 193–209. Springer, Cham (2019). https://doi.org/10.1007/978-3-030-28954-6_10

59. Montavon, G., Kauffmann, J., Samek, W., Müller, K.R.: Explaining the predictions of unsupervised learning models. In: Holzinger, A., et al. (eds.) xxAI 2020. LNAI, vol. 13200, pp. 117–138. Springer, Cham (2022)

60. Montavon, G., Lapuschkin, S., Binder, A., Samek, W., Müller, K.R.: Explaining nonlinear classification decisions with deep Taylor decomposition. Pattern Recogn. **65**, 211–222 (2017)

61. Montavon, G., Samek, W., Müller, K.R.: Methods for interpreting and understanding deep neural networks. Digit. Signal Process. **73**, 1–15 (2018)

62. Morch, N.J., et al.: Visualization of neural networks using saliency maps. In: Proceedings of ICNN 1995-International Conference on Neural Networks, vol. 4, pp. 2085–2090 (1995)

63. O'Sullivan, S., et al.: Legal, regulatory, and ethical frameworks for development of standards in artificial intelligence (AI) and autonomous robotic surgery. Int. J. Med. Robot. Comput. Assisted Surg. **15**(1), e1968 (2019)

64. Pearl, J.: The limitations of opaque learning machines. In: Brockman, J. (ed.) Possible Minds: 25 Ways of Looking at AI, pp. 13–19. Penguin, New York (2019)

65. Pearl, J.: The seven tools of causal inference, with reflections on machine learning. Commun. ACM **62**(3), 54–60 (2019)

66. Pearl, J., Mackenzie, D.: The Book of Why. Basic Books, New York (2018)

67. Ribeiro, M.T., Singh, S., Guestrin, C.: Why should I trust you?: explaining the predictions of any classifier. In: 22nd ACM SIGKDD International Conference on Knowledge Discovery and Data Mining (KDD 2016), pp. 1135–1144. ACM (2016)
68. Ribeiro, M.T., Singh, S., Guestrin, C.: Anchors: high-precision model-agnostic explanations. In: Proceedings of the AAAI Conference on Artificial Intelligence, vol. 32, no. 1 (2018)
69. Robnik-Šikonja, M., Kononenko, I.: Explaining classifications for individual instances. IEEE Trans. Knowl. Data Eng. **20**(5), 589–600 (2008)
70. Samek, W., Montavon, G., Lapuschkin, S., Anders, C.J., Müller, K.R.: Explaining deep neural networks and beyond: a review of methods and applications. Proc. IEEE **109**(3), 247–278 (2021)
71. Saranti, A., Taraghi, B., Ebner, M., Holzinger, A.: Insights into learning competence through probabilistic graphical models. In: Holzinger, A., Kieseberg, P., Tjoa, A.M., Weippl, E. (eds.) CD-MAKE 2019. LNCS, vol. 11713, pp. 250–271. Springer, Cham (2019). https://doi.org/10.1007/978-3-030-29726-8_16
72. Schnake, T., et al.: XAI for graphs: explaining graph neural network predictions by identifying relevant walks. arXiv preprint arXiv:2006.03589 (2020)
73. Schneeberger, D., Stöger, K., Holzinger, A.: The European legal framework for medical AI. In: Holzinger, A., Kieseberg, P., Tjoa, A.M., Weippl, E. (eds.) CD-MAKE 2020. LNCS, vol. 12279, pp. 209–226. Springer, Cham (2020). https://doi.org/10.1007/978-3-030-57321-8_12
74. Schoelkopf, B.: Causality for machine learning. arXiv preprint arXiv:1911.10500 (2019)
75. Simonyan, K., Vedaldi, A., Zisserman, A.: Deep inside convolutional networks: visualising image classification models and saliency maps. arXiv preprint arXiv:1312.6034 (2013)
76. Stoeger, K., Schneeberger, D., Kieseberg, P., Holzinger, A.: Legal aspects of data cleansing in medical AI. Comput. Law Secur. Rev. **42**, 105587 (2021)
77. Student: The probable error of a mean. Biometrika, pp. 1–25 (1908)
78. Sturm, I., Lapuschkin, S., Samek, W., Müller, K.R.: Interpretable deep neural networks for single-trial EEG classification. J. Neurosci. Methods **274**, 141–145 (2016)
79. Sun, J., Lapuschkin, S., Samek, W., Binder, A.: Explain and improve: LRP-inference fine tuning for image captioning models. Inf. Fusion **77**, 233–246 (2022)
80. Sundararajan, M., Taly, A., Yan, Q.: Axiomatic attribution for deep networks. In: Proceedings of the 34th International Conference on Machine Learning. Proceedings of Machine Learning Research, vol. 70, pp. 3319–3328. PMLR, 06–11 August 2017
81. Sutton, R.S., Barto, A.G.: Reinforcement Learning: An Introduction. MIT Press, Cambridge (2018)
82. Vedantam, R., Lawrence Zitnick, C., Parikh, D.: Cider: consensus-based image description evaluation. In: Proceedings of the IEEE Conference on Computer Vision and Pattern Recognition, pp. 4566–4575 (2015)
83. Vu, M., Thai, M.T.: PGM-explainer: probabilistic graphical model explanations for graph neural networks. In: Advances in Neural Information Processing Systems, vol. 33, pp. 12225–12235 (2020)
84. Wang, J., Wiens, J., Lundberg, S.: Shapley flow: a graph-based approach to interpreting model predictions. In: 24th International Conference on Artificial Intelligence and Statistics (AISTATS). Proceedings of Machine Learning Research, vol. 130, pp. 721–729. PMLR (2021)

85. Williams, R.J.: Simple statistical gradient-following algorithms for connectionist reinforcement learning. Mach. Learn. **8**(3), 229–256 (1992)
86. Xu, J., Xue, M., Picek, S.: Explainability-based backdoor attacks against graph neural networks. In: Proceedings of the 3rd ACM Workshop on Wireless Security and Machine Learning, pp. 31–36 (2021)
87. Yeom, S.K., et al.: Pruning by explaining: a novel criterion for deep neural network pruning. Pattern Recogn. **115**, 107899 (2021)
88. Yuan, H., Tang, J., Hu, X., Ji, S.: XGNN: towards model-level explanations of graph neural networks. In: Proceedings of the 26th ACM SIGKDD International Conference on Knowledge Discovery & Data Mining, pp. 430–438 (2020)
89. Zeiler, M.D., Fergus, R.: Visualizing and understanding convolutional networks. In: Fleet, D., Pajdla, T., Schiele, B., Tuytelaars, T. (eds.) ECCV 2014. LNCS, vol. 8689, pp. 818–833. Springer, Cham (2014). https://doi.org/10.1007/978-3-319-10590-1_53
90. Zhang, A., Lipton, Z.C., Li, M., Smola, A.J.: Dive into deep learning. Release 0.17.0, Open Source (2021)
91. Zintgraf, L.M., Cohen, T.S., Adel, T., Welling, M.: Visualizing deep neural network decisions: prediction difference analysis. arXiv preprint arXiv:1702.04595 (2017)

General Pitfalls of Model-Agnostic Interpretation Methods for Machine Learning Models

Christoph Molnar[1,7]([⊠]) [iD], Gunnar König[1,4] [iD], Julia Herbinger[1] [iD],
Timo Freiesleben[2,3] [iD], Susanne Dandl[1] [iD], Christian A. Scholbeck[1] [iD],
Giuseppe Casalicchio[1] [iD], Moritz Grosse-Wentrup[4,5,6] [iD], and Bernd Bischl[1] [iD]

[1] Department of Statistics, LMU Munich, Munich, Germany
`christoph.molnar.ai@gmail.com`
[2] Munich Center for Mathematical Philosophy, LMU Munich, Munich, Germany
[3] Graduate School of Systemic Neurosciences, LMU Munich, Munich, Germany
[4] Research Group Neuroinformatics, Faculty for Computer Science,
University of Vienna, Vienna, Austria
[5] Research Platform Data Science @ Uni Vienna, Vienna, Austria
[6] Vienna Cognitive Science Hub, Vienna, Austria
[7] Leibniz Institute for Prevention Research and Epidemiology - BIPS GmbH,
Bremen, Germany

Abstract. An increasing number of model-agnostic interpretation techniques for machine learning (ML) models such as partial dependence plots (PDP), permutation feature importance (PFI) and Shapley values provide insightful model interpretations, but can lead to wrong conclusions if applied incorrectly. We highlight many general pitfalls of ML model interpretation, such as using interpretation techniques in the wrong context, interpreting models that do not generalize well, ignoring feature dependencies, interactions, uncertainty estimates and issues in high-dimensional settings, or making unjustified causal interpretations, and illustrate them with examples. We focus on pitfalls for global methods that describe the average model behavior, but many pitfalls also apply to local methods that explain individual predictions. Our paper addresses ML practitioners by raising awareness of pitfalls and identifying solutions for correct model interpretation, but also addresses ML researchers by discussing open issues for further research.

Keywords: Interpretable machine learning · Explainable AI

This work is funded by the Bavarian State Ministry of Science and the Arts (coordinated by the Bavarian Research Institute for Digital Transformation (bidt)), by the German Federal Ministry of Education and Research (BMBF) under Grant No. 01IS18036A, by the German Research Foundation (DFG), Emmy Noether Grant 437611051, and by the Graduate School of Systemic Neurosciences (GSN) Munich. The authors of this work take full responsibilities for its content.

A. Holzinger et al. (Eds.): xxAI 2020, LNAI 13200, pp. 39–68, 2022.
https://doi.org/10.1007/978-3-031-04083-2_4

1 Introduction

In recent years, both industry and academia have increasingly shifted away from parametric models, such as generalized linear models, and towards non-parametric and non-linear machine learning (ML) models such as random forests, gradient boosting, or neural networks. The major driving force behind this development has been a considerable outperformance of ML over traditional models on many prediction tasks [32]. In part, this is because most ML models handle interactions and non-linear effects automatically. While classical statistical models – such as generalized additive models (GAMs) – also support the inclusion of interactions and non-linear effects, they come with the increased cost of having to (manually) specify and evaluate these modeling options. The benefits of many ML models are partly offset by their lack of interpretability, which is of major importance in many applications. For certain model classes (e.g. linear models), feature effects or importance scores can be directly inferred from the learned parameters and the model structure. In contrast, it is more difficult to extract such information from complex non-linear ML models that, for instance, do not have intelligible parameters and are hence often considered black boxes. However, model-agnostic interpretation methods allow us to harness the predictive power of ML models while gaining insights into the black-box model. These interpretation methods are already applied in many different fields. Applications of interpretable machine learning (IML) include understanding pre-evacuation decision-making [124] with partial dependence plots [36], inferring behavior from smartphone usage [105, 106] with the help of permutation feature importance [107] and accumulated local effect plots [3], or understanding the relation between critical illness and health records [70] using Shapley additive explanations (SHAP) [78]. Given the widespread application of interpretable machine learning, it is crucial to highlight potential pitfalls, that, in the worst case, can produce incorrect conclusions.

This paper focuses on pitfalls for model-agnostic IML methods, i.e. methods that can be applied to any predictive model. Model-specific methods, in contrast, are tied to a certain model class (e.g. saliency maps [57] for gradient-based models, such as neural networks), and are mainly considered out-of-scope for this work. We focus on pitfalls for global interpretation methods, which describe the expected behavior of the entire model with respect to the whole data distribution. However, many of the pitfalls also apply to local explanation methods, which explain individual predictions or classifications. Global methods include the partial dependence plot (PDP) [36], partial importance (PI) [19], accumulated local affects (ALE) [3], or the permutation feature importance (PFI) [12, 19, 33]. Local methods include the individual conditional expectation (ICE) curves [38], individual conditional importance (ICI) [19], local interpretable model-agnostic explanations (LIME) [94], Shapley values [108] and SHapley Additive exPlanations (SHAP) [77, 78] or counterfactual explanations [26, 115]. Furthermore, we distinguish between feature effect and feature importance methods. A feature effect indicates the direction and magnitude of a change in predicted outcome due to changes in feature values. Effect methods include

		Local	Global
Feature	**Effects**	ICE LIME Counterfactuals Shapley Values SHAP	PDP ALE
	Importance	ICI	PI PFI SAGE

Fig. 1. Selection of popular model-agnostic interpretation techniques, classified as local or global, and as effect or importance methods.

Shapley values, SHAP, LIME, ICE, PDP, or ALE. Feature importance methods quantify the contribution of a feature to the model performance (e.g. via a loss function) or to the variance of the prediction function. Importance methods include the PFI, ICI, PI, or SAGE. See Fig. 1 for a visual summary.

The interpretation of ML models can have subtle pitfalls. Since many of the interpretation methods work by similar principles of manipulating data and "probing" the model [100], they also share many pitfalls. The sources of these pitfalls can be broadly divided into three categories: (1) application of an unsuitable ML model which does not reflect the underlying data generating process very well, (2) inherent limitations of the applied IML method, and (3) wrong application of an IML method. Typical pitfalls for (1) are bad model generalization or the unnecessary use of complex ML models. Applying an IML method in a wrong way (3) often results from the users' lack of knowledge of the inherent limitations of the chosen IML method (2). For example, if feature dependencies and interactions are present, potential extrapolations might lead to misleading interpretations for perturbation-based IML methods (inherent limitation). In such cases, methods like PFI might be a wrong choice to quantify feature importance.

Table 1. Categorization of the pitfalls by source.

Sources of pitfall	Sections
Unsuitable ML model	3, 4
Limitation of IML method	5.1, 6.1, 6.2, 9.1, 9.2
Wrong application of IML method	2, 5.2, 5.3, 7, 8, 9.3, 10

Contributions: We uncover and review general pitfalls of model-agnostic interpretation techniques. The categorization of these pitfalls into different sources is provided in Table 1. Each section describes and illustrates a pitfall, reviews possible solutions for practitioners to circumvent the pitfall, and discusses open issues that require further research. The pitfalls are accompanied by illustrative examples for which the code can be found in this repository: https://github.com/ compstat-lmu/code_pitfalls_iml.git. In addition to reproducing our examples, we invite readers to use this code as a starting point for their own experiments and explorations.

Related Work: Rudin et al. [96] present principles for interpretability and discuss challenges for model interpretation with a focus on inherently interpretable models. Das et al. [27] survey methods for explainable AI and discuss challenges with a focus on saliency maps for neural networks. A general warning about using and explaining ML models for high stakes decisions has been brought forward by Rudin [95], in which the author argues against model-agnostic techniques in favor of inherently interpretable models. Krishnan [64] criticizes the general conceptual foundation of interpretability, but does not dispute the usefulness of available methods. Likewise, Lipton [73] criticizes interpretable ML for its lack of causal conclusions, trust, and insights, but the author does not discuss any pitfalls in detail. Specific pitfalls due to dependent features are discussed by Hooker [54] for PDPs and functional ANOVA as well as by Hooker and Mentch [55] for feature importance computations. Hall [47] discusses recommendations for the application of particular interpretation methods but does not address general pitfalls.

2   Assuming One-Fits-All Interpretability

Pitfall: Assuming that a single IML method fits in all interpretation contexts can lead to dangerous misinterpretation. IML methods condense the complexity of ML models into human-intelligible descriptions that only provide insight into specific aspects of the model and data. The vast number of interpretation methods make it difficult for practitioners to choose an interpretation method that can answer their question. Due to the wide range of goals that are pursued under the umbrella term "interpretability", the methods differ in which aspects of the model and data they describe.

For example, there are several ways to quantify or rank the features according to their relevance. The relevance measured by PFI can be very different from the relevance measured by the SHAP importance. If a practitioner aims to gain insight into the relevance of a feature regarding the model's generalization error, a loss-based method (on unseen test data) such as PFI should be used. If we aim to expose which features the model relies on for its prediction or classification – irrespective of whether they aid the model's generalization performance – PFI on test data is misleading. In such scenarios, one should quantify the relevance of a feature regarding the model's prediction (and not the model's generalization error) using methods like the SHAP importance [76].

We illustrate the difference in Fig. 2. We simulated a data-generating process where the target is completely independent of all features. Hence, the features are just noise and should not contribute to the model's generalization error. Consequently, the features are not considered relevant by PFI on test data.

However, the model mechanistically relies on a number of spuriously correlated features. This reliance is exposed by marginal global SHAP importance.

As the example demonstrates, it would be misleading to view the PFI computed on test data or global SHAP as one-fits-all feature importance techniques. Like any IML method, they can only provide insight into certain aspects of model and data.

Many pitfalls in this paper arise from situations where an IML method that was designed for one purpose is applied in an unsuitable context. For example, extrapolation (Sect. 5.1) can be problematic when we aim to study how the model behaves under realistic data but simultaneously can be the correct choice if we want to study the sensitivity to a feature outside the data distribution.

For some IML techniques – especially local methods – even the same method can provide very different explanations, depending on the choice of hyperparameters: For counterfactuals, explanation goals are encoded in their optimization metrics [26,34] such as sparsity and data faithfulness; The scope and meaning of LIME explanations depend on the kernel width and the notion of complexity [8,37].

Solution: The suitability of an IML method cannot be evaluated with respect to one-fits-all interpretability but must be motivated and assessed with respect to well-defined interpretation goals. Similarly, practitioners must tailor the choice of the IML method and its respective hyperparameters to the interpretation context. This implies that these goals need to be clearly stated in a detailed manner *before* any analysis – which is still often not the case.

Open Issues: Since IML methods themselves are subject to interpretation, practitioners must be informed about which conclusions can or cannot be drawn given different choices of IML technique. In general, there are three aspects to be considered: (a) an intuitively understandable and plausible algorithmic construction of the IML method to achieve an explanation; (b) a clear mathematical axiomatization of interpretation goals and properties, which are linked by proofs and theoretical considerations to IML methods, and properties of models and data characteristics; (c) a practical translation for practitioners of the axioms from (b) in terms of what an IML method provides and what not, ideally with implementable guidelines and diagnostic checks for violated assumptions to guarantee correct interpretations. While (a) is nearly always given for any published method, much work remains for (b) and (c).

3   Bad Model Generalization

Pitfall: Under- or overfitting models can result in misleading interpretations with respect to the true feature effects and importance scores, as the model does not match the underlying data-generating process well [39]. Formally, most IML methods are designed to interpret the model instead of drawing inferences about

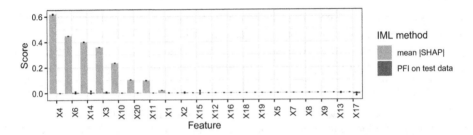

Fig. 2. Assuming one-fits-all interpretability. A default `xgboost` regression model that minimizes the mean squared error (MSE) was fitted on 20 independently and uniformly distributed features to predict another independent, uniformly sampled target. In this setting, predicting the (unconditional) mean $\mathbb{E}[Y]$ in a constant model is optimal. The learner overfits due to a small training data size. Mean marginal SHAP (red, error bars indicate 0.05 and 0.95 quantiles) exposes all mechanistically used features. In contrast, PFI on test data (blue, error bars indicate 0.05 and 0.95 quantiles) considers all features to be irrelevant, since no feature contributes to the generalization performance.

the data-generating process. In practice, however, the latter is often the goal of the analysis, and then an interpretation can only be as good as its underlying model. If a model approximates the data-generating process well enough, its interpretation should reveal insights into the underlying process.

Solution: In-sample evaluation (i.e. on training data) should not be used to assess the performance of ML models due to the risk of overfitting on the training data, which will lead to overly optimistic performance estimates. We must resort to out-of-sample validation based on resampling procedures such as hold-out for larger datasets or cross-validation, or even repeated cross-validation for small sample size scenarios. These resampling procedures are readily available in software [67,89], and well-studied in theory as well as practice [4,11,104], although rigorous analysis of cross-validation is still considered an open problem [103]. Nested resampling is necessary, when computational model selection and hyperparameter tuning are involved [10]. This is important, as the Bayes error for most practical situations is unknown, and we cannot make absolute statements about whether a model already optimally fits the data.

Figure 3 shows the mean squared errors for a simulated example on both training and test data for a support vector machine (SVM), a random forest, and a linear model. Additionally, PDPs for all models are displayed, which show to what extent each model's effect estimates deviate from the ground truth. The linear model is unable to represent the non-linear relationship, which is reflected in a high error on both test and training data and the linear PDPs. In contrast, the random forest has a low training error but a much higher test error, which indicates overfitting. Also, the PDPs for the random forest display overfitting behavior, as the curves are quite noisy, especially at the lower and upper value

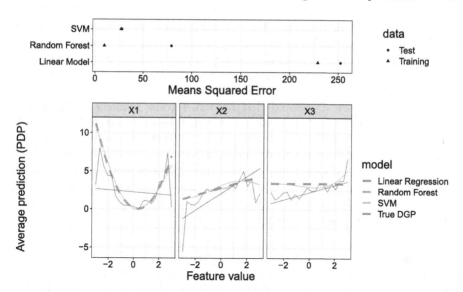

Fig. 3. Bad model generalization. Top: Performance estimates on training and test data for a linear regression model (underfitting), a random forest (overfitting) and a support vector machine with radial basis kernel (good fit). The three features are drawn from a uniform distribution, and the target was generated as $Y = X_1^2 + X_2 - 5X_1X_2 + \epsilon$, with $\epsilon \sim N(0,5)$.**Bottom:** PDPs for the data-generating process (DGP) – which is the ground truth – and for the three models.

ranges of each feature. The SVM with both low training and test error comes closest to the true PDPs.

4   Unnecessary Use of Complex Models

Pitfall: A common mistake is to use an opaque, complex ML model when an interpretable model would have been sufficient, i.e. when the performance of interpretable models is only negligibly worse – or maybe the same or even better – than that of the ML model. Although model-agnostic methods can shed light on the behavior of complex ML models, inherently interpretable models still offer a higher degree of transparency [95] and considering them increases the chance of discovering the true data-generating function [23]. What constitutes an interpretable model is highly dependent on the situation and target audience, as even a linear model might be difficult to interpret when many features and interactions are involved.

It is commonly believed that complex ML models always outperform more interpretable models in terms of accuracy and should thus be preferred. However, there are several examples where interpretable models have proven to be serious competitors: More than 15 years ago, Hand [49] demonstrated that simple models often achieve more than 90% of the predictive power of potentially highly complex models across the UCI benchmark data repository and concluded that such models often should be preferred due to their inherent interpretability; Makridakis et al. [79] systematically compared various ML models (including long-short-term-memory models and multi-layer neural networks) to statistical models (e.g. damped exponential smoothing and the Theta method) in time series forecasting tasks and found that the latter consistently show greater predictive accuracy; Kuhle et al. [65] found that random forests, gradient boosting and neural networks did not outperform logistic regression in predicting fetal growth abnormalities; Similarly, Wu et al. [120] have shown that a logistic regression model performs as well as AdaBoost and even better than an SVM in predicting heart disease from electronic health record data; Baesens et al. [7] showed that simple interpretable classifiers perform competitively for credit scoring, and in an update to the study the authors note that "the complexity and/or recency of a classifier are misleading indicators of its prediction performance" [71].

Solution: We recommend starting with simple, interpretable models such as linear regression models and decision trees. Generalized additive models (GAM) [50] can serve as a gradual transition between simple linear models and more complex machine learning models. GAMs have the desirable property that they can additively model smooth, non-linear effects and provide PDPs out-of-the-box, but without the potential pitfall of masking interactions (see Sect. 6). The additive model structure of a GAM is specified before fitting the model so that only the pre-specified feature or interaction effects are estimated. Interactions between features can be added manually or algorithmically (e.g. via a forward greedy search) [18]. GAMs can be fitted with component-wise boosting [99]. The boosting approach allows to smoothly increase model complexity, from sparse linear models to more complex GAMs with non-linear effects and interactions. This smooth transition provides insight into the tradeoffs between model simplicity and performance gains. Furthermore, component-wise boosting has an in-built feature selection mechanism as the model is build incrementally, which is especially useful in high-dimensional settings (see Sect. 9.1). The predictive performance of models of different complexity should be carefully measured and compared. Complex models should only be favored if the additional performance gain is both significant and relevant – a judgment call that the practitioner must ultimately make. Starting with simple models is considered best practice in data science, independent of the question of interpretability [23]. The comparison of predictive performance between model classes of different complexity can add further insights for interpretation.

Open Issues: Measures of model complexity allow quantifying the trade-off between complexity and performance and to automatically optimize for multiple objectives beyond performance. Some steps have been made towards quantifying model complexity, such as using functional decomposition and quantifying the complexity of the components [82] or measuring the stability of predictions [92]. However, further research is required, as there is no single perfect definition of interpretability, but rather multiple depending on the context [30, 95].

5   Ignoring Feature Dependence

5.1   Interpretation with Extrapolation

Pitfall: When features are dependent, perturbation-based IML methods such as PFI, PDP, LIME, and Shapley values extrapolate in areas where the model was trained with little or no training data, which can cause misleading interpretations [55]. This is especially true if the ML model relies on feature interactions [45] – which is often the case. Perturbations produce artificial data points that are used for model predictions, which in turn are aggregated to produce global or local interpretations [100]. Feature values can be perturbed by replacing original values with values from an equidistant grid of that feature, with permuted or randomly subsampled values [19], or with quantiles. We highlight two major issues: First, if features are dependent, all three perturbation approaches produce unrealistic data points, i.e. the new data points are located outside of the multivariate joint distribution of the data (see Fig. 4). Second, even if features are independent, using an equidistant grid can produce unrealistic values for the feature of interest. Consider a feature that follows a skewed distribution with outliers. An equidistant grid would generate many values between outliers and non-outliers. In contrast to the grid-based approach, the other two approaches maintain the marginal distribution of the feature of interest.

Both issues can result in misleading interpretations (illustrative examples are given in [55,84]), since the model is evaluated in areas of the feature space with few or no observed real data points, where model uncertainty can be expected to be very high. This issue is aggravated if interpretation methods integrate over such points with the same weight and confidence as for much more realistic samples with high model confidence.

Solution: Before applying interpretation methods, practitioners should check for dependencies between features in the data, e.g. via descriptive statistics or measures of dependence (see Sect. 5.2). When it is unavoidable to include dependent features in the model (which is usually the case in ML scenarios), additional information regarding the strength and shape of the dependence structure should be provided. Sometimes, alternative interpretation methods can be used as a workaround or to provide additional information. Accumulated local effect plots (ALE) [3] can be applied when features are dependent, but can produce non-intuitive effect plots for simple linear models with interactions [45]. For other methods such as the PFI, conditional variants exist [17,84,107]. In the case of LIME, it was suggested to focus in sampling on realistic (i.e. close to the data manifold) [97] and relevant areas (e.g. close to the decision boundary) [69]. Note, however, that conditional interpretations are often different and should not be used as a substitute for unconditional interpretations (see Sect. 5.3). Furthermore, dependent features should not be interpreted separately but rather jointly. This can be achieved by visualizing e.g. a 2-dimensional ALE plot of two dependent features, which, admittedly, only works for very low-dimensional combinations. Especially in high-dimensional settings where dependent features

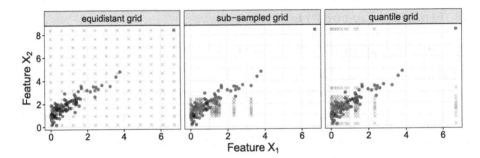

Fig. 4. Interpretation with extrapolation. Illustration of artificial data points generated by three different perturbation approaches. The black dots refer to observed data points and the red crosses to the artificial data points.

can be grouped in a meaningful way, grouped interpretation methods might be more reasonable (see Sect. 9.1).

We recommend using quantiles or randomly subsampled values over equidistant grids. By default, many implementations of interpretability methods use an equidistant grid to perturb feature values [41,81,89], although some also allow using user-defined values.

Open Issues: A comprehensive comparison of strategies addressing extrapolation and how they affect an interpretation method is currently missing. This also includes studying interpretation methods and their conditional variants when they are applied to data with different dependence structures.

5.2   Confusing Linear Correlation with General Dependence

Pitfall: Features with a Pearson correlation coefficient (PCC) close to zero can still be dependent and cause misleading model interpretations (see Fig. 5). While independence between two features implies that the PCC is zero, the converse is generally false. The PCC, which is often used to analyze dependence, only tracks linear correlations and has other shortcomings such as sensitivity to outliers [113]. Any type of dependence between features can have a strong impact on the interpretation of the results of IML methods (see Sect. 5.1). Thus, knowledge about the (possibly non-linear) dependencies between features is crucial for an informed use of IML methods.

Solution: Low-dimensional data can be visualized to detect dependence (e.g. scatter plots) [80]. For high-dimensional data, several other measures of dependence in addition to PCC can be used. If dependence is monotonic, Spearman's rank correlation coefficient [72] can be a simple, robust alternative to PCC. For categorical or mixed features, separate dependence measures have been proposed, such as Kendall's rank correlation coefficient for ordinal features, or the phi coefficient and Goodman & Kruskal's lambda for nominal features [59].

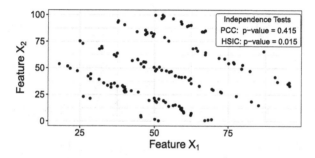

Fig. 5. Confusing linear correlation with dependence. Highly dependent features X_1 and X_2 that have a correlation close to zero. A test (H_0: Features are independent) using Pearson correlation is not significant, but for HSIC, the H_0-hypothesis gets rejected. Data from [80].

Studying non-linear dependencies is more difficult since a vast variety of possible associations have to be checked. Nevertheless, several non-linear association measures with sound statistical properties exist. Kernel-based measures, such as kernel canonical correlation analysis (KCCA) [6] or the Hilbert-Schmidt independence criterion (HSIC) [44], are commonly used. They have a solid theoretical foundation, are computationally feasible, and robust [113]. In addition, there are information-theoretical measures, such as (conditional) mutual information [24] or the maximal information coefficient (MIC) [93], that can however be difficult to estimate [9,116]. Other important measures are e.g. the distance correlation [111], the randomized dependence coefficient (RDC) [74], or the alternating conditional expectations (ACE) algorithm [14]. In addition to using PCC, we recommend using at least one measure that detects non-linear dependencies (e.g. HSIC).

5.3 Misunderstanding Conditional Interpretation

Pitfall: Conditional variants of interpretation techniques avoid extrapolation but require a different interpretation. Interpretation methods that perturb features independently of others will extrapolate under dependent features but provide insight into the model's mechanism [56,61]. Therefore, these methods are said to be true to the model but not true to the data [21].

For feature effect methods such as the PDP, the plot can be interpreted as the isolated, average effect the feature has on the prediction. For the PFI, the importance can be interpreted as the drop in performance when the feature's information is "destroyed" (by perturbing it). Marginal SHAP value functions [78] quantify a feature's contribution to a specific prediction, and marginal SAGE value functions [25] quantify a feature's contribution to the overall prediction performance. All the aforementioned methods extrapolate under dependent features (see also Sect. 5.1), but satisfy sensitivity, i.e. are zero if a feature is not used by the model [25,56,61,110].

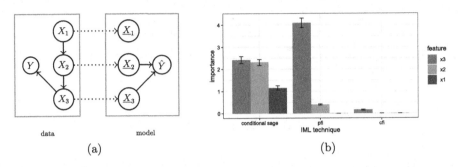

(a)   (b)

Fig. 6. Misunderstanding conditional interpretation. A linear model was fitted on the data-generating process modeled using a linear Gaussian structural causal model. The entailed directed acyclic graph is depicted on the left. For illustrative purposes, the original model coefficients were updated such that not only feature X_3, but also feature X_2 is used by the model. PFI on test data considers both X_3 and X_2 to be relevant. In contrast, conditional feature importance variants either only consider X_3 to be relevant (CFI) or consider all features to be relevant (conditional SAGE value function).

Conditional variants of these interpretation methods do not replace feature values independently of other features, but in such a way that they conform to the conditional distribution. This changes the interpretation as the effects of all dependent features become entangled. Depending on the method, conditional sampling leads to a more or less restrictive notion of relevance.

For example, for dependent features, the Conditional Feature Importance (CFI) [17,84,107,117] answers the question: "How much does the model performance drop if we permute a feature, *but given that we know the values of the other features?*" [63,84,107].[1] Two highly dependent features might be individually important (based on the unconditional PFI), but have a very low conditional importance score because the information of one feature is contained in the other and vice versa.

In contrast, the conditional variant of PDP, called marginal plot or M-plot [3], violates sensitivity, i.e. may even show an effect for features that are not used by the model. This is because for M-plots, the feature of interest is not sampled conditionally on the remaining features, but rather the remaining features are sampled conditionally on the feature of interest. As a consequence, the distribution of dependent covariates varies with the value of the feature of interest. Similarly, conditional SAGE and conditional SHAP value functions sample the remaining features conditional on the feature of interest and therefore violate sensitivity [25,56,61,109].

We demonstrate the difference between PFI, CFI, and conditional SAGE value functions on a simulated example (Fig. 6) where the data-generating mech-

---

[1] While for CFI the conditional independence of the feature of interest X_j with the target Y given the remaining features X_{-j} ($Y \perp X_j | X_{-j}$) is already a sufficient condition for zero importance, the corresponding PFI may still be nonzero [63].

anism is known. While PFI only considers features to be relevant if they are actually used by the model, SAGE value functions may also consider a feature to be important that is not directly used by the model if it contains information that the model exploits. CFI only considers a feature to be relevant if it is both mechanistically used by the model and contributes unique information about Y.

Solution: When features are highly dependent and conditional effects and importance scores are used, the practitioner must be aware of the distinct interpretation. Recent work formalizes the implications of marginal and conditional interpretation techniques [21,25,56,61,63]. While marginal methods provide insight into the model's mechanism but are not true to the data, their conditional variants are not true to the model but provide insight into the associations in the data.

If joint insight into model and data is required, designated methods must be used. ALE plots [3] provide interval-wise unconditional interpretations that are true to the data. They have been criticized to produce non-intuitive results for certain data-generating mechanisms [45]. Molnar et al. [84] propose a subgroup-based conditional sampling technique that allows for group-wise marginal interpretations that are true to model and data and that can be applied to feature importance and feature effects methods such as conditional PDPs and CFI. For feature importance, the DEDACT framework [61] allows to decompose conditional importance measures such as SAGE value functions into their marginal contributions and vice versa, thereby allowing global insight into both: the sources of prediction-relevant information in the data as well as into the feature pathways by which the information enters the model.

Open Issues: The quality of conditional IML techniques depends on the goodness of the conditional sampler. Especially in continuous, high-dimensional settings, conditional sampling is challenging. More research on the robustness of interpretation techniques regarding the quality of the sample is required.

6   Misleading Interpretations Due to Feature Interactions

6.1   Misleading Feature Effects Due to Aggregation

Pitfall: Global interpretation methods, such as PDP or ALE plots, visualize the average effect of a feature on a model's prediction. However, they can produce misleading interpretations when features interact. Figure 7 A and B show the marginal effect of features X_1 and X_2 of the below-stated simulation example. While the PDP of the non-interacting feature X_1 seems to capture the true underlying effect of X_1 on the target quite well (A), the global aggregated effect of the interacting feature X_2 (B) shows almost no influence on the target, although an effect is clearly there by construction.

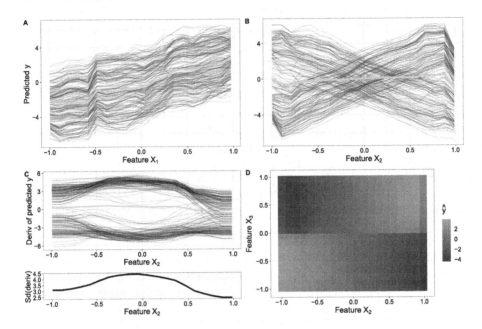

Fig. 7. Misleading effect due to interactions. Simulation example with interactions: $Y = 3X_1 - 6X_2 + 12X_2 \mathbb{1}_{(X_3 \geq 0)} + \epsilon$ with $X_1, X_2, X_3 \overset{i.i.d.}{\sim} U[-1, 1]$ and $\epsilon \overset{i.i.d.}{\sim} N(0, 0.3)$. A random forest with 500 trees is fitted on 1000 observations. Effects are calculated on 200 randomly sampled (training) observations. **A, B:** PDP (yellow) and ICE curves of X_1 and X_2; **C:** Derivative ICE curves and their standard deviation of X_2; **D:** 2-dimensional PDP of X_2 and X_3.

Solution: For the PDP, we recommend to additionally consider the corresponding ICE curves [38]. While PDP and ALE average out interaction effects, ICE curves directly show the heterogeneity between individual predictions. Figure 7 A illustrates that the individual marginal effect curves all follow an upward trend with only small variations. Hence, by aggregating these ICE curves to a global marginal effect curve such as the PDP, we do not lose much information. However, when the regarded feature interacts with other features, such as feature X_2 with feature X_3 in this example, then marginal effect curves of different observations might not show similar effects on the target. Hence, ICE curves become very heterogeneous, as shown in Fig. 7 B. In this case, the influence of feature X_2 is not well represented by the global average marginal effect. Particularly for continuous interactions where ICE curves start at different intercepts, we recommend the use of derivative or centered ICE curves, which eliminate differences in intercepts and leave only differences due to interactions [38]. Derivative ICE curves also point out the regions of highest interaction with other features. For example, Fig. 7 C indicates that predictions for X_2 taking values close to 0 strongly depend on other features' values. While these methods show that interactions are present with regards to the feature of interest but do not reveal other features with which it interacts, the 2-dimensional PDP or ALE plot are options to visualize 2-way interaction effects. The 2-dimensional PDP in Fig. 7 D shows that predictions with regards to feature X_2 highly depend on the feature values of feature X_3.

Other methods that aim to gain more insights into these visualizations are based on clustering homogeneous ICE curves, such as visual interaction effects (VINE) [16] or [122]. As an example, in Fig. 7 B, it would be more meaningful to average over the upward and downward proceeding ICE curves separately and hence show that the average influence of feature X_2 on the target depends on an interacting feature (here: X_3). Work by Zon et al. [125] followed a similar idea by proposing an interactive visualization tool to group Shapley values with regards to interacting features that need to be defined by the user.

Open Issues: The introduced visualization methods are not able to illustrate the type of the underlying interaction and most of them are also not applicable to higher-order interactions.

6.2 Failing to Separate Main from Interaction Effects

Pitfall: Many interpretation methods that quantify a feature's importance or effect cannot separate an interaction from main effects. The PFI, for example, includes both the importance of a feature and the importance of all its interactions with other features [19]. Also local explanation methods such as LIME and Shapley values only provide additive explanations without separation of main effects and interactions [40].

Solution: Functional ANOVA introduced by [53] is probably the most popular approach to decompose the joint distribution into main and interaction effects. Using the same idea, the H-Statistic [35] quantifies the interaction strength between two features or between one feature and all others by decomposing the 2-dimensional PDP into its univariate components. The H-Statistic is based on the fact that, in the case of non-interacting features, the 2-dimensional partial dependence function equals the sum of the two underlying univariate partial dependence functions. Another similar interaction score based on partial dependencies is defined by [42]. Instead of decomposing the partial dependence function, [87] uses the predictive performance to measure interaction strength. Based on Shapley values, Lundberg et al. [77] proposed SHAP interaction values, and Casalicchio et al. [19] proposed a fair attribution of the importance of interactions to the individual features.

Furthermore, Hooker [54] considers dependent features and decomposes the predictions in main and interaction effects. A way to identify higher-order interactions is shown in [53].

Open Issues: Most methods that quantify interactions are not able to identify higher-order interactions and interactions of dependent features. Furthermore, the presented solutions usually lack automatic detection and ranking of all inter-actions of a model. Identifying a suitable shape or form of the modeled inter-action is not straightforward as interactions can be very different and complex, e.g., they can be a simple product of features (multiplicative interaction) or can have a complex joint non-linear effect such as smooth spline surface.

7   Ignoring Model and Approximation Uncertainty

Pitfall: Many interpretation methods only provide a mean estimate but do not quantify uncertainty. Both the model training and the computation of interpre-tation are subject to uncertainty. The model is trained on (random) data, and therefore should be regarded as a random variable. Similarly, LIME's surrogate model relies on perturbed and reweighted samples of the data to approximate the prediction function locally [94]. Other interpretation methods are often defined in terms of expectations over the data (PFI, PDP, Shapley values, ...), but are approximated using Monte Carlo integration. Ignoring uncertainty can result in the interpretation of noise and non-robust results. The true effect of a feature may be flat, but – purely by chance, especially on smaller datasets – the Shap-ley value might show an effect. This effect could cancel out once averaged over multiple model fits.

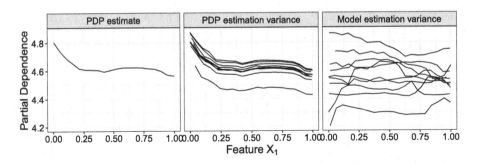

Fig. 8. Ignoring model and approximation uncertainty. PDP for X_1 with $Y = 0 \cdot X_1 + \sum_{j=2}^{10} X_j + \epsilon_i$ with $X_1, \ldots, X_{10} \sim U[0,1]$ and $\epsilon_i \sim N(0, 0.9)$. **Left:** PDP for X_1 of a random forest trained on 100 data points. **Middle:** Multiple PDPs (10x) for the model from left plots, but with different samples (each n=100) for PDP estimation. **Right:** Repeated (10x) data samples of n=100 and newly fitted random forest.

Figure 8 shows that a single PDP (first plot) can be misleading because it does not show the variance due to PDP estimation (second plot) and model fitting (third plot). If we are not interested in learning about a specific model, but rather about the relationship between feature X_1 and the target (in this case), we should consider the model variance.

Solution: By repeatedly computing PDP and PFI with a given model, but with different permutations or bootstrap samples, the uncertainty of the estimate can be quantified, for example in the form of confidence intervals. For PFI, frameworks for confidence intervals and hypothesis tests exist [2,117], but they assume a fixed model. If the practitioner wants to condition the analysis on the modeling process and capture the process' variance instead of conditioning on a fixed model, PDP and PFI should be computed on multiple model fits [83].

Open Issues: While Moosbauer et al. [85] derived confidence bands for PDPs for probabilistic ML models that cover the model's uncertainty, a general model-agnostic uncertainty measure for feature effect methods such as ALE [3] and PDP [36] has (to the best of our knowledge) not been introduced yet.

8 Ignoring the Rashomon Effect

Pitfall: Sometimes different models explain the data-generating process equally well, but contradict each other. This phenomenon is called the Rashomon effect, named after the movie "Rashomon" from the year 1950. Breiman formalized it for predictive models in 2001 [13]: Different prediction models might perform equally well (Rashomon set), but construct the prediction function in a different way (e.g. relying on different features). This can result in conflicting interpretations and conclusions about the data. Even small differences in the training data can cause one model to be preferred over another.

For example, Dong and Rudin [29] identified a Rashomon set of equally well performing models for the COMPAS dataset. They showed that the models differed greatly in the importance they put on certain features. Specifically, if criminal history was identified as less important, race was more important and vice versa. Cherry-picking one model and its underlying explanation might not be sufficient to draw conclusions about the data-generating process. As Hancox-Li [48] states "just because race happens to be an unimportant variable in that one explanation does not mean that it is objectively an unimportant variable".

The Rashomon effect can also occur at the level of the interpretation method itself. Differing hyperparameters or interpretation goals can be one reason (see Sect. 2). But even if the hyperparameters are fixed, we could still obtain contradicting explanations by an interpretation method, e.g., due to a different data sample or initial seed.

A concrete example of the Rashomon effect is counterfactual explanations. Different counterfactuals may all alter the prediction in the desired way, but point to different feature changes required for that change. If a person is deemed uncreditworthy, one corresponding counterfactual explaining this decision may point to a scenario in which the person had asked for a shorter loan duration and amount, while another counterfactual may point to a scenario in which the person had a higher income and more stable job. Focusing on only one counterfactual explanation in such cases strongly limits the possible epistemic access.

Solution: If multiple, equally good models exist, their interpretations should be compared. Variable importance clouds [29] is a method for exploring variable importance scores for equally good models within one model class. If the interpretations are in conflict, conclusions must be drawn carefully. Domain experts or further constraints (e.g. fairness or sparsity) could help to pick a suitable model. Semenova et al. [102] also hypothesized that a large Rashomon set could contain simpler or more interpretable models, which should be preferred according to Sect. 4.

In the case of counterfactual explanations, multiple, equally good explanations exist. Here, methods that return a set of explanations rather than a single one should be used – for example, the method by Dandl et al. [26] or Mothilal et al. [86].

Open Issues: Numerous very different counterfactual explanations are overwhelming for users. Methods for aggregating or combining explanations are still a matter of future research.

9   Failure to Scale to High-Dimensional Settings

9.1   Human-Intelligibility of High-Dimensional IML Output

Pitfall: Applying IML methods naively to high-dimensional datasets (e.g. visualizing feature effects or computing importance scores on feature level) leads to an overwhelming and high-dimensional IML output, which impedes human analysis. Especially interpretation methods that are based on visualizations make it difficult for practitioners in high-dimensional settings to focus on the most important insights.

Solution: A natural approach is to reduce the dimensionality before applying any IML methods. Whether this facilitates understanding or not depends on the possible semantic interpretability of the resulting, reduced feature space – as features can either be selected or dimensionality can be reduced by linear or non-linear transformations. Assuming that users would like to interpret in the original feature space, many feature selection techniques can be used [46], resulting in much sparser and consequently easier to interpret models. Wrapper selection approaches are model-agnostic and algorithms like greedy forward selection or subset selection procedures [5,60], which start from an empty model and iteratively add relevant (subsets of) features if needed, even allow to measure the relevance of features for predictive performance. An alternative is to directly use models that implicitly perform feature selection such as LASSO [112] or component-wise boosting [99] as they can produce sparse models with fewer features. In the case of LIME or other interpretation methods based on surrogate models, the aforementioned techniques could be applied to the surrogate model.

When features can be meaningfully grouped in a data-driven or knowledge-driven way [51], applying IML methods directly to grouped features instead of single features is usually more time-efficient to compute and often leads to more appropriate interpretations. Examples where features can naturally be grouped include the grouping of sensor data [20], time-lagged features [75], or one-hot-encoded categorical features and interaction terms [43]. Before a model is fitted, groupings could already be exploited for dimensionality reduction, for example by selecting groups of features by the group LASSO [121].

For model interpretation, various papers extended feature importance methods from single features to groups of features [5,43,114,119]. In the case of grouped PFI, this means that we perturb the entire group of features at once and measure the performance drop compared to the unperturbed dataset. Compared to standard PFI, the grouped PFI does not break the association to the other features of the group, but to features of other groups and the target. This is especially useful when features within the same group are highly correlated (e.g. time-lagged features), but between-group dependencies are rather low. Hence, this might also be a possible solution for the extrapolation pitfall described in Sect. 5.1.

We consider the PhoneStudy in [106] as an illustration. The PhoneStudy dataset contains 1821 features to analyze the link between human behavior based on smartphone data and participants' personalities. Interpreting the results in this use case seems to be challenging since features were dependent and single feature effects were either small or non-linear [106]. The features have been grouped in behavior-specific categories such as app-usage, music consumption, or overall phone usage. Au et al. [5] calculated various grouped importance scores on the feature groups to measure their influence on a specific personality trait (e.g. conscientiousness). Furthermore, the authors applied a greedy forward subset selection procedure via repeated subsampling on the feature groups and showed that combining app-usage features and overall phone usage features were most of the times sufficient for the given prediction task.

Open Issues: The quality of a grouping-based interpretation strongly depends on the human intelligibility and meaningfulness of the grouping. If the grouping structure is not naturally given, then data-driven methods can be used. However, if feature groups are not meaningful (e.g. if they cannot be described by a super-feature such as app-usage), then subsequent interpretations of these groups are purposeless. One solution could be to combine feature selection strategies with interpretation methods. For example, LIME's surrogate model could be a LASSO model. However, beyond surrogate models, the integration of feature selection strategies remains an open issue that requires further research.

Existing research on grouped interpretation methods mainly focused on quantifying grouped feature importance, but the question of "how a group of features influences a model's prediction" remains almost unanswered. Only recently, [5,15,101] attempted to answer this question by using dimension-reduction techniques (such as PCA) before applying the interpretation method. However, this is also a matter of further research.

9.2 Computational Effort

Pitfall: Some interpretation methods do not scale linearly with the number of features. For example, for the computation of exact Shapley values the number of possible coalitions [25,78], or for a (full) functional ANOVA decomposition the number of components (main effects plus all interactions) scales with $\mathcal{O}(2^p)$ [54].[2]

Solution: For the functional ANOVA, a common solution is to keep the analysis to the main effects and selected 2-way interactions (similar for PDP and ALE). Interesting 2-way interactions can be selected by another method such as the H-statistic [35]. However, the selection of 2-way interactions requires additional computational effort. Interaction strength usually decreases quickly with increasing interaction size, and one should only consider d-way interactions when all their $(d-1)$-way interactions were significant [53]. For Shapley-based methods, an efficient approximation exists that is based on randomly sampling and evaluating feature orderings until the estimates converge. The variance of the estimates reduces in $\mathcal{O}(\frac{1}{m})$, where m is the number of evaluated orderings [25,78].

9.3 Ignoring Multiple Comparison Problem

Pitfall: Simultaneously testing the importance of multiple features will result in false-positive interpretations if the multiple comparisons problem (MCP) is ignored. The MCP is well known in significance tests for linear models and exists similarly in testing for feature importance in ML. For example, suppose we simultaneously test the importance of 50 features (with the H_0-hypothesis of zero importance) at the significance level $\alpha = 0.05$. Even if all features are unimportant, the probability of observing that at least one feature is significantly important is $1 - \mathbb{P}(\text{'no feature important'}) = 1 - (1 - 0.05)^{50} \approx 0.923$. Multiple comparisons become even more problematic the higher the dimension of the dataset.

Solution: Methods such as Model-X knockoffs [17] directly control for the false discovery rate (FDR). For all other methods that provide p-values or confidence intervals, such as PIMP (Permutation IMPortance) [2], which is a testing approach for PFI, MCP is often ignored in practice to the best of our knowledge, with some exceptions[105,117]. One of the most popular MCP adjustment methods is the Bonferroni correction [31], which rejects a null hypothesis if its p-value is smaller than α/p, with p as the number of tests. It has the disadvantage that it increases the probability of false negatives [90]. Since MCP is well known in statistics, we refer the practitioner to [28] for an overview and discussion of alternative adjustment methods, such as the Bonferroni-Holm method [52].

---

[2] Similar to the PDP or ALE plots, the functional ANOVA components describe individual feature effects and interactions.

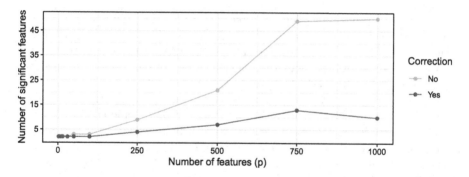

Fig. 9. Failure to scale to high-dimensional settings. Comparison of the number of features with significant importance - once with and once without Bonferroni-corrected significance levels for a varying number of added noise variables. Datasets were sampled from $Y = 2X_1 + 2X_2^2 + \epsilon$ with $X_1, X_2, \epsilon \sim N(0,1)$. $X_3, X_4, ..., X_p \sim N(0,1)$ are additional noise variables with p ranging between 2 and 1000. For each p, we sampled two datasets from this data-generating process – one to train a random forest with 500 trees on and one to test whether feature importances differed from 0 using PIMP. In all experiments, X_1 and X_2 were correctly identified as important.

As an example, in Fig. 9 we compare the number of features with significant importance measured by PIMP once with and once without Bonferroni-adjusted significance levels ($\alpha = 0.05$ vs. $\alpha = 0.05/p$). Without correcting for multi-comparisons, the number of features mistakenly evaluated as important grows considerably with increasing dimension, whereas Bonferroni correction results in only a modest increase.

10   Unjustified Causal Interpretation

Pitfall: Practitioners are often interested in causal insights into the underlying data-generating mechanisms, which IML methods do not generally provide. Common causal questions include the identification of causes and effects, predicting the effects of interventions, and answering counterfactual questions [88]. For example, a medical researcher might want to identify risk factors or predict average and individual treatment effects [66]. In search of answers, a researcher can therefore be tempted to interpret the result of IML methods from a causal perspective.

However, a causal interpretation of predictive models is often not possible. Standard supervised ML models are not designed to model causal relationships but to merely exploit associations. A model may therefore rely on causes and effects of the target variable as well as on variables that help to reconstruct unobserved influences on Y, e.g. causes of effects [118]. Consequently, the question of whether a variable is relevant to a predictive model (indicated e.g. by PFI > 0) does not directly indicate whether a variable is a cause, an effect, or does not stand in any causal relation to the target variable. Furthermore, even if a model would rely solely on direct causes for the prediction, the causal structure between features must be taken into account. Intervening on a variable in the real world may affect not only Y but also other variables in the feature set. Without assumptions about the underlying causal structure, IML methods cannot account for these adaptions and guide action [58,62].

As an example, we constructed a dataset by sampling from a structural causal model (SCM), for which the corresponding causal graph is depicted in Fig. 10. All relationships are linear Gaussian with variance 1 and coefficients 1. For a linear model fitted on the dataset, all features were considered to be relevant based on the model coefficients ($\hat{y} = 0.329x_1 + 0.323x_2 - 0.327x_3 + 0.342x_4 + 0.334x_5$, $R^2 = 0.943$), although x_3, x_4 and x_5 do not cause Y.

Solution: The practitioner must carefully assess whether sufficient assumptions can be made about the underlying data-generating process, the learned model, and the interpretation technique. If these assumptions are met, a causal interpretation may be possible. The PDP between a feature and the target can be interpreted as the respective average causal effect if the model performs well and the set of remaining variables is a valid adjustment set [123]. When it is known whether a model is deployed in a causal or anti-causal setting – i.e. whether the model attempts to predict an effect from its causes or the other way round – a partial identification of the causal roles based on feature relevance is possible (under strong and non-testable assumptions) [118]. Designated tools and approaches are available for causal discovery and inference [91].

Open Issues: The challenge of causal discovery and inference remains an open key issue in the field of ML. Careful research is required to make explicit under which assumptions what insight about the underlying data-generating mechanism can be gained by interpreting an ML model.

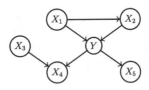

Fig. 10. Causal graph

11   Discussion

In this paper, we have reviewed numerous pitfalls of local and global model-agnostic interpretation techniques, e.g. in the case of bad model generalization, dependent features, interactions between features, or causal interpretations. We have not attempted to provide an exhaustive list of all potential pitfalls in ML model interpretation, but have instead focused on common pitfalls that apply to various model-agnostic IML methods and pose a particularly high risk.

We have omitted pitfalls that are more specific to one IML method type: For local methods, the vague notions of neighborhood and distance can lead to misinterpretations [68,69], and common distance metrics (such as the Euclidean distance) are prone to the curse of dimensionality [1]; Surrogate methods such as LIME may not be entirely faithful to the original model they replace in interpretation. Moreover, we have not addressed pitfalls associated with certain data types (like the definition of superpixels in image data [98]), nor those related to human cognitive biases (e.g. the illusion of model understanding [22]).

Many pitfalls in the paper are strongly linked with axioms that encode desiderata of model interpretation. For example, pitfall Sect. 5.3 (misunderstanding conditional interpretations) is related to violations of sensitivity [56,110]. As such, axioms can help to make the strengths and limitations of methods explicit. Therefore, we encourage an axiomatic evaluation of interpretation methods.

We hope to promote a more cautious approach when interpreting ML models in practice, to point practitioners to already (partially) available solutions, and to stimulate further research on these issues. The stakes are high: ML algorithms are increasingly used for socially relevant decisions, and model interpretations play an important role in every empirical science. Therefore, we believe that users can benefit from concrete guidance on properties, dangers, and problems of IML techniques – especially as the field is advancing at high speed. We need to strive towards a recommended, well-understood set of tools, which will in turn require much more careful research. This especially concerns the meta-issues of comparisons of IML techniques, IML diagnostic tools to warn against misleading interpretations, and tools for analyzing multiple dependent or interacting features.

References

1. Aggarwal, C.C., Hinneburg, A., Keim, D.A.: On the surprising behavior of distance metrics in high dimensional space. In: Van den Bussche, J., Vianu, V. (eds.) ICDT 2001. LNCS, vol. 1973, pp. 420–434. Springer, Heidelberg (2001). https://doi.org/10.1007/3-540-44503-X_27
2. Altmann, A., Toloşi, L., Sander, O., Lengauer, T.: Permutation importance: a corrected feature importance measure. Bioinformatics **26**(10), 1340–1347 (2010). https://doi.org/10.1093/bioinformatics/btq134
3. Apley, D.W., Zhu, J.: Visualizing the effects of predictor variables in black box supervised learning models. J. R. Stat. Soc.: Ser. B (Stat. Methodol.) **82**(4), 1059–1086 (2020). https://doi.org/10.1111/rssb.12377
4. Arlot, S., Celisse, A.: A survey of cross-validation procedures for model selection. Statist. Surv. **4**, 40–79 (2010). https://doi.org/10.1214/09-SS054
5. Au, Q., Herbinger, J., Stachl, C., Bischl, B., Casalicchio, G.: Grouped feature importance and combined features effect plot. arXiv preprint arXiv:2104.11688 (2021)
6. Bach, F.R., Jordan, M.I.: Kernel independent component analysis. J. Mach. Learn. Res. **3**(Jul), 1–48 (2002)

7. Baesens, B., Van Gestel, T., Viaene, S., Stepanova, M., Suykens, J., Vanthienen, J.: Benchmarking state-of-the-art classification algorithms for credit scoring. J. Oper. Res. Soc. **54**(6), 627–635 (2003). https://doi.org/10.1057/palgrave.jors. 2601545

8. Bansal, N., Agarwal, C., Nguyen, A.: SAM: the sensitivity of attribution methods to hyperparameters. In: Proceedings of the IEEE/CVF Conference on Computer Vision and Pattern Recognition, pp. 8673–8683 (2020)

9. Belghazi, M.I., et al.: Mutual information neural estimation. In: International Conference on Machine Learning, pp. 531–540 (2018)

10. Bischl, B., et al.: Hyperparameter optimization: foundations, algorithms, best practices and open challenges. arXiv preprint arXiv:2107.05847 (2021)

11. Bischl, B., Mersmann, O., Trautmann, H., Weihs, C.: Resampling methods for meta-model validation with recommendations for evolutionary computation. Evol. Comput. **20**(2), 249–275 (2012). https://doi.org/10.1162/EVCO_a_00069

12. Breiman, L.: Random forests. Mach. Learn. **45**(1), 5–32 (2001). https://doi.org/10.1023/A:1010933404324

13. Breiman, L.: Statistical modeling: the two cultures (with comments and a rejoinder by the author). Stat. Sci. **16**(3), 199–231 (2001). https://doi.org/10.1214/ss/1009213726

14. Breiman, L., Friedman, J.H.: Estimating optimal transformations for multiple regression and correlation. J. Am. Stat. Assoc. **80**(391), 580–598 (1985). https://doi.org/10.1080/01621459.1985.10478157

15. Brenning, A.: Transforming feature space to interpret machine learning models. arXiv:2104.04295 (2021)

16. Britton, M.: Vine: visualizing statistical interactions in black box models. arXiv preprint arXiv:1904.00561 (2019)

17. Candes, E., Fan, Y., Janson, L., Lv, J.: Panning for gold:'model-x'knockoffs for high dimensional controlled variable selection. J. R. Stat. Soc.: Ser. B (Stat. Methodol.) **80**(3), 551–577 (2018). https://doi.org/10.1111/rssb.12265

18. Caruana, R., Lou, Y., Gehrke, J., Koch, P., Sturm, M., Elhadad, N.: Intelligible models for healthcare: predicting pneumonia risk and hospital 30-day readmission. In: Proceedings of the 21th ACM SIGKDD International Conference on Knowledge Discovery and Data Mining, pp. 1721–1730 (2015). https://doi.org/10.1145/2783258.2788613

19. Casalicchio, G., Molnar, C., Bischl, B.: Visualizing the feature importance for black box models. In: Berlingerio, M., Bonchi, F., Gärtner, T., Hurley, N., Ifrim, G. (eds.) ECML PKDD 2018. LNCS (LNAI), vol. 11051, pp. 655–670. Springer, Cham (2019). https://doi.org/10.1007/978-3-030-10925-7_40

20. Chakraborty, D., Pal, N.R.: Selecting useful groups of features in a connectionist framework. IEEE Trans. Neural Netw. **19**(3), 381–396 (2008). https://doi.org/10.1109/TNN.2007.910730

21. Chen, H., Janizek, J.D., Lundberg, S., Lee, S.I.: True to the model or true to the data? arXiv preprint arXiv:2006.16234 (2020)

22. Chromik, M., Eiband, M., Buchner, F., Krüger, A., Butz, A.: I think I get your point, AI! the illusion of explanatory depth in explainable AI. In: 26th International Conference on Intelligent User Interfaces, IUI 2021, pp. 307–317. Association for Computing Machinery, New York (2021). https://doi.org/10.1145/3397481.3450644

23. Claeskens, G., Hjort, N.L., et al.: Model Selection and Model Averaging. Cambridge Books (2008). https://doi.org/10.1017/CBO9780511790485

24. Cover, T.M., Thomas, J.A.: Elements of Information Theory. Wiley (2012). https://doi.org/10.1002/047174882X
25. Covert, I., Lundberg, S.M., Lee, S.I.: Understanding global feature contributions with additive importance measures. In: Larochelle, H., Ranzato, M., Hadsell, R., Balcan, M.F., Lin, H. (eds.) Advances in Neural Information Processing Systems, vol. 33, pp. 17212–17223. Curran Associates, Inc. (2020)
26. Dandl, S., Molnar, C., Binder, M., Bischl, B.: Multi-objective counterfactual explanations. In: Bäck, T., et al. (eds.) PPSN 2020. LNCS, vol. 12269, pp. 448–469. Springer, Cham (2020). https://doi.org/10.1007/978-3-030-58112-1_31
27. Das, A., Rad, P.: Opportunities and challenges in explainable artificial intelligence (XAI): a survey. arXiv preprint arXiv:2006.11371 (2020)
28. Dickhaus, T.: Simultaneous Statistical Inference. Springer, Heidelberg (2014). https://doi.org/10.1007/978-3-642-45182-9
29. Dong, J., Rudin, C.: Exploring the cloud of variable importance for the set of all good models. Nat. Mach. Intell. 2(12), 810–824 (2020). https://doi.org/10.1038/s42256-020-00264-0
30. Doshi-Velez, F., Kim, B.: Towards a rigorous science of interpretable machine learning. arXiv preprint arXiv:1702.08608 (2017)
31. Dunn, O.J.: Multiple comparisons among means. J. Am. Stat. Assoc. 56(293), 52–64 (1961). https://doi.org/10.1080/01621459.1961.10482090
32. Fernández-Delgado, M., Cernadas, E., Barro, S., Amorim, D.: Do we need hundreds of classifiers to solve real world classification problems. J. Mach. Learn. Res. 15(1), 3133–3181 (2014). https://doi.org/10.5555/2627435.2697065
33. Fisher, A., Rudin, C., Dominici, F.: All models are wrong, but many are useful: learning a variable's importance by studying an entire class of prediction models simultaneously. J. Mach. Learn. Res. 20(177), 1–81 (2019)
34. Freiesleben, T.: Counterfactual explanations & adversarial examples-common grounds, essential differences, and potential transfers. arXiv preprint arXiv:2009.05487 (2020)
35. Friedman, J.H., Popescu, B.E.: Predictive learning via rule ensembles. Ann. Appl. Stat. 2(3), 916–954 (2008). https://doi.org/10.1214/07-AOAS148
36. Friedman, J.H., et al.: Multivariate adaptive regression splines. Ann. Stat. 19(1), 1–67 (1991). https://doi.org/10.1214/aos/1176347963
37. Garreau, D., von Luxburg, U.: Looking deeper into tabular lime. arXiv preprint arXiv:2008.11092 (2020)
38. Goldstein, A., Kapelner, A., Bleich, J., Pitkin, E.: Peeking inside the black box: visualizing statistical learning with plots of individual conditional expectation. J. Comput. Graph. Stat. 24(1), 44–65 (2015). https://doi.org/10.1080/10618600.2014.907095
39. Good, P.I., Hardin, J.W.: Common Errors in Statistics (and How to Avoid Them). Wiley (2012). https://doi.org/10.1002/9781118360125
40. Gosiewska, A., Biecek, P.: Do not trust additive explanations. arXiv preprint arXiv:1903.11420 (2019)
41. Greenwell, B.M.: PDP: an R package for constructing partial dependence plots. R J. 9(1), 421–436 (2017). https://doi.org/10.32614/RJ-2017-016
42. Greenwell, B.M., Boehmke, B.C., McCarthy, A.J.: A simple and effective model-based variable importance measure. arXiv:1805.04755 (2018)
43. Gregorutti, B., Michel, B., Saint-Pierre, P.: Grouped variable importance with random forests and application to multiple functional data analysis. Comput. Stat. Data Anal. 90, 15–35 (2015). https://doi.org/10.1016/j.csda.2015.04.002

44. Gretton, A., Bousquet, O., Smola, A., Schölkopf, B.: Measuring statistical dependence with Hilbert-Schmidt norms. In: Jain, S., Simon, H.U., Tomita, E. (eds.) ALT 2005. LNCS (LNAI), vol. 3734, pp. 63–77. Springer, Heidelberg (2005). https://doi.org/10.1007/11564089_7
45. Grömping, U.: Model-agnostic effects plots for interpreting machine learning models. Reports in Mathematics, Physics and Chemistry Report 1/2020 (2020)
46. Guyon, I., Elisseeff, A.: An introduction to variable and feature selection. J. Mach. Learn. Res. **3**(Mar), 1157–1182 (2003)
47. Hall, P.: On the art and science of machine learning explanations. arXiv preprint arXiv:1810.02909 (2018)
48. Hancox-Li, L.: Robustness in machine learning explanations: does it matter? In: Proceedings of the 2020 Conference on Fairness, Accountability, and Transparency, FAT* 2020, pp. 640–647. Association for Computing Machinery, New York (2020). https://doi.org/10.1145/3351095.3372836
49. Hand, D.J.: Classifier technology and the illusion of progress. Stat. Sci. **21**(1), 1–14 (2006). https://doi.org/10.1214/088342306000000060
50. Hastie, T., Tibshirani, R.: Generalized additive models. Stat. Sci. **1**(3), 297–310 (1986). https://doi.org/10.1214/ss/1177013604
51. He, Z., Yu, W.: Stable feature selection for biomarker discovery. Comput. Biol. Chem. **34**(4), 215–225 (2010). https://doi.org/10.1016/j.compbiolchem.2010.07.002
52. Holm, S.: A simple sequentially rejective multiple test procedure. Scand. J. Stat. **6**(2), 65–70 (1979)
53. Hooker, G.: Discovering additive structure in black box functions. In: Proceedings of the Tenth ACM SIGKDD International Conference on Knowledge Discovery and Data Mining, KDD 2004, pp. 575–580. Association for Computing Machinery, New York (2004). https://doi.org/10.1145/1014052.1014122
54. Hooker, G.: Generalized functional ANOVA diagnostics for high-dimensional functions of dependent variables. J. Comput. Graph. Stat. **16**(3), 709–732 (2007). https://doi.org/10.1198/106186007X237892
55. Hooker, G., Mentch, L.: Please stop permuting features: an explanation and alternatives. arXiv preprint arXiv:1905.03151 (2019)
56. Janzing, D., Minorics, L., Blöbaum, P.: Feature relevance quantification in explainable AI: a causality problem. arXiv preprint arXiv:1910.13413 (2019)
57. Kadir, T., Brady, M.: Saliency, scale and image description. Int. J. Comput. Vis. **45**(2), 83–105 (2001). https://doi.org/10.1023/A:1012460413855
58. Karimi, A.H., Schölkopf, B., Valera, I.: Algorithmic recourse: from counterfactual explanations to interventions. arXiv:2002.06278 (2020)
59. Khamis, H.: Measures of association: how to choose? J. Diagn. Med. Sonography **24**(3), 155–162 (2008). https://doi.org/10.1177/8756479308317006
60. Kohavi, R., John, G.H.: Wrappers for feature subset selection. Artif. Intell. **97**(1–2), 273–324 (1997)
61. König, G., Freiesleben, T., Bischl, B., Casalicchio, G., Grosse-Wentrup, M.: Decomposition of global feature importance into direct and associative components (DEDACT). arXiv preprint arXiv:2106.08086 (2021)
62. König, G., Freiesleben, T., Grosse-Wentrup, M.: A causal perspective on meaningful and robust algorithmic recourse. arXiv preprint arXiv:2107.07853 (2021)
63. König, G., Molnar, C., Bischl, B., Grosse-Wentrup, M.: Relative feature importance. In: 2020 25th International Conference on Pattern Recognition (ICPR), pp. 9318–9325. IEEE (2021). https://doi.org/10.1109/ICPR48806.2021.9413090

64. Krishnan, M.: Against interpretability: a critical examination of the interpretability problem in machine learning. Philos. Technol. **33**(3), 487–502 (2019). https://doi.org/10.1007/s13347-019-00372-9

65. Kuhle, S., et al.: Comparison of logistic regression with machine learning methods for the prediction of fetal growth abnormalities: a retrospective cohort study. BMC Pregnancy Childbirth **18**(1), 1–9 (2018). https://doi.org/10.1186/s12884-018-1971-2

66. König, G., Grosse-Wentrup, M.: A Causal Perspective on Challenges for AI in Precision Medicine (2019)

67. Lang, M., et al.: MLR3: a modern object-oriented machine learning framework in R. J. Open Source Softw. (2019). https://doi.org/10.21105/joss.01903

68. Laugel, T., Lesot, M.J., Marsala, C., Renard, X., Detyniecki, M.: The dangers of post-hoc interpretability: unjustified counterfactual explanations. In: Twenty-Eighth International Joint Conference on Artificial Intelligence, IJCAI 2019, pp. 2801–2807. International Joint Conferences on Artificial Intelligence Organization (2019)

69. Laugel, T., Renard, X., Lesot, M.J., Marsala, C., Detyniecki, M.: Defining locality for surrogates in post-hoc interpretablity. arXiv preprint arXiv:1806.07498 (2018)

70. Lauritsen, S.M., et al.: Explainable artificial intelligence model to predict acute critical illness from electronic health records. Nat. Commun. **11**(1), 1–11 (2020). https://doi.org/10.1038/s41467-020-17431-x

71. Lessmann, S., Baesens, B., Seow, H.V., Thomas, L.C.: Benchmarking state-of-the-art classification algorithms for credit scoring: an update of research. Eur. J. Oper. Res. **247**(1), 124–136 (2015). https://doi.org/10.1016/j.ejor.2015.05.030

72. Liebetrau, A.: Measures of Association. No. Bd. 32; Bd. 1983 in 07, SAGE Publications (1983)

73. Lipton, Z.C.: The mythos of model interpretability. Queue **16**(3), 31–57 (2018). https://doi.org/10.1145/3236386.3241340

74. Lopez-Paz, D., Hennig, P., Schölkopf, B.: The randomized dependence coefficient. In: Advances in Neural Information Processing Systems, pp. 1–9 (2013). https://doi.org/10.5555/2999611.2999612

75. Lozano, A.C., Abe, N., Liu, Y., Rosset, S.: Grouped graphical granger modeling for gene expression regulatory networks discovery. Bioinformatics **25**(12), i110–i118 (2009). https://doi.org/10.1093/bioinformatics/btp199

76. Lundberg, S.M., et al.: From local explanations to global understanding with explainable AI for trees. Nat. Mach. Intell. **2**(1), 56–67 (2020). https://doi.org/10.1038/s42256-019-0138-9

77. Lundberg, S.M., Erion, G.G., Lee, S.I.: Consistent individualized feature attribution for tree ensembles. arXiv preprint arXiv:1802.03888 (2018)

78. Lundberg, S.M., Lee, S.I.: A unified approach to interpreting model predictions. In: NIPS, vol. 30, pp. 4765–4774. Curran Associates, Inc. (2017). https://doi.org/10.5555/3295222.3295230

79. Makridakis, S., Spiliotis, E., Assimakopoulos, V.: Statistical and machine learning forecasting methods: concerns and ways forward. PloS One **13**(3) (2018). https://doi.org/10.1371/journal.pone.0194889

80. Matejka, J., Fitzmaurice, G.: Same stats, different graphs: generating datasets with varied appearance and identical statistics through simulated annealing. In: Proceedings of the 2017 CHI Conference on Human Factors in Computing Systems, pp. 1290–1294 (2017). https://doi.org/10.1145/3025453.3025912

81. Molnar, C., Casalicchio, G., Bischl, B.: IML: an R package for interpretable machine learning. J. Open Source Softw. **3**(26), 786 (2018). https://doi.org/10.21105/joss.00786
82. Molnar, C., Casalicchio, G., Bischl, B.: Quantifying model complexity via functional decomposition for better post-hoc interpretability. In: Cellier, P., Driessens, K. (eds.) ECML PKDD 2019. CCIS, vol. 1167, pp. 193–204. Springer, Cham (2020). https://doi.org/10.1007/978-3-030-43823-4_17
83. Molnar, C., Freiesleben, T., König, G., Casalicchio, G., Wright, M.N., Bischl, B.: Relating the partial dependence plot and permutation feature importance to the data generating process. arXiv preprint arXiv:2109.01433 (2021)
84. Molnar, C., König, G., Bischl, B., Casalicchio, G.: Model-agnostic feature importance and effects with dependent features-a conditional subgroup approach. arXiv preprint arXiv:2006.04628 (2020)
85. Moosbauer, J., Herbinger, J., Casalicchio, G., Lindauer, M., Bischl, B.: Towards explaining hyperparameter optimization via partial dependence plots. In: 8th ICML Workshop on Automated Machine Learning (AutoML) (2020)
86. Mothilal, R.K., Sharma, A., Tan, C.: Explaining machine learning classifiers through diverse counterfactual explanations. CoRR abs/1905.07697 (2019). http://arxiv.org/abs/1905.07697
87. Oh, S.: Feature interaction in terms of prediction performance. Appl. Sci. **9**(23) (2019). https://doi.org/10.3390/app9235191
88. Pearl, J., Mackenzie, D.: The Ladder of Causation. The Book of Why: The New Science of Cause and Effect, pp. 23–52. Basic Books, New York (2018). https://doi.org/10.1080/14697688.2019.1655928
89. Pedregosa, F., et al.: Scikit-learn: machine learning in Python. J. Mach. Learn. Res. **12**, 2825–2830 (2011). https://doi.org/10.5555/1953048.2078195
90. Perneger, T.V.: What's wrong with Bonferroni adjustments. BMJ **316**(7139), 1236–1238 (1998). https://doi.org/10.1136/bmj.316.7139.1236
91. Peters, J., Janzing, D., Scholkopf, B.: Elements of Causal Inference - Foundations and Learning Algorithms. The MIT Press (2017). https://doi.org/10.5555/3202377
92. Philipp, M., Rusch, T., Hornik, K., Strobl, C.: Measuring the stability of results from supervised statistical learning. J. Comput. Graph. Stat. **27**(4), 685–700 (2018). https://doi.org/10.1080/10618600.2018.1473779
93. Reshef, D.N., et al.: Detecting novel associations in large data sets. Science **334**(6062), 1518–1524 (2011). https://doi.org/10.1126/science.1205438
94. Ribeiro, M.T., Singh, S., Guestrin, C.: Why should I trust you?: explaining the predictions of any classifier. In: Proceedings of the 22nd ACM SIGKDD International Conference on Knowledge Discovery and Data Mining, pp. 1135–1144. ACM (2016). https://doi.org/10.1145/2939672.2939778
95. Rudin, C.: Stop explaining black box machine learning models for high stakes decisions and use interpretable models instead. Nat. Mach. Intell. **1**(5), 206–215 (2019). https://doi.org/10.1038/s42256-019-0048-x
96. Rudin, C., Chen, C., Chen, Z., Huang, H., Semenova, L., Zhong, C.: Interpretable machine learning: fundamental principles and 10 grand challenges. arXiv preprint arXiv:2103.11251 (2021)
97. Saito, S., Chua, E., Capel, N., Hu, R.: Improving lime robustness with smarter locality sampling. arXiv preprint arXiv:2006.12302 (2020)
98. Schallner, L., Rabold, J., Scholz, O., Schmid, U.: Effect of superpixel aggregation on explanations in lime-a case study with biological data. arXiv preprint arXiv:1910.07856 (2019)

99. Schmid, M., Hothorn, T.: Boosting additive models using component-wise p-splines. Comput. Stat. Data Anal. **53**(2), 298–311 (2008). https://doi.org/10.1016/j.csda.2008.09.009

100. Scholbeck, C.A., Molnar, C., Heumann, C., Bischl, B., Casalicchio, G.: Sampling, intervention, prediction, aggregation: a generalized framework for model-agnostic interpretations. In: Cellier, P., Driessens, K. (eds.) ECML PKDD 2019. CCIS, vol. 1167, pp. 205–216. Springer, Cham (2020). https://doi.org/10.1007/978-3-030-43823-4_18

101. Seedorff, N., Brown, G.: Totalvis: a principal components approach to visualizing total effects in black box models. SN Comput. Sci. **2**(3), 1–12 (2021). https://doi.org/10.1007/s42979-021-00560-5

102. Semenova, L., Rudin, C., Parr, R.: A study in Rashomon curves and volumes: a new perspective on generalization and model simplicity in machine learning. arXiv preprint arXiv:1908.01755 (2021)

103. Shalev-Shwartz, S., Ben-David, S.: Understanding Machine Learning: From Theory to Algorithms. Cambridge University Press, Cambridge (2014)

104. Simon, R.: Resampling strategies for model assessment and selection. In: Dubitzky, W., Granzow, M., Berrar, D. (eds.) Fundamentals of Data Mining in Genomics and Proteomics, pp. 173–186. Springer, Cham (2007). https://doi.org/10.1007/978-0-387-47509-7_8

105. Stachl, C., et al.: Behavioral patterns in smartphone usage predict big five personality traits. PsyArXiv (2019). https://doi.org/10.31234/osf.io/ks4vd

106. Stachl, C., et al.: Predicting personality from patterns of behavior collected with smartphones. Proc. Natl. Acad. Sci. (2020). https://doi.org/10.1073/pnas.1920484117

107. Strobl, C., Boulesteix, A.L., Kneib, T., Augustin, T., Zeileis, A.: Conditional variable importance for random forests. BMC Bioinform. **9**(1), 307 (2008). https://doi.org/10.1186/1471-2105-9-307

108. Štrumbelj, E., Kononenko, I.: Explaining prediction models and individual predictions with feature contributions. Knowl. Inf. Syst. **41**(3), 647–665 (2013). https://doi.org/10.1007/s10115-013-0679-x

109. Sundararajan, M., Najmi, A.: The many Shapley values for model explanation. arXiv preprint arXiv:1908.08474 (2019)

110. Sundararajan, M., Taly, A., Yan, Q.: Axiomatic attribution for deep networks. In: International Conference on Machine Learning, pp. 3319–3328. PMLR (2017)

111. Székely, G.J., Rizzo, M.L., Bakirov, N.K., et al.: Measuring and testing dependence by correlation of distances. Ann. Stat. **35**(6), 2769–2794 (2007). https://doi.org/10.1214/009053607000000505

112. Tibshirani, R.: Regression shrinkage and selection via the lasso. J. Roy. Stat. Soc.: Ser. B (Methodol.) **58**(1), 267–288 (1996). https://doi.org/10.1111/j.1467-9868.2011.00771.x

113. Tjøstheim, D., Otneim, H., Støve, B.: Statistical dependence: beyond pearson's p. arXiv preprint arXiv:1809.10455 (2018)

114. Valentin, S., Harkotte, M., Popov, T.: Interpreting neural decoding models using grouped model reliance. PLoS Comput. Biol. **16**(1), e1007148 (2020). https://doi.org/10.1371/journal.pcbi.1007148

115. Wachter, S., Mittelstadt, B., Russell, C.: Counterfactual explanations without opening the black box: automated decisions and the GDPR. Harv. JL Tech. **31**, 841 (2017). https://doi.org/10.2139/ssrn.3063289

116. Walters-Williams, J., Li, Y.: Estimation of mutual information: a survey. In: Wen, P., Li, Y., Polkowski, L., Yao, Y., Tsumoto, S., Wang, G. (eds.) RSKT 2009. LNCS (LNAI), vol. 5589, pp. 389–396. Springer, Heidelberg (2009). https://doi.org/10.1007/978-3-642-02962-2_49
117. Watson, D.S., Wright, M.N.: Testing conditional independence in supervised learning algorithms. arXiv preprint arXiv:1901.09917 (2019)
118. Weichwald, S., Meyer, T., Özdenizci, O., Schölkopf, B., Ball, T., Grosse-Wentrup, M.: Causal interpretation rules for encoding and decoding models in neuroimaging. Neuroimage **110**, 48–59 (2015). https://doi.org/10.1016/j.neuroimage.2015.01.036
119. Williamson, B.D., Gilbert, P.B., Simon, N.R., Carone, M.: A unified approach for inference on algorithm-agnostic variable importance. arXiv:2004.03683 (2020)
120. Wu, J., Roy, J., Stewart, W.F.: Prediction modeling using EHR data: challenges, strategies, and a comparison of machine learning approaches. Med. Care S106–S113 (2010). https://doi.org/10.1097/MLR.0b013e3181de9e17
121. Yuan, M., Lin, Y.: Model selection and estimation in regression with grouped variables. J. R. Stat. Soc.: Ser. B (Statistical Methodology) **68**(1), 49–67 (2006). https://doi.org/10.1111/j.1467-9868.2005.00532.x
122. Zhang, X., Wang, Y., Li, Z.: Interpreting the black box of supervised learning models: visualizing the impacts of features on prediction. Appl. Intell. **51**(10), 7151–7165 (2021). https://doi.org/10.1007/s10489-021-02255-z
123. Zhao, Q., Hastie, T.: Causal interpretations of black-box models. J. Bus. Econ. Stat. 1–10 (2019). https://doi.org/10.1080/07350015.2019.1624293
124. Zhao, X., Lovreglio, R., Nilsson, D.: Modelling and interpreting pre-evacuation decision-making using machine learning. Autom. Constr. **113**, 103140 (2020). https://doi.org/10.1016/j.autcon.2020.103140
125. van der Zon, S.B., Duivesteijn, W., van Ipenburg, W., Veldsink, J., Pechenizkiy, M.: ICIE 1.0: a novel tool for interactive contextual interaction explanations. In: Alzate, C., et al. (eds.) MIDAS/PAP -2018. LNCS (LNAI), vol. 11054, pp. 81–94. Springer, Cham (2019). https://doi.org/10.1007/978-3-030-13463-1_6

CLEVR-X: A Visual Reasoning Dataset for Natural Language Explanations

Leonard Salewski[1]([⊠]) [iD], A. Sophia Koepke[1] [iD], Hendrik P. A. Lensch[1] [iD],
and Zeynep Akata[1,2,3] [iD]

[1] University of Tübingen, Tübingen, Germany
{leonard.salewski,a-sophia.koepke,hendrik.lensch,
zeynep.akata}@uni-tuebingen.de
[2] MPI for Informatics, Saarbrücken, Germany
[3] MPI for Intelligent Systems, Tübingen, Germany

Abstract. Providing explanations in the context of Visual Question Answering (VQA) presents a fundamental problem in machine learning. To obtain detailed insights into the process of generating natural language explanations for VQA, we introduce the large-scale CLEVR-X dataset that extends the CLEVR dataset with natural language explanations. For each image-question pair in the CLEVR dataset, CLEVR-X contains multiple structured textual explanations which are derived from the original scene graphs. By construction, the CLEVR-X explanations are correct and describe the reasoning and visual information that is necessary to answer a given question. We conducted a user study to confirm that the ground-truth explanations in our proposed dataset are indeed complete and relevant. We present baseline results for generating natural language explanations in the context of VQA using two state-of-the-art frameworks on the CLEVR-X dataset. Furthermore, we provide a detailed analysis of the explanation generation quality for different question and answer types. Additionally, we study the influence of using different numbers of ground-truth explanations on the convergence of natural language generation (NLG) metrics. The CLEVR-X dataset is publicly available at https://github.com/ExplainableML/CLEVR-X.

Keywords: Visual question answering · Natural language explanations

1 Introduction

Explanations for automatic decisions form a crucial step towards increasing transparency and human trust in deep learning systems. In this work, we focus on natural language explanations in the context of vision-language tasks.

In particular, we consider the vision-language task of Visual Question Answering (VQA) which consists of answering a question about an image. This requires multiple skills, such as visual perception, text understanding, and cross-modal reasoning in the visual and language domains. A natural language explanation for a given answer allows a better understanding of the reasoning process for answering the question and adds transparency. However, it is challenging to formulate what comprises a good textual explanation in the context of VQA involving natural images.

A. Holzinger et al. (Eds.): xxAI 2020, LNAI 13200, pp. 69–88, 2022.
https://doi.org/10.1007/978-3-031-04083-2_5

VQA-X
Question: Does this scene look like it could be from the early 1950s?

e-SNLI-VE
Hypothesis: A woman is holding a child.

CLEVR-X
Question: There is a purple metallic ball; what number of cyan objects are right of it?

Answer | Explanation:
Yes | The photo is in black and white and the cars are all classic designs from the 1950s

Answer | Explanation:
Entailment | If a woman holds a child she is holding a child.

Answer | Explanation:
1 | There is a cyan cylinder which is on the right side of the purple metallic ball.

Fig. 1. Comparing examples from the VQA-X (left), e-SNLI-VE (middle), and CLEVR-X (right) datasets. The explanation in VQA-X requires prior knowledge (about cars from the 1950s), e-SNLI-VE argues with a tautology, and our CLEVR-X only uses abstract visual reasoning.

Explanation datasets commonly used in the context of VQA, such as the VQA-X dataset [26] or the e-SNLI-VE dataset [13,29] for visual entailment, contain explanations of widely varying quality since they are generated by humans. The ground-truth explanations in VQA-X and e-SNLI-VE can range from statements that merely describe an image to explaining the reasoning about the question and image involving prior information, such as common knowledge. One example for a ground-truth explanation in VQA-X that requires prior knowledge about car designs from the 1950s can be seen in Fig. 1. The e-SNLI-VE dataset contains numerous explanation samples which consist of repeated statements ("x because x"). Since existing explanation datasets for vision-language tasks contain immensely varied explanations, it is challenging to perform a structured analysis of strengths and weaknesses of existing explanation generation methods.

In order to fill this gap, we propose the novel, diagnostic CLEVR-X dataset for visual reasoning with natural language explanations. It extends the synthetic CLEVR [27] dataset through the addition of structured natural language explanations for each question-image pair. An example for our proposed CLEVR-X dataset is shown in Fig. 1. The synthetic nature of the CLEVR-X dataset results in several advantages over datasets that use human explanations. Since the explanations are synthetically constructed from the underlying scene graph, the explanations are *correct* and do not require auxiliary prior knowledge. The synthetic textual explanations do not suffer from errors that get introduced with human explanations. Nevertheless, the explanations in the CLEVR-X dataset are human parsable as demonstrated in the human user study that we conducted. Furthermore, the explanations contain all the information that is necessary to answer a given question about an image without seeing the image. This means that the explanations are *complete* with respect to the question about the image.

The CLEVR-X dataset allows for detailed diagnostics of natural language explanation generation methods in the context of VQA. For instance, it contains a wider range of question types than other related datasets. We provide baseline performances on the CLEVR-X dataset using recent frameworks for natural language explanations in the context of VQA. Those frameworks are jointly trained to answer the question and provide a textual explanation. Since the question family, question complexity (number of reasoning steps required), and the answer type (binary, counting, attributes) is known for each question and answer, the results can be analyzed and split according to these groups. In particular, the challenging counting problem [48], which is not well-represented in the VQA-X dataset, can be studied in detail on CLEVR-X. Furthermore, our dataset contains multiple ground-truth explanations for each image-question pair. These capture a large portion of the space of correct explanations which allows for a thorough analysis of the influence of the number of ground-truth explanations used on the evaluation metrics. Our approach of constructing textual explanations from a scene graph yields a great resource which could be extended to other datasets that are based on scene graphs, such as the CLEVR-CoGenT dataset.

To summarize, we make the following four contributions: (1) We introduce the CLEVR-X dataset with natural language explanations for Visual Question Answering; (2) We confirm that the CLEVR-X dataset consists of correct explanations that contain sufficient relevant information to answer a posed question by conducting a user study; (3) We provide baseline performances with two state-of-the-art methods that were proposed for generating textual explanations in the context of VQA; (4) We use the CLEVR-X dataset for a detailed analysis of the explanation generation performance for different subsets of the dataset and to better understand the metrics used for evaluation.

2   Related Work

In this section, we discuss several themes in the literature that relate to our work, namely *Visual Question Answering*, *Natural language explanations (for vision-language tasks)*, and the *CLEVR dataset*.

Visual Question Answering (VQA). The VQA [5] task has been addressed by several works that apply attention mechanisms to text and image features [16,45,55,56,60]. However, recent works observed that the question-answer bias in common VQA datasets can be exploited in order to answer questions without leveraging any visual information [1,2,27,59]. This has been further investigated in more controlled dataset settings, such as the CLEVR [27], VQA-CP [2], and GQA [25] datasets. In addition to a controlled dataset setting, our proposed CLEVR-X dataset contains natural language explanations that enable a more detailed analysis of the reasoning in the context of VQA.

Natural Language Explanations. Decisions made by neural networks can be visually explained with visual attribution that is determined by introspecting trained networks and their features [8,43,46,57,58], by using input perturbations [14,15,42], or by training a probabilistic feature attribution model along with a task-specific CNN [30]. Complementary to visual explanations methods that tend to not help users distinguish between correct and incorrect predictions [32], natural language explanations have been investigated for a variety of tasks, such as fine-grained visual object classification [20,21], or self-driving car models [31]. The requirement to ground language explanations in the input image can prevent shortcuts, such as relying on dataset statistics or referring to instance attributes that are not present in the image. For a comprehensive overview of research on explainability and interpretability, we refer to recent surveys [7, 10, 17].

Natural Language Explanations for Vision-Language Tasks. Multiple datasets for natural language explanations in the context of vision-language tasks have been proposed, such as the VQA-X [26], VQA-E [35], and e-SNLI-VE datasets [29]. VQA-X [26] augments a small subset of the VQA v2 [18] dataset for the Visual Question Answering task with human explanations. Similarly, the VQA-E dataset [35] extends the VQA v2 dataset by sourcing explanations from image captions. However, the VQA-E explanations resemble image descriptions and do not provide satisfactory justifications whenever prior knowledge is required [35]. The e-SNLI-VE [13,29] dataset combines human explanations from e-SNLI [11] and the image-sentence pairs for the Visual Entailment task from SNLI-VE [54]. In contrast to the VQA-E, VQA-X, and e-SNLI-VE datasets which consist of human explanations or image captions, our proposed dataset contains systematically constructed explanations derived from the associated scene graphs. Recently, several works have aimed at generating natural language explanations for vision-language tasks [26,29,38,40,52,53]. In particular, we use the PJ-X [26] and FM [53] frameworks to obtain baseline results on our proposed CLEVR-X dataset.

The CLEVR Dataset. The CLEVR dataset [27] was proposed as a diagnostic dataset to inspect the visual reasoning of VQA models. Multiple frameworks have been proposed to address the CLEVR task [23,24,28,41,44,47]. To add explainability, the XNM model [44] adopts the scene graph as an inductive bias which enables the visualization of the reasoning based on the attention on the nodes of the graph. There have been numerous dataset extensions for the CLEVR dataset, for instance to measure the generalization capabilities of models pre-trained on CLEVR (CLOSURE [51]), to evaluate object detection and segmentation (CLEVR-Ref+ [37]), or to benchmark visual dialog models (CLEVR dialog [34]). The Compositional Reasoning Under Uncertainty (CURI) benchmark uses the CLEVR renderer to construct a test bed for compositional and relational learning under uncertainty [49]. [22] provide an extensive survey of further experimental diagnostic benchmarks for analyzing explainable machine learning frameworks along with proposing the KandinskyPATTERNS benchmark that contains synthetic images with simple 2-dimensional objects. It can be used for testing the quality of explanations and concept learning. Additionally, [6] proposed the CLEVR-XAI-simple and CLEVR-XAI-complex datasets which provide ground-truth segmentation information for heatmap-based visual explanations. Our CLEVR-X augments the existing CLEVR dataset with explanations, but in contrast to (heatmap-based) visual explanations, we focus on natural language explanations.

3   The CLEVR-X Dataset

In this section, we introduce the CLEVR-X dataset that consists of natural language explanations in the context of VQA. The CLEVR-X dataset extends the CLEVR dataset with 3.6 million natural language explanations for 850k question-image pairs. In Sect. 3.1, we briefly describe the CLEVR dataset, which forms the base for our proposed dataset. Next, we present an overview of the CLEVR-X dataset by describing how the natural language explanations were obtained in Sect. 3.2, and by providing a comprehensive analysis of the CLEVR-X dataset in Sect. 3.3. Finally, in Sect. 3.4, we present results for a user study on the CLEVR-X dataset.

3.1 The CLEVR Dataset

The CLEVR dataset consists of images with corresponding full scene graph annotations which contain information about all objects in a given scene (as nodes in the graph) along with spatial relationships for all object pairs. The synthetic images in the CLEVR dataset contain three to ten (at least partially visible) objects in each scene, where each object has the four distinct properties size, color, material, and shape. There are three shapes (box, sphere, cylinder), eight colors (gray, red, blue, green, brown, purple, cyan, yellow), two sizes (large, small), and two materials (rubber, metallic). This allows for up to 96 different combinations of properties.

There are a total of 90 different question families in the dataset which are grouped into 9 different question types. Each type contains questions from between 5 and 28 question families. In the following, we describe the 9 question types in more detail.

***Hop* Questions:** The *zero hop, one hop, two hop,* and *three hop* question types contain up to three relational reasoning steps, e.g. "What color is the cube to the left of the ball?" is a *one hop* question.

***Compare* and *Relate* Questions:** The *compare integer, same relate,* and *comparison* question types require the understanding and comparison of multiple objects in a scene. Questions of the *compare integer* type compare counts corresponding to two independent clauses (e.g. "Are there more cubes than red balls?"). *Same relate* questions reason about objects that have the same attribute as another previously specified object (e.g. "What is the color of the cube that has the same size as the ball?"). In contrast, *comparison* question types compare the attributes of two objects (e.g. "Is the color of the cube the same as the ball?").

***Single and/or* Questions:** *Single or* questions identify objects that satisfy an exclusive disjunction condition (e.g. "How many objects are either red or blue?"). Similarly, *single and* questions apply multiple relations and filters to find an object that satisfies all conditions (e.g. "How many objects are red and to the left of the cube.").

Each CLEVR question can be represented by a corresponding functional program and its natural language realization. A functional program is composed of basic functions that resemble elementary visual reasoning operations, such as *filtering* objects by one or more properties, *relating* objects to each other, or *querying* object properties. Furthermore, logical operations like *and* and *or*, as well as counting operations like *count, less, more,* and *equal* are used to build complex questions. Executing the functional program associated with the question against the scene graph yields the correct answer to the question. We can distinguish between three different answer types: Binary answers (yes or no), counting answers (integers from 0 to 10), and attribute answers (any of the possible values of shape, color, size, or material).

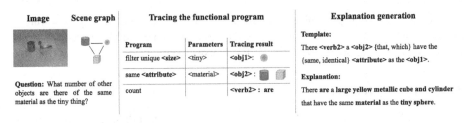

Fig. 2. CLEVR-X dataset generation: Generating a natural language explanation for a sample from the CLEVR dataset. Based on the question, the functional program for answering the question is executed on the scene graph and traced. A language template is used to cast the gathered information into a natural language explanation.

3.2   Dataset Generation

Here, we describe the process for generating natural language explanations for the CLEVR-X dataset. In contrast to image captions, the CLEVR-X explanations only describe image elements that are relevant to a specific input question. The explanation generation process for a given question-image pair is illustrated in Fig. 2. It consists of three steps: Tracing the functional program, relevance filtering (not shown in the figure), and explanation generation. In the following, we will describe those steps in detail.

Tracing the Functional Program. Given a question-image pair from the CLEVR dataset, we trace the execution of the functional program (that corresponds to the question) on the scene graph (which is associated with the image). The generation of the CLEVR dataset uses the same step to obtain a question-answer pair. When executing the basic functions that comprise the functional program, we record their outputs in order to collect all the information required for explaining a ground-truth answer.

In particular, we trace the *filter*, *relate* and *same-property* functions and record the returned objects and their properties, such as shape, size etc. As a result, the tracing omits objects in the scene that are not relevant for the question. As we are aiming for complete explanations for all question types, each explanation has to mention all the objects that were needed to answer the question, i.e. all the evidence that was obtained during tracing. For example, for *counting* questions, all objects that match the *filter* function preceding the *counting* step are recorded during tracing. For *and* questions, we merge the tracing results of the preceding functions which results in short and readable explanations. In summary, the tracing produces a *complete* and *correct* understanding of the objects and relevant properties which contributed to an answer.

Relevance Filtering. To keep the explanation at a reasonable length, we filter the object attributes that are mentioned in the explanation according to their relevance. For example, the color of an object is not relevant for a given question that asks about the material of said object. We deem all properties that were listed in the question to be relevant. This makes it easier to recognize the same referenced object in both the question and explanation. As the shape property also serves as a noun in CLEVR, our explanations always mention the shape to avoid using generic shape descriptions like "object" or "thing". We distinguish between objects which are used to build the question (e.g. "[...] that is left of the *cube*?") and those that are the subject of the posed question (e.g. "What color is the *sphere* that is left of the cube?"). For the former, we do not mention any additional properties, and for the latter, we mention the queried property (e.g. `color`) for question types yielding attribute answers.

Explanation Generation. To obtain the final natural language explanations, each question type is equipped with one or more natural language templates with variations in terms of the wording used. Each template contains placeholders which are filled with the output of the previous steps, i.e. the tracing of the functional program and subsequent filtering for relevance. As mentioned above, our explanations use the same property descriptions that appeared in the question. This is done to ensure that the wording of the explanation is consistent with the given question, e.g. for the question "Is there a small object?" we generate the explanation "Yes there is a small cube."[1] . We randomly sample synonyms for describing the properties of objects that do not appear in the question. If multiple objects are mentioned in the explanation, we randomize their order. If the tracing step returned an empty set, e.g. if no object exists that matches the given filtering function for an *existence* or *counting* question, we state that no relevant object is contained in the scene (e.g. "There is no red cube.").

In order to decrease the overall sentence length and to increase the readability, we aggregate repetitive descriptions (e.g. "There is a red cube and a red cube") using numerals (e.g. "There are two red cubes."). In addition, if a function of the functional program merely restricts the output set of a preceding function, we only mention the outputs of the later function. For instance, if a `same-color` function yields a large and a small cube, and a subsequent `filter-large` function restricts the output to only the large cube, we do not mention the output of `same-color`, as the output of the following `filter-large` causes natural language redundancies[2] .

The selection of different language templates, random sampling of synonyms and randomization of the object order (if possible) results in multiple different explanations. We uniformly sample up to 10 different explanations per question for our dataset.

Dataset Split. We provide explanations for the CLEVR training and validation sets, skipping only a negligible subset (less than 0.04‰) of questions due to malformed question programs from the CLEVR dataset, e.g. due to disjoint parts of their abstract syntax trees. In total, this affected 25 CLEVR training and 4 validation questions.

As the scene graphs and question functional programs are not publicly available for the CLEVR test set, we use the original CLEVR validation subset as the CLEVR-X test set. 20% of the CLEVR training set serve as the CLEVR-X validation set. We perform this split on the image-level to avoid any overlap between images in the CLEVR-X training and validation sets. Furthermore, we verified that the relative proportion of

---

[1] The explanation could have used the synonym "box" instead of "cube". In contrast, "tiny" and "small" are also synonyms in CLEVR, but the explanation would not have been consistent with the question which used "small".

[2] E.g. for the question: "How many large objects have the same color as the cube?", we do not generate the explanation "There are a small and a large cube that have the same color as the red cylinder of which only the large cube is large." but instead only write "There is a large cube that has the same color as the red cylinder.".

samples from each question and answer type in the CLEVR-X training and validation sets is similar, such that there are no biases towards specific question or answer types.

Code for generating the CLEVR-X dataset and the dataset itself are publicly available at https://github.com/ExplainableML/CLEVR-X.

3.3   Dataset Analysis

Table 1. Statistics of the CLEVR-X dataset compared to the VQA-X, and e-SNLI-VE datasets. We show the total number of images, questions, and explanations, vocabulary size, and the average number of explanations per question, the average number of words per explanation, and the average number of words per question. Note that subsets do not necessarily add up to the Total since some subsets have overlaps (e.g. for the vocabulary).

Dataset	Subset	Total #				Average #		
		Images	Questions	Explanations	Vocabulary	Explanations	Expl. Words	Quest. Words
VQA-X	Train	24,876	29,549	31,536	9,423	1.07	10.55	7.50
	Val	1,431	1,459	4,377	3,373	3.00	10.88	7.56
	Test	1,921	1,921	5,904	3,703	3.07	10.93	7.31
	Total	28,180	32,886	41,817	10,315	1.48	10.64	7.49
e-SNLI-VE	Train	29,779	401,672	401,672	36,778	1.00	13.62	8.23
	Val	1,000	14,339	14,339	8,311	1.00	14.67	8.10
	Test	998	14,712	14,712	8,334	1.00	14.59	8.20
	Total	31,777	430,723	430,723	38,208	1.00	13.69	8.23
CLEVR-X	Train	56,000	559,969	2,401,275	96	4.29	21.52	21.61
	Val	14,000	139,995	599,711	96	4.28	21.54	21.62
	Test	15,000	149,984	644,151	96	4.29	21.54	21.62
	Total	85,000	849,948	3,645,137	96	4.29	21.53	21.61

We compare the CLEVR-X dataset to the related VQA-X and e-SNLI-VE datasets in Table 1. Similar to CLEVR-X, VQA-X contains natural language explanations for the VQA task. However, different to the natural images and human explanations in VQA-X, CLEVR-X consists of synthetic images and explanations. The e-SNLI-VE dataset provides explanations for the visual entailment (VE) task. VE consists of classifying an input image-hypothesis pair into entailment / neutral / contradiction categories.

The CLEVR-X dataset is significantly larger than the VQA-X and e-SNLI-VE datasets in terms of the number of images, questions, and explanations. In contrast to the two other datasets, CLEVR-X provides (on average) multiple explanations for each question-image pair in the train set. Additionally, the average number of words per explanation is also higher. Since the explanations are built so that they explain each component mentioned in the question, long questions require longer explanations than short questions. Nevertheless, by design, there are no unnecessary redundancies. The explanation length in CLEVR-X is very strongly correlated with the length of the corresponding question (Spearman's correlation coefficient between the number of words in the explanations and questions is 0.89).

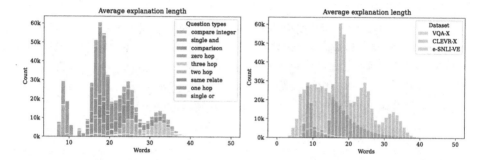

Fig. 3. Stacked histogram of the average explanation lengths measured in words for the nine question types for the CLEVR-X training set (left). Explanation length distribution for the CLEVR-X, VQA-X, and e-SNLI-VE training sets (right). The long tail of the e-SNLI-VE distribution (125 words) was cropped out for better readability.

Figure 3 (left) shows the explanation length distribution in the CLEVR-X dataset for the nine question types. The shortest explanation consists of 7 words, and the longest one has 53 words. On average, the explanations contain 21.53 words. In Fig. 3 (right) and Table 1, we can observe that explanations in CLEVR-X tend to be longer than the explanations in the VQA-X dataset. Furthermore, VQA-X has significantly fewer samples overall than the CLEVR-X dataset. The e-SNLI-VE dataset also contains longer explanations (that are up to 125 words long), but the CLEVR-X dataset is significantly larger than the e-SNLI-VE dataset. However, due to the synthetic nature and limited domain of CLEVR, the vocabulary of CLEVR-X is very small with only 96 different words. Unfortunately, VQA-X and e-SNLI-VE contain spelling errors, resulting in multiple versions of the same words. Models trained on CLEVR-X circumvent those aforementioned challenges and can purely focus on visual reasoning and explanations for the same. Therefore, Natural Language Generation (NLG) metrics applied to CLEVR-X indeed capture the factual correctness and completeness of an explanation.

3.4   User Study on Explanation Completeness and Relevance

In this section, we describe our user study for evaluating the completeness and relevance of the generated ground-truth explanations in the CLEVR-X dataset. We wanted to verify whether humans are successfully able to parse the synthetically generated textual explanations and to select complete and relevant explanations. While this is obvious for easier explanations like "There is a blue sphere.", it is less trivial for more complex explanations such as "There are two red cylinders in front of the green cube that is to the right of the tiny ball." Thus, strong human performance in the user study indicates that the sentences are parsable by humans.

We performed our user study using Amazon Mechanical Turk (MTurk). It consisted of two types of Human Intelligence Tasks (HITs). Each HIT was made up of (1) An explanation of the task; (2) A non-trivial example, where the correct answers are already selected; (3) A CAPTCHA [3] to verify that the user is human; (4) The problem definition consisting of a question and an image; (5) A user qualification step, for which the user has to correctly answer a question about an image. This ensures that the user is

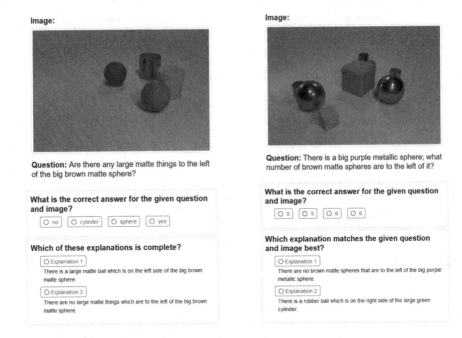

Fig. 4. Two examples from our user study to evaluate the completeness (left) and relevance (right) of natural language explanations in the CLEVR-X dataset.

able to answer the question in the first place, a necessary condition to participate in our user study; (6) Two explanations from which the user needs to choose one. Example screenshots of the user interface for the user study are shown in Fig. 4.

For the two different HIT types, we randomly sampled 100 explanations from each of the 9 question types, resulting in a total of 1800 samples for the completeness and relevance tasks. For each task sample, we requested 3 different MTurk workers based in the US (with high acceptance rate of > 95% and over 5000 accepted HITs). A total of 78 workers participated in the completeness HITs. They took on average 144.83 s per HIT. The relevance task was carried out by 101 workers which took on average 120.46 s per HIT. In total, 134 people participated in our user study. In the following, we describe our findings regarding the completeness and relevance of the CLEVR-X explanations in more detail.

Explanation Completeness. In the first part of the user study, we evaluated whether human users are able to determine if the ground-truth explanations in the CLEVR-X dataset are complete (and also correct). We presented the MTurk workers with an image, a question, and two explanations. As can be seen in Fig. 4 (left), a user had to first select the correct answer (*yes*) before deciding which of the two given explanations was complete. By design, one of the explanations presented to the user was the complete one from the CLEVR-X dataset and the other one was a modified version for which at least one necessary object had been removed. As simply deleting an object from a textual explanation could lead to grammar errors, we re-generated the explanations after removing objects from the tracing results. This resulted in incomplete, albeit grammatically correct, explanations.

Table 2. Results for the user study evaluating the accuracy for the completeness and relevance tasks for the nine question types in the CLEVR-X dataset.

	Zero hop	One hop	Two hop	Three hop	Same relate	Comparison	Compare integer	Single or	Single and	All
Completeness	100.00	98.00	98.67	94.00	100.00	83.67	77.00	84.00	94.33	92.19
Relevance	99.67	99.00	95.67	89.00	95.67	87.33	83.67	90.67	92.00	92.52

To evaluate the ability to determine the completeness of explanations, we measured the accuracy of selecting the complete explanation. The human participants obtained an average accuracy of 92.19%, confirming that complete explanations which mention all objects necessary to answer a given question were preferred over incomplete ones. The performance was weaker for complex question types, such as *compare-integer* and *comparison* with accuracies of only 77.00% and 83.67% respectively, compared to the easier *zero-hop* and *one-hop* questions with accuracies of 100% and 98.00% respectively.

Additionally, there were huge variations in performance across different participants of the completeness study (Fig. 5 (top left)), with the majority performing very well (>97% answering accuracy) for most question types. For the *compare-integer*, *comparison* and *single or* question types, some workers exhibited a much weaker performance with answering accuracies as low as 0%. The average turnaround time shown in Fig. 5 (bottom left) confirms that complex question types required less time to be solved than more complex question types, such as *three hop* and *compare integer* questions. Similar to the performance, the work time varied greatly between different users.

Explanation Relevance. In the second part of our user study, we analyzed if humans are able to identify explanations which are relevant for a given image. For a given question-image pair, the users had to first select the correct answer. Furthermore, they were provided with a correct explanation and another randomly chosen explanation from the same question family (that did not match the image). The task consisted of selecting the correct explanation that matched the image and question content. Explanation 1 in the example user interface shown in Fig. 4 (right) was the relevant one, since Explanation 2 does not match the question and image.

The participants of our user study were able to determine which explanation matched the given question-image example with an average accuracy of 92.52%. Again, the performance for complex question types was weaker than for easier questions. The difficulty of the question influences the accuracy of detecting the relevant explanation, since this task first requires understanding the question. Furthermore, complex questions tend to be correlated with complex scenes that contain many objects which makes the user's task more challenging. The accuracy for *three-hop* questions was 89.00% compared to 99.67% for *zero-hop* questions. For *compare-integer* and *comparison* questions, the users obtained accuracies of 83.67% and 87.33% respectively, which is significantly lower than the overall average accuracy.

We analyzed the answering accuracy per worker in Fig. 5 (top). The performance varies greatly between workers, with the majority performing very well (>90% answering accuracy) for most question types. Some workers showed much weaker perfor-

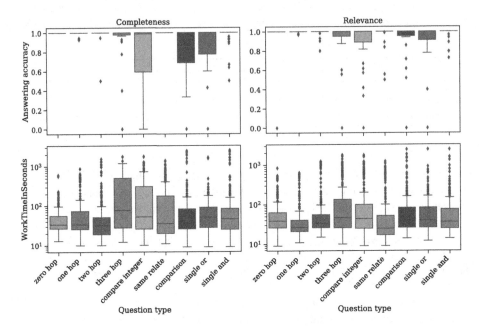

Fig. 5. Average answering accuracies for each worker (top) and average work time (bottom) for the user study (left: completeness, right: relevance). The boxes indicate the mean as well as lower and upper quartiles, the lines extend 1.5 interquartile ranges of the lower and upper quartile. All other values are plotted as diamonds.

mance with answering accuracies as low as 0% (e.g. for *compare-integer* and *single or* questions). Furthermore, the distribution of work time for the relevance task is shown in Fig. 5 (bottom right). The turnaround times for each worker exhibit greater variation on the completeness task (bottom left) compared to the relevance task (bottom right). This might be due to the nature of the different tasks. For the completeness task, the users need to check if the explanation contains all the elements that are necessary to answer the given question. The relevance task, on the other hand, can be solved by detecting a single non-relevant object to discard the wrong explanation.

Our user study confirmed that humans are able to parse the synthetically generated natural language explanations in the CLEVR-X dataset. Furthermore, the results have shown that users prefer complete and relevant explanations in our dataset over corrupted samples.

4   Experiments

We describe the experimental setup for establishing baselines on our proposed CLEVR-X dataset in Sect. 4.1. In Sect. 4.2, we present quantitative results on the CLEVR-X dataset. Additionally, we analyze the generated explanations for the CLEVR-X dataset in relation to the question and answer types in Sect. 4.3. Furthermore, we study the behavior of the NLG metrics when using different numbers of ground-truth explanations for testing in Sect. 4.4. Finally, we present qualitative explanation generation results on the CLEVR-X dataset in Sect. 4.5.

4.1 Experimental Setup

In this section, we provide details about the datasets and models used to establish baselines for our CLEVR-X dataset and about their training details. Furthermore, we explain the metrics for evaluating the explanation generation performance.

Datasets. In the following, we summarize the datasets that were used for our experiments. In addition to providing baseline results on CLEVR-X, we also report experimental results on the VQA-X and e-SNLI-VE datasets. Details about our proposed **CLEVR-X** dataset can be found in Sect. 3. The **VQA-X** dataset [26] is a subset of the VQA v2 dataset with a single human-generated textual explanation per question-image pair in the training set and 3 explanations for each sample in the validation and test sets. The **e-SNLI-VE** dataset [13,29] is a large-scale dataset with natural language explanations for the visual entailment task.

Methods. We used multiple frameworks to provide baselines on our proposed CLEVR-X dataset. For the **random words** baseline, we sample random word sequences of length w for the answer and explanation words for each test sample. The full vocabulary corresponding to a given dataset is used as the sampling pool, and w denotes the average number of words forming an answer and explanation in a given dataset. For the **random explanations** baseline, we randomly sample an answer-explanation pair from the training set and use this as the prediction. The explanations from this baseline are well-formed sentences. However, the answers and explanations most likely do not match the question or the image. For the random-words and random-explanations baselines, we report the NLG metrics for all samples in the test set (instead of only considering the correctly answered samples, since the random sampling of the answer does not influence the explanation). The Pointing and Justification model **PJ-X** [26] provides text-based post-hoc justifications for the VQA task. It combines a modified MCB [16] framework, pre-trained on the VQA v2 dataset, with a visual pointing and textual justification module. The Faithful Multimodal (**FM**) model [53] aims at grounding parts of generated explanations in the input image to provide explanations that are *faithful* to the input image. It is based on the Up-Down VQA model [4]. In addition, FM contains an explanation module which enforces consistency between the predicted answer, explanation and the attention of the VQA model. The implementations for the PJ-X and FM models are based on those provided by the authors of [29].

Implementation and Training Details. We extracted $14 \times 14 \times 1024$ grid features for the images in the CLEVR-X dataset using a ResNet-101 [19], pre-trained on ImageNet [12]. These grid features served as inputs to the FM [53] and PJ-X [26] frameworks. The CLEVR-X explanations are lower case and punctuation is removed from the sentences. We selected the best model on the CLEVR-X validation set based on the highest mean of the four NLG metrics, where explanations for incorrect answers were set to an empty string. This metric accounts for the answering performance as well as for the explanation quality. The final models were evaluated on the CLEVR-X test set. For PJ-X, our best model was trained for 52 epochs, using the Adam optimizer [33] with a learning rate of 0.0002 and a batch size of 256. We did not use gradient clipping for PJ-X. Our strongest FM model was trained for 30 epochs, using the Adam optimizer with a learning rate of 0.0002, a batch size of 128, and gradient clipping of 0.1. All other hyperparameters were taken from [26,53].

Evaluation Metrics. To evaluate the quality of the generated explanations, we use the standard natural language generation metrics BLEU [39], METEOR [9], ROUGE-L [36] and CIDEr [50]. By design, there is no correct explanation that can justify a wrong answer. We follow [29] and report the quality of the generated explanations for the subset of correctly answered questions.

4.2 Evaluating Explanations Generated by State-of-the-Art Methods

In this section, we present quantitative results for generating explanations for the CLEVR-X dataset (Table 3). The random words baseline exhibits weak explanation performance for all NLG metrics on CLEVR-X. Additionally, the random answering accuracy is very low at 3.6%. The results are similar on VQA-X and e-SNLI-VE. The random explanations baseline achieves stronger explanation results on all three datasets, but is still significantly worse than the trained models. This confirms that, even with a medium-sized answer space (28 options) and a small vocabulary (96 words), it is not possible to achieve good scores on our dataset using a trivial approach.

We observed that the PJ-X model yields a significantly stronger performance on CLEVR-X in terms of the NLG metrics for the generated explanations compared to the FM model, with METEOR scores of 58.9 and 52.5 for PJ-X and FM respectively. Across all explanation metrics, the scores on the VQA-X and e-SNLI-VE datasets are in a lower range than those on CLEVR-X. For PJ-X, we obtain a CIDEr score of 639.8 on CLEVR-X and 82.7 and 72.5 on VQA-X and e-SNLI-VE. This can be attributed to the smaller vocabulary and longer sentences, which allow n-gram based metrics (e.g. BLEU) to match parts of sentences more easily.

In contrast to the explanation generation performance, the FM model is better at answering questions than PJ-X on CLEVR-X with an answering accuracy of 80.3% for FM compared to 63.0% for PJ-X. Compared to recent models tuned to the CLEVR task, the answering performances of PJ-X and FM do not seem very strong. However, the PJ-X backbone MCB [16] (which is crucial for the answering performance) preceded the publication of the CLEVR dataset. A version of the MCB backbone (CNN+LSTM+MCB in the CLEVR publication [27]) achieved an answering accuracy of 51.4% on CLEVR [27], whereas PJ-X is able to correctly answer 63% of the questions. The strongest model discussed in the initial CLEVR publication (CNN+LSTM+SA in [27]) achieved an answering accuracy of 68.5%.

4.3 Analyzing Results on CLEVR-X by Question and Answer Types

In Fig. 6 (left and middle), we present the performance for PJ-X on CLEVR-X for the nine question and three answer types. The explanation results for samples which require counting abilities (counting answers) are lower than those for attribute answers (57.3 vs. 63.3). This is in line with prior findings that VQA models struggle with counting problems [48]. The explanation quality for binary questions is even lower with a METEOR score of only 55.6. The generated explanations are of higher quality for easier question types; *zero-hop* questions yield a METEOR score of 64.9 compared to 62.1 for

Table 3. Explanation generation results on the CLEVR-X, VQA-X, and e-SNLI-VE test sets using BLEU-4 (B4), METEOR (M), ROUGE-L (RL), CIDEr (C), and answer accuracy (*Acc*). Higher is better for all reported metrics. For the random baselines, *Acc* corresponds to $100/\text{\# answers}$ for CLEVR-X and e-SNLI-VE, and to the VQA answer score for VQA-X. (Rnd. words: random words, Rnd. expl: Random explanations)

Model	CLEVR-X					VQA-X					e-SNLI-VE				
	B4	M	RL	C	Acc	B4	M	RL	C	Acc	B4	M	RL	C	Acc
Rnd. words	0.0	8.4	11.4	5.9	3.6	0.0	1.2	0.7	0.1	0.1	0.0	0.3	0.0	0.0	33.3
Rnd. expl	10.9	16.6	35.3	30.4	3.6	0.9	6.5	18.4	21.6	0.2	0.4	5.4	9.9	2.6	33.3
FM [53]	78.8	52.5	85.8	566.8	80.3	23.1	20.4	47.1	87.0	75.5	8.2	15.6	29.9	83.6	58.5
PJ-X [26]	87.4	58.9	93.4	639.8	63.0	22.7	19.7	46.0	82.7	76.4	7.3	14.7	28.6	72.5	69.2

three-hop questions. It can also be seen that *single-or* questions are harder to explain than *single-and* questions. These trends can be observed across all NLG explanation metrics.

4.4 Influence of Using Different Numbers of Ground-Truth Explanations

In this section, we study the influence of using multiple ground-truth explanations for evaluation on the behavior of the NLG metrics. This gives insights about whether the metrics can correctly rate a model's performance with a limited number of ground-truth explanations. We set an upper bound k on the number of explanations used and randomly sample k explanations if a test sample has more than k explanations for $k \in \{1, 2, \ldots, 10\}$. Figure 6 (right) shows the NLG metrics (normalized with the maximum value for each metric on the test set for all ground-truth explanations) for the PJ-X model depending on the average number of ground-truth references used on the test set.

Out of the four metrics, BLEU-4 converges the slowest, requiring close to 3 ground-truth explanations to obtain a relative metric value of 95%. Hence, BLEU-4 might not be able to reliably predict the explanation quality on the e-SNLI-VE dataset which has only one explanation for each test sample. CIDEr converges faster than ROUGE and METEOR, and achieves 95.7% of its final value with only one ground-truth explanation. This could be caused by the fact, that CIDEr utilizes a tf-idf weighting scheme for different words, which is built from all reference sentences in the subset that the metric is computed on. This allows CIDEr to be more sensitive to important words (e.g. attributes and shapes) and to give less weight, for instance, to stopwords, such as "the". The VQA-X and e-SNLI-VE datasets contain much lower average numbers of explanations for each dataset sample (1.4 and 1.0). Since there could be many more possible explanations for samples in those datasets that describe different aspects than those mentioned in the ground truth, automated metric may not be able to correctly judge a prediction even if it is correct and faithful w.r.t. to the image and question.

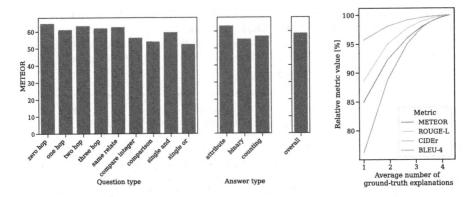

Fig. 6. Explanation generation results for PJ-X on the CLEVR-X test set according to question (left) and answer (middle) types compared to the overall explanation quality. Easier types yield higher METEOR scores. NLG metrics using different numbers of ground-truth explanations on the CLEVR-X test set (right). CIDEr converges faster than the other NLG metrics.

4.5 Qualitative Explanation Generation Results

We show examples for explanations generated with the PJ-X framework on CLEVR-X in Fig. 7. As can be seen across the three examples presented, PJ-X generates high-quality explanations which closely match the ground-truth explanations.

In the left-most example in Fig. 7, we can observe slight variations in grammar when comparing the generated explanation to the ground-truth explanation. However, the content of the generated explanation corresponds to the ground truth. Furthermore, some predicted explanations differ from the ground-truth explanation in the use of another synonym for a predicted attribute. For instance, in the middle example in Fig. 7, the ground-truth explanation describes the size of the cylinder as "small", whereas the predicted explanation uses the equivalent attribute "tiny". In contrast to other datasets, the set of ground-truth explanations for each sample in CLEVR-X contains these variations. Therefore, the automated NLG metrics do not decrease when such variations are found in the predictions. For the first and second example, PJ-X obtains the highest possible explanation score (100.0) in terms of the BLEU-4, METEOR, and ROUGE-L metrics.

We show a failure case where PJ-X predicted the wrong answer in Fig. 7 (right). The generated answer-explanation pair shows that the predicted explanation is consistent with the wrong answer prediction and does not match the input question-image pair. The NLG metrics for this case are significantly weaker with a BLEU-4 score of 0.0, as there are no matching 4-grams between the prediction and the ground truth.

Question: How many tiny red things are the same material as the big sphere?

Question: The cylinder has what size?

Question: Are there any small matte cubes?

GT Answer | Explanation:
1 | The tiny red metal block has the same material as a big sphere.
Pred. Answer | Expl.
1 | There is the tiny red metal block which has the identical material as a big sphere.
B4 / M / RL / C:
100.0 / 100.0 / 100.0 / 744.0

GT Answer | Explanation:
Small | The cylinder is small.
Pred. Answer | Expl.
Small | The cylinder is tiny.

B4 / M / RL / C:
100.0 / 100.0 / 100.0 / 462.4

GT Answer | Explanation:
No | There are no small matte cubes.
Pred. Answer | Expl.
Yes | There is a small matte cube.

B4 / M / RL / C:
0.0 / 76.9 / 57.1 / 157.1

Fig. 7. Examples for answers and explanations generated with the PJ-X framework on the CLEVR-X dataset, showing correct answer predictions (left, middle) and a failure case (right). The NLG metrics obtained with the explanations for the correctly predicted answers are high compared to those for the explanation corresponding to the wrong answer prediction.

5 Conclusion

We introduced the novel CLEVR-X dataset which contains natural language explanations for the VQA task on the CLEVR dataset. Our user study confirms that the explanations in the CLEVR-X dataset are complete and match the questions and images. Furthermore, we have provided baseline performances using the PJ-X and FM frameworks on the CLEVR-X dataset. The structured nature of our proposed dataset allowed the detailed evaluation of the explanation generation quality according to answer and question types. We observed that the generated explanations were of higher quality for easier answer and question categories. One of our findings is, that explanations for counting problems are worse than for other answer types, suggesting that further research into this direction is needed. Additionally, we find that the four NLG metrics used to evaluate the quality of the generated explanations exhibit different convergence patterns depending on the number of available ground-truth references.

Since this work only considered two natural language generation methods for VQA as baselines, the natural next step will be the benchmarking and closer investigation of additional recent frameworks for textual explanations in the context of VQA on the CLEVR-X dataset. We hope that our proposed CLEVR-X benchmark will facilitate further research to improve the generation of natural language explanations in the context of vision-language tasks.

Acknowledgements. The authors thank the Amazon Mechanical Turk workers that participated in the user study. This work was supported by the DFG - EXC number 2064/1 - project number 390727645, by the DFG: SFB 1233, Robust Vision: Inference Principles and Neural

Mechanisms - project number: 276693517, by the ERC (853489 - DEXIM), and by the BMBF (FKZ: 01IS18039A). L. Salewski thanks the International Max Planck Research School for Intelligent Systems (IMPRS-IS) for support.

References

1. Agrawal, A., Batra, D., Parikh, D.: Analyzing the behavior of visual question answering models. In: EMNLP, pp. 1955–1960. Association for Computational Linguistics (2016)
2. Agrawal, A., Batra, D., Parikh, D., Kembhavi, A.: Don't just assume; look and answer: overcoming priors for visual question answering. In: CVPR, pp. 4971–4980 (2018)
3. Ahn, L.V., Blum, M., Hopper, N.J., Langford, J.: CAPTCHA: using hard AI problems for security: In: Biham, E. (eds.) EUROCRYPT 2003. LNCS, vol. 2656, pp. 294–311. Springer, Heidelberg (2003). https://doi.org/10.1007/3-540-39200-9_18
4. Anderson, P., et al.: Bottom-up and top-down attention for image captioning and visual question answering. In: CVPR, pp. 6077–6086 (2018)
5. Antol, S., et al.: VQA: Visual Question Answering. In: ICCV, pp. 2425–2433 (2015)
6. Arras, L., Osman, A., Samek, W.: CLEVR-XAI: a benchmark dataset for the ground truth evaluation of neural network explanations. Inform. Fusion 81, 14–40 (2022)
7. Arrieta, A.B., et al.: Explainable artificial intelligence (xAI): concepts, taxonomies, opportunities and challenges toward responsible AI. Inf. Fusion 58, 82–115 (2020)
8. Bach, S., Binder, A., Montavon, G., Klauschen, F., Müller, K.R., Samek, W.: On pixel-wise explanations for non-linear classifier decisions by layer-wise relevance propagation. PLoS ONE 10(7), e0130140 (2015)
9. Banerjee, S., Lavie, A.: METEOR: an automatic metric for MT evaluation with improved correlation with human judgments. In: ACL Workshop, pp. 65–72 (2005)
10. Brundage, M., et al.: Toward trustworthy AI development: mechanisms for supporting verifiable claims. arXiv preprint arXiv:2004.07213 (2020)
11. Camburu, O.M., Rocktäschel, T., Lukasiewicz, T., Blunsom, P.: e-SNLI: Natural language inference with natural language explanations. In: NeurIPS (2018)
12. Deng, J., Dong, W., Socher, R., Li, L.J., Li, K., Fei-Fei, L.: Imagenet: a large-scale hierarchical image database. In: CVPR, pp. 248–255 (2009)
13. Do, V., Camburu, O.M., Akata, Z., Lukasiewicz, T.: e-SNLI-VE-2.0: corrected visual-textual entailment with natural language explanations. arXiv preprint arXiv:2004.03744 (2020)
14. Fong, R.C., Patrick, M., Vedaldi, A.: Understanding deep networks via extremal perturbations and smooth masks. In: ICCV, pp. 2950–2958 (2019)
15. Fong, R.C., Vedaldi, A.: Interpretable explanations of black boxes by meaningful perturbation. In: ICCV, pp. 3429–3437 (2017)
16. Fukui, A., Park, D.H., Yang, D., Rohrbach, A., Darrell, T., Rohrbach, M.: Multimodal compact bilinear pooling for visual question answering and visual grounding. In: EMNLP, pp. 457–468 (2016)
17. Gilpin, L.H., Bau, D., Yuan, B.Z., Bajwa, A., Specter, M., Kagal, L.: Explaining explanations: an overview of interpretability of machine learning. In: IEEE DSAA, pp. 80–89 (2018)
18. Goyal, Y., Khot, T., Summers-Stay, D., Batra, D., Parikh, D.: Making the V in VQA matter: elevating the role of image understanding in Visual Question Answering. In: CVPR, pp. 6904–6913 (2017)
19. He, K., Zhang, X., Ren, S., Sun, J.: Deep residual learning for image recognition. In: CVPR, pp. 770–778 (2016)
20. Hendricks, L.A., Hu, R., Darrell, T., Akata, Z.: Generating counterfactual explanations with natural language. arXiv preprint arXiv:1806.09809 (2018)

21. Hendricks, L.A., Hu, R., Darrell, T., Akata, Z.: Grounding visual explanations. In: ECCV, pp. 264–279 (2018)
22. Holzinger, A., Saranti, A., Mueller, H.: KANDINSKYpatterns - an experimental exploration environment for pattern analysis and machine intelligence. arXiv preprint arXiv:2103.00519 (2021)
23. Hudson, D., Manning, C.D.: Learning by abstraction: the neural state machine. In: NeurIPS (2019)
24. Hudson, D.A., Manning, C.D.: Compositional attention networks for machine reasoning. In: ICLR (2018)
25. Hudson, D.A., Manning, C.D.: GQA: a new dataset for real-world visual reasoning and compositional question answering. In: CVPR, pp. 6693–6702 (2019)
26. Park, D.H., et al.: Multimodal explanations: justifying decisions and pointing to the evidence. In: CVPR, pp. 8779–8788 (2018)
27. Johnson, J., Hariharan, B., van der Maaten, L., Fei-Fei, L., Zitnick, C.L., Girshick, R.B.: Clevr: a diagnostic dataset for compositional language and elementary visual reasoning. In: CVPR, pp. 2901–2910 (2017)
28. Johnson, J., et al.: Inferring and executing programs for visual reasoning. In: ICCV, pp. 2989–2998 (2017)
29. Kayser, M., Camburu, O.M., Salewski, L., Emde, C., Do, V., Akata, Z., Lukasiewicz, T.: e-VIL: a dataset and benchmark for natural language explanations in vision-language tasks. In: ICCV, pp. 1244–1254 (2021)
30. Kim, J.M., Choe, J., Akata, Z., Oh, S.J.: Keep calm and improve visual feature attribution. In: ICCV, pp. 8350–8360 (2021)
31. Kim, J., Rohrbach, A., Darrell, T., Canny, J., Akata, Z.: Textual explanations for self-driving vehicles. In: ECCV, pp. 563–578 (2018)
32. Kim, S.S., Meister, N., Ramaswamy, V.V., Fong, R., Russakovsky, O.: Hive: evaluating the human interpretability of visual explanations. arXiv preprint arXiv:2112.03184 (2021)
33. Kingma, D.P., Ba, J.: Adam: a method for stochastic optimization. In: ICLR (2015)
34. Kottur, S., Moura, J.M., Parikh, D., Batra, D., Rohrbach, M.: CLEVR-dialog: a diagnostic dataset for multi-round reasoning in visual dialog. In: NAACL, pp. 582–595 (2019)
35. Li, Q., Tao, Q., Joty, S., Cai, J., Luo, J.: VQA-E: explaining, elaborating, and enhancing your answers for visual questions. In: ECCV, pp. 552–567 (2018)
36. Lin, C.Y.: ROUGE: a package for automatic evaluation of summaries. In: ACL, pp. 74–81 (2004)
37. Liu, R., Liu, C., Bai, Y., Yuille, A.L.: CLEVR-Ref+: diagnosing visual reasoning with referring expressions. In: CVPR, pp. 4185–4194 (2019)
38. Marasović, A., Bhagavatula, C., Park, J.s., Le Bras, R., Smith, N.A., Choi, Y.: Natural language rationales with full-stack visual reasoning: from pixels to semantic frames to commonsense graphs. In: EMNLP, pp. 2810–2829 (2020)
39. Papineni, K., Roukos, S., Ward, T., Zhu, W.J.: Bleu: a method for automatic evaluation of machine translation. In: ACL, pp. 311–318 (2002)
40. Patro, B., Patel, S., Namboodiri, V.: Robust explanations for visual question answering. In: WACV, pp. 1577–1586 (2020)
41. Perez, E., Strub, F., De Vries, H., Dumoulin, V., Courville, A.: FiLM: visual Reasoning with a general conditioning layer. In: AAAI, vol. 32 (2018)
42. Petsiuk, V., Das, A., Saenko, K.: Rise: randomized input sampling for explanation of black-box models. In: BMVC, p. 151 (2018)
43. Selvaraju, R.R., Cogswell, M., Das, A., Vedantam, R., Parikh, D., Batra, D.: Grad-cam: visual explanations from deep networks via gradient-based localization. In: ICCV, pp. 618–626 (2017)

44. Shi, J., Zhang, H., Li, J.: Explainable and explicit visual reasoning over scene graphs. In: CVPR, pp. 8376–8384 (2019)
45. Shih, K.J., Singh, S., Hoiem, D.: Where to look: Focus regions for visual question answering. In: CVPR. pp. 4613–4621 (2016)
46. Simonyan, K., Vedaldi, A., Zisserman, A.: Deep inside convolutional networks: Visualising image classification models and saliency maps. In: ICLR Workshop (2014)
47. Suarez, J., Johnson, J., Li, F.F.: Ddrprog: a clevr differentiable dynamic reasoning programmer. arXiv preprint arXiv:1803.11361 (2018)
48. Trott, A., Xiong, C., Socher, R.: Interpretable counting for visual question answering. In: ICLR (2018)
49. Vedantam, R., Szlam, A., Nickel, M., Morcos, A., Lake, B.M.: CURI: a benchmark for productive concept learning under uncertainty. In: ICML, pp. 10519–10529 (2021)
50. Vedantam, R., Zitnick, C.L., Parikh, D.: Cider: consensus-based image description evaluation. In: CVPR, pp. 4566–4575 (2015)
51. de Vries, H., Bahdanau, D., Murty, S., Courville, A.C., Beaudoin, P.: CLOSURE: assessing systematic generalization of CLEVR models. In: NeurIPS Workshop (2019)
52. Wu, J., Chen, L., Mooney, R.: Improving VQA and its explanations by comparing competing explanations. In: AAAI Workshop (2021)
53. Wu, J., Mooney, R.: Faithful multimodal explanation for visual question answering. In: ACL Workshop, pp. 103–112 (2019)
54. Xie, N., Lai, F., Doran, D., Kadav, A.: Visual entailment: a novel task for fine-grained image understanding. arXiv preprint arXiv:1901.06706 (2019)
55. Xu, H., Saenko, K.: Ask, attend and answer: exploring question-guided spatial attention for visual question answering. In: ECCV, pp. 451–466 (2016)
56. Yang, Z., He, X., Gao, J., Deng, L., Smola, A.: Stacked attention networks for image question answering. In: CVPR, pp. 21–29 (2016)
57. Zeiler, M.D., Fergus, R.: Visualizing and understanding convolutional networks. In: Fleet, D., Pajdla, T., Schiele, B., Tuytelaars, T. (eds.) ECCV 2014. LNCS, vol. 8689, pp. 818–833. Springer, Cham (2014). https://doi.org/10.1007/978-3-319-10590-1_53
58. Zhang, J., Bargal, S.A., Lin, Z., Brandt, J., Shen, X., Sclaroff, S.: Top-down neural attention by excitation backprop. Int. J. Comput. Vision **126**(10), 1084–1102 (2018). https://doi.org/10.1007/s11263-017-1059-x
59. Zhang, P., Goyal, Y., Summers-Stay, D., Batra, D., Parikh, D.: Yin and yang: balancing and answering binary visual questions. In: CVPR, pp. 5014–5022 (2016)
60. Zhu, Y., Groth, O., Bernstein, M., Fei-Fei, L.: Visual7w: grounded question answering in images. In: CVPR, pp. 4995–5004 (2016)

New Developments in Explainable AI

A Rate-Distortion Framework
for Explaining Black-Box Model Decisions

Stefan Kolek[1]([✉]), Duc Anh Nguyen[1], Ron Levie[3], Joan Bruna[2], and Gitta Kutyniok[1]

[1] Department of Mathematics, LMU Munich, Munich, Germany
kolek@math.lmu.de
[2] Courant Institute of Mathematical Sciences, NYU, New York, USA
[3] Faculty of Mathematics, Technion - Israel Institute of Technology, Haifa, Israel
levieron@technion.ac.il

Abstract. We present the *Rate-Distortion Explanation* (RDE) framework, a mathematically well-founded method for explaining black-box model decisions. The framework is based on perturbations of the target input signal and applies to any differentiable pre-trained model such as neural networks. Our experiments demonstrate the framework's adaptability to diverse data modalities, particularly images, audio, and physical simulations of urban environments.

1 Introduction

Powerful machine learning models such as deep neural networks are inherently opaque, which has motivated numerous explanation methods that the research community developed over the last decade [1,2,7,15,16,20,26,29]. The meaning and validity of an explanation depends on the underlying principle of the explanation framework. Therefore, a trustworthy explanation framework must align intuition with mathematical rigor while maintaining maximal flexibility and applicability. We believe the *Rate-Distortion Explanation* (RDE) framework, first proposed by [16], then extended by [9], as well as the similar framework in [2], meets the desired qualities. In this chapter, we aim to present the RDE framework in a revised and holistic manner. Our generalized RDE framework can be applied to any model (not just classification tasks), supports in-distribution interpretability (by leveraging in-painting GANs), and admits interpretation queries (by considering suitable input signal representations).

The typical setting of a (local) explanation method is given by a pre-trained model $\Phi : \mathbb{R}^n \to \mathbb{R}^m$, and a data instance $x \in \mathbb{R}^n$. The model Φ can be either a classification task with m class labels or a regression task with m-dimensional model output. The model decision $\Phi(x)$ is to be explained. In the original RDE framework [16], an explanation for $\Phi(x)$ is a set of feature components $S \subset \{1, \ldots, n\}$ in x that are deemed relevant for the decision $\Phi(x)$. The core principle behind the RDE framework is that a set $S \subset \{1, \ldots, n\}$ contains all the relevant components if $\Phi(x)$ remains (approximately) unchanged after modifying x_{S^c}, i.e.,

© The Author(s) 2022
A. Holzinger et al. (Eds.): xxAI 2020, LNAI 13200, pp. 91–115, 2022.
https://doi.org/10.1007/978-3-031-04083-2_6

the components in x that are not deemed relevant. In other words, S contains all relevant features if they are sufficient for producing the output $\Phi(x)$. To convey concise explanatory information, one aims to find the *minimal set* $S \subset \{1, \ldots, n\}$ with all the relevant components. As demonstrated in [16] and [31], the minimal relevant set $S \subset \{1, \ldots, n\}$ cannot be found combinatorially in an efficient manner for large input sizes. A meaningful approximation can nevertheless be found by optimizing a sparse continuous mask $s \in [0, 1]^n$ that has no significant effect on the output $\Phi(x)$ in the sense that $\Phi(x) \approx \Phi(x \odot s + (1 - s) \odot v)$ should hold for appropriate perturbations $v \in \mathbb{R}^n$, where $\odot$ denotes the componentwise multiplication. Suppose $d(\Phi(x), \Phi(y))$ is a measure of distortion (e.g. the ℓ_2-norm) between the model outputs for $x, y \in \mathbb{R}^n$ and $\mathcal{V}$ is a distribution over appropriate perturbations $v \sim \mathcal{V}$. An explanation in the RDE framework can be found as a solution mask s^* to the following minimization problem:

$$s^* := \operatorname*{arg\,min}_{s \in [0,1]^n} \mathbb{E}_{v \sim \mathcal{V}} \left[d\Big(\Phi(x), \Phi(x \odot s + (1 - s) \odot v)\Big) \right] + \lambda \|s\|_1,$$

where $\lambda > 0$ is a hyperparameter controlling the sparsity of the mask.

We further generalize the RDE framework to abstract input signal representations $x = f(h)$, where f is a data representation function with input h. The philosophy of the generalized RDE framework is that an explanation for generic input signals $x = f(h)$ should be some simplified version of the signal, which is interpretable to humans. This is achieved by demanding sparsity in a suitable representation system h, which ideally optimally represents the class of explanations that are desirable for the underlying domain and interpretation query. This philosophy underpins our experiments on image classification in the wavelet domain, on audio signal classification in the Fourier domain, and on radio map estimation in an urban environment domain. Therein we demonstrate the versatility of our generalized RDE framework.

2   Related Works

To our knowledge, the explanation principle of optimizing a mask $s \in [0, 1]^n$ has been first proposed in [7]. Fong et al. [7] explained image classification decisions by considering one of the two "deletion games": (1) optimizing for the smallest deletion mask that causes the class score to drop significantly or (2) optimizing for the largest deletion mask that has no significant effect on the class score. The original RDE approach [16] is based on the second deletion game and connects the deletion principle to rate-distortion-theory, which studies lossy data compression. Deleted entries in [7] were replaced with either constants, noise, or blurring and deleted entries in [16] were replaced with noise.

Explanation methods introduced before the "deletion games" principle from [7] were typically based upon gradient-based methods [26,29], propagation of activations in neurons [1,25], surrogate models [20], and game-theory [15].

Gradient-based methods such as smoothgrad [26] suffer from a lacking principle of relevance beyond local sensitivity. Reference-based methods such as Integrated Gradients [29] and DeepLIFT [25] depend on a reference value, which has no clear optimal choice. DeepLIFT and LRP assign relevance by propagating neuron activations, which makes them dependent on the implementation of Φ. LIME [20] uses an interpretable surrogate model that approximates Φ in a neighborhood around x. Surrogate model explanations are inherently limited for complex models Φ (such as image classifiers) as they only admit very local approximations. Generally, explanations that only depend on the model behavior on a small neighborhood U_x of x offer limited insight. Lastly, Shapley values-based explanations [15] are grounded in Shapley values from game-theory. They assign relevance scores as weighted averages of marginal contributions of respective features. Though Shapley values are mathematically well-founded, relevance scores cannot be computed exactly for common input sizes such as $n \geq 50$, since one exact relevance score generally requires $O(2^n)$ evaluations of Φ [30].

A notable difference between the RDE method and additive feature explanations [15] is that the values in the mask s^* do not add up to the model output. The additive property as in [15] takes the view that features individually contribute to the model output and relevance should be reflected by their contributions. We emphasize that the RDE method is designed to look for a *set* of relevant features and *not* an estimate of individual relative contributions. This is particularly desirable when only groups of features are interpretable, as for example in image classification tasks, where individual pixels do not carry any interpretable meaning. Similarly to Shapley values, the explanation in the RDE framework cannot be computed exactly, as it requires solving a non-convex minimization problem. However, the RDE method can take full advantage of modern optimization techniques. Furthermore, the RDE method is a model-agnostic explanation technique, with a mathematically principled and intuitive notion of relevance as well as enough flexibility to incorporate the model behavior on meaningful input regions of Φ.

The meaning of an explanation based on deletion masks $s \in [0,1]^n$ depends on the nature of the perturbations that replace the deleted regions. Random [7,16] or blurred [7] replacements $v \in \mathbb{R}^n$ may result in a data point $x \odot s + (1-s) \odot v$ that falls out of the natural data manifold on which Φ was trained on. This is a subtle though important problem, since such an explanation may depend on evaluations of Φ on data points from undeveloped decision regions. The latter motivates *in-distribution interpretability*, which considers meaningful perturbations that keep $x \odot s + (1-s) \odot v$ in the data manifold. [2] was the first work that suggested to use an inpainting-GAN to generate meaningful perturbations to the "deletion games". The authors of [9] then applied in-distribution interpretability to the RDE method in the challenging modalities *music* and *physical simulations of urban environments*. Moreover, they demonstrated that the RDE method in [16] can be extended to answer so-called "*interpretation queries*". For example, the RDE method was applied in [9] to an instrument classifier to answer the global interpretation query "*Is magnitude or phase in the signal more important for the classifier?*". Most recently, in [11], we introduced CartoonX as a novel explanation method for image classifiers, answering the interpretation query *"What is the relevant piece-wise smooth part of an image?"* by applying RDE in the wavelet basis of images.

3   Rate-Distortion Explanation Framework

Based on the original RDE approach from [16], in this section, we present a general formulation of the RDE framework and discuss several implementations. While [16] focuses merely on image classification with explanations in pixel representation, we will apply the RDE framework not only to more challenging domains but also to different input signal representations. Not surprisingly, the combinatorical optimization problem in the RDE framework, even in simpler form, is extremely hard to solve [16,31]. This motivates heuristic solution strategies, which will be discussed in Subsect. 3.2.

3.1   General Formulation

It is well-known that in practice there are different ways to describe a signal $x \in \mathbb{R}^n$. Generally speaking, x can be represented by a data representation function $f : \prod_{i=1}^{k} \mathbb{R}^{d_i} \to \mathbb{R}^n$,

$$x = f(h_1, \ldots, h_k), \tag{1}$$

for some inputs $h_i \in \mathbb{R}^{d_i}$, $d_i \in \mathbb{N}$, $i \in \{1, \ldots, k\}$, $k \in \mathbb{N}$. Note, we do not restrict ourselves to linear data representation functions f. To briefly illustrate the generality of this abstract representation, we consider the following examples.

Example 1 (Pixel representation). An arbitrary (vectorized) image $x \in \mathbb{R}^n$ can be simply represented pixelwise

$$x = \begin{bmatrix} x_1 \\ \vdots \\ x_n \end{bmatrix} = f(h_1, \ldots, h_n),$$

with $h_i := x_i$ being the individual pixel values and $f \colon \mathbb{R}^n \to \mathbb{R}^n$ being the identity transform.

Due to its simplicity, this standard basis representation is a reasonable choice when explaining image classification models. However, in many other applications, one requires more sophisticated representations of the signals, such as through a possibly redundant dictionary.

Example 2. Let $\{\psi_j\}_{j=1}^{k}$, $k \in \mathbb{N}$, be a dictionary in $\mathbb{R}^n$, e.g., a basis. A signal $x \in \mathbb{R}^n$ is represented as

$$x = \sum_{j=1}^{k} h_j \psi_j,$$

where $h_j \in \mathbb{R}$, $j \in \{1, \ldots, k\}$, are appropriate coefficients. In terms of the abstract representation (1), we have $d_j = 1$ for $j \in \{1, \ldots, k\}$ and f is the function that yields the weighted sum over ψ_j. Note that Example 1 can be seen as a special case of this representation.

The following gives an example of a non-linear representation function f.

Example 3. Consider the discrete inverse Fourier transform, defined as

$$f : \prod_{j=1}^{n} \mathbb{R}_+ \times \prod_{j=1}^{n} [0, 2\pi] \to \mathbb{C}^n,$$

$$\left[f(m_1, \ldots, m_n, \omega_1, \ldots, \omega_n) \right]_l := \frac{1}{n} \sum_{j=1}^{n} \underbrace{m_j e^{i\omega_j}}_{:=c_j \in \mathbb{C}} e^{i2\pi l(j-1)/n}, \ l \in \{1, \ldots, n\},$$

where m_j and ω_j are respectively the magnitude and the phase of the j-th discrete Fourier coefficient c_j. Thus every signal $x \in \mathbb{R}^n \subseteq \mathbb{C}^n$ can be represented in terms of (1) with f being the discrete inverse Fourier transform while h_j, $j = 1, \ldots, k$ (with $k = 2n$) being specified as $m_{j'}$ and $\omega_{j'}$, $j' = 1, \ldots, n$.

Further examples of dictionaries $\{\psi_j\}_{j=1}^{k}$ include the discrete wavelet [21], cosine [19] or shearlet [12] representation systems and many more. In these cases, the coefficients h_i are given by the forward transform and f is referred to as the backward transform. Note that in the above examples we have $d_i = 1$, i.e., the input vectors h_i are real-valued. In many situations, one is also interested in representations $x = f(h_1, \ldots, h_k)$ with $h_i \in \mathbb{R}^{d_i}$ where $d_i > 1$.

Example 4. Let $k = 2$ and define f again as the discrete inverse Fourier transform, but as a function of two components: (1) the entire magnitude spectrum and (2) the entire frequency spectrum, namely

$$f : \mathbb{R}_+^n \times [0, 2\pi]^n,$$

$$\left[f(m, \omega) \right]_l := \frac{1}{n} \sum_{j=1}^{n} \underbrace{m_j e^{i\omega_j}}_{:=c_n \in \mathbb{C}} e^{i2\pi l(j-1)/n}, \ l \in \{1, \ldots, n\}.$$

Similarly, instead of individual pixel values, one can consider patches of pixels in an image $x \in \mathbb{R}^n$ from Example 1 as the input vectors h_i to the identity transform f. We will come back to these examples in the experiments in Sect. 4.

Finally, we would like to remark that our abstract representation

$$x = f(h_1, \ldots, h_k)$$

also covers the cases where the signal is the output of a decoder or generative model f with inputs $h_1, \ldots, h_k$ as the code or the latent variables.

As was discussed in previous sections, the main idea of the RDE framework is to extract the relevant features of the signal based on the optimization over its perturbations defined through masks. The ingredients of this idea are formally defined below.

Definition 1 (Obfuscations and expected distortion). *Let $\Phi : \mathbb{R}^n \to \mathbb{R}^m$ be a model and $x \in \mathbb{R}^n$ a data point with a data representation $x = f(h_1, ..., h_k)$ as discussed above. For every mask $s \in [0,1]^k$, let $\mathcal{V}_s$ be a probability distribution over $\prod_{i=1}^{k} \mathbb{R}^{d_i}$. Then the* obfuscation *of x with respect to s and $\mathcal{V}_s$ is defined as the random vector*

$$y := f(s \odot h + (1-s) \odot v),$$

where $v \sim \mathcal{V}_s$, $(s \odot h)_i = s_i h_i \in \mathbb{R}^{d_i}$ and $((1-s) \odot v)_i = (1-s_i)v_i \in \mathbb{R}^{d_i}$ for $i \in \{1, \dots, k\}$. Furthermore, the expected distortion *of x with respect to the mask s and the perturbation distribution $\mathcal{V}_s$ is defined as*

$$D(x, s, \mathcal{V}_s, \Phi) := \underset{v \sim \mathcal{V}_s}{\mathbb{E}} \left[d\Big(\Phi(x), \Phi(y)\Big) \right],$$

where $d : \mathbb{R}^m \times \mathbb{R}^m \to \mathbb{R}_+$ is a measure of distortion between two model outputs.

In the RDE framework, the explanation is given by a mask that minimizes distortion while remaining relatively sparse. The rate-distortion-explanation mask is defined in the following.

Definition 2 (The RDE mask). *In the setting of Definition 1 we define the RDE mask as a solution $s^*(\ell)$ to the minimization problem*

$$\min_{s \in \{0,1\}^k} D(x, s, \mathcal{V}_s, \Phi) \quad s.t. \quad \|s\|_0 \leq \ell, \tag{2}$$

where $\ell \in \{1, \dots, k\}$ is the desired level of sparsity.

Here, the RDE mask is defined as the binary mask that minimizes the expected distortion while keeping the sparsity smaller than a certain threshold. Besides this, one could obviously also define the RDE mask as the sparsest binary mask that keeps the distortion lower than a given threshold, as defined in [16]. Geometrically, one can interpret the RDE mask as a subspace that is stable under Φ. If $x = f(h)$ is the input signal and s is the RDE mask for $\Phi(x)$ on the coefficients h, then the associated subspace $R_\Phi(s)$ is defined as the space of feasible obfuscations of x with s under $\mathcal{V}_s$, i.e.,

$$R_\Phi(s) := \{f(s \odot h + (1-s) \odot v) \mid v \in \mathrm{supp}\mathcal{V}_s\},$$

where $\mathrm{supp}\mathcal{V}_s$ denotes the support of the distribution $\mathcal{V}_s$. The model Φ will act similarly on signals in $R_\Phi(s)$ due to the low expected distortion $D(x, s, \mathcal{V}_s, \Phi)$— making the subspace stable under Φ. Note that RDE directly optimizes towards a subspace that is stable under Φ. If, instead, one would choose the mask s based on information of the gradient $\nabla\Phi(x)$ and Hessian $\nabla^2\Phi(x)$, then only a local neighborhood around x would tend to be stable under Φ due to the local nature of the gradient and Hessian. Before discussing practical algorithms to approximate the RDE mask in Subsect. 3.2, we will review frequently used obfuscation strategies, i.e., the distribution $\mathcal{V}_s$, and measures of distortion.

3.1.1   Obfuscation Strategies and in-Distribution Interpretability

The meaning of an explanation in RDE depends greatly on the nature of the perturbations $v \sim \mathcal{V}_s$. A particular choice of $\mathcal{V}_s$ defines an *obfuscation strategy*. Obfuscations are either *in-distribution*, i.e., if the obfuscation $f(s \odot h + (1-s) \odot v)$ lies on the natural data manifold that Φ was trained on, or *out-of-distribution* otherwise. Out-of-distribution obfuscations pose the following problem. The RDE mask (see Definition 2) depends on evaluations of Φ on obfuscations $f(s \odot h + (1-s) \odot v)$. If $f(s \odot h + (1-s) \odot v)$ is not on the natural data manifold that Φ was trained on, then it may lie in undeveloped regions of Φ. In practice, we are interested in explaining the behavior of Φ on realistic data and an explanation can be corrupted if Φ did not develop the region of out-of distribution points $f(s \odot h + (1-s) \odot v)$. One can guard against this by choosing $\mathcal{V}_s$ so that $f(s \odot h + (1-s) \odot v)$ is in-distribution. Choosing $\mathcal{V}_s$ in-distribution boils down to modeling the conditional data distribution – a non-trivial task.

Example 5 (In-distribution obfuscation strategy). In light of the recent success of generative adversarial networks (GANs) in generative modeling [8], one can train an in-painting GAN [32]

$$G(h, s, z) \in \prod_{i=1}^{k} \mathbb{R}^{d_i},$$

where z are random latent variables of the GAN, such that the obfuscation $f(s \odot h + (1-s) \odot G(h, s, z))$ lies on the natural data manifold (see also [2]). In other words, one can choose $\mathcal{V}_s$ as the distribution of $v := G(h, s, z)$, where the randomness comes from the random latent variables z.

Example 6 (Out-of-distribution obfuscation strategies). A very simple obfuscation strategy is Gaussian noise. In that case, one defines $\mathcal{V}_s$ for every $s \in [0,1]^k$ as $\mathcal{V}_s := \mathcal{N}(\mu, \Sigma)$, where μ and Σ denote a pre-defined mean vector and covariance matrix. In Sect. 4.1, we give an example of a reasonable choice for μ and Σ for image data. Alternatively, for images with pixel representation (see Example 1) one can mask out the deleted pixels by blurred inputs, $v = K * x$, where K is a suitable blur kernel.

Table 1. Common obfuscation strategies with their perturbation formulas.

Obfuscation strategy	Perturbation formula	In-distribution
Constant	$v \in \mathbb{R}^d$	–
Noise	$v \sim \mathcal{N}(\mu, \Sigma)$	–
Blurring	$v = K * x$	–
Inpainting-GAN	$v = G(h, s, z)$	✓

We summarize common obfuscation strategies for a given target signal in Table 1.

3.1.2   Measure of Distortion

Various options exist for the measure $d \colon \mathbb{R}^m \times \mathbb{R}^m \to \mathbb{R}$ of the distortion between model outputs. The measure of distortion should be chosen according to the task of the model $\Phi \colon \mathbb{R}^n \to \mathbb{R}^m$ and the objective of the explanation.

Example 7 (Measure of distortion for classification task). Consider a classification model $\Phi \colon \mathbb{R}^n \to \mathbb{R}^m$ and a target input signal $x \in \mathbb{R}^n$. The model Φ assigns to each class $j \in \{1, \dots, m\}$ a (pre-softmax) score $\Phi_j(x)$ and the predicted label is given by $j^* := \arg\max_{j \in \{1,\dots,m\}} \Phi_j(x)$. One commonly used measure of the distortion between the outputs at x and another data point $y \in \mathbb{R}^n$ is given as

$$d_1\big(\Phi(x), \Phi(y)\big) := \big(\Phi_{j^*}(x) - \Phi_{j^*}(y)\big)^2.$$

On the other hand, the vector $[\Phi_j(x)]_{j=1}^m$ is usually normalized to a probability vector $[\tilde{\Phi}_j(x)]_{j=1}^m$ by applying the softmax function, namely $\tilde{\Phi}_j(x) := \exp\Phi_j(x)/\sum_{i=1}^m \exp\Phi_i(x)$. This, in turn, gives another measure of the distortion between $\Phi(x), \Phi(y) \in \mathbb{R}^m$, namely

$$d_2\big(\Phi(x), \Phi(y)\big) := \big(\tilde{\Phi}_{j^*}(x) - \tilde{\Phi}_{j^*}(y)\big)^2,$$

where $j^* := \arg\max_{j \in \{1,\dots,m\}} \Phi_j(x) = \arg\max_{j \in \{1,\dots,m\}} \tilde{\Phi}_j(x)$. An important property of the softmax function is the invariance under translation by a vector $[c, \dots, c]^\top \in \mathbb{R}^m$, where $c \in \mathbb{R}$ is a constant. By definition, only d_2 respects this invariance while d_1 does not.

Example 8 (Measure of distortion for regression task). Consider a regression model $\Phi \colon \mathbb{R}^n \to \mathbb{R}^m$ and an input signal $x \in \mathbb{R}^n$. One can then define the measure of distortion between the outputs of x and another data point $y \in \mathbb{R}^n$ as

$$d_3\big((\Phi(x), \Phi(y)\big) := \|\Phi(x) - \Phi(y)\|_2^2.$$

Sometimes it is reasonable to consider a certain subset of components $J \subseteq \{1, \dots, m\}$ of the output vectors instead of all m entries. Denoting the vector formed by corresponding entries by $\Phi_J(x)$, the measure of distortion between the outputs can be defined as

$$d_4\big((\Phi(x), \Phi(y)\big) := \|\Phi_J(x) - \Phi_J(y)\|_2^2.$$

The measure d_4 will be used in our experiments for radio maps in Subsect. 4.3.

3.2   Implementation

The RDE mask from Definition 2 was defined as a solution to

$$\min_{s \in \{0,1\}^k} \quad D(x, s, \mathcal{V}_s, \Phi) \quad \text{s.t.} \quad \|s\|_0 \le \ell.$$

In practice, we need to relax this problem. We offer the following three approaches.

3.2.1 ℓ_1-relaxation with Lagrange Multiplier

The RDE mask can be approximately computed by finding an approximate solution to the following relaxed minimization problem:

$$\min_{s \in [0,1]^k} \quad D(x, s, \mathcal{V}_s, \Phi) + \lambda \|s\|_1, \qquad (\mathcal{P}_1)$$

where $\lambda > 0$ is a hyperparameter for the sparsity level. Note that the optimization problem is not necessarily convex, thus the solution might not be unique.

The expected distortion $D(x, s, \mathcal{V}_s, \Phi)$ can typically be approximated with simple Monte-Carlo estimates, i.e., by averaging i.i.d. samples from $\mathcal{V}_s$. After estimating $D(x, s, \mathcal{V}_s, \Phi)$, one can optimize the mask s with stochastic gradient descent (SGD) to solve the optimization problem $(\mathcal{P}_1)$.

3.2.2 Bernoulli Relaxation

By viewing the binary mask as Bernoulli random variables $s \sim \text{Ber}(\theta)$ and optimizing over θ, one can guarantee that the expected distortion $D(x, s, \mathcal{V}_s, \Phi)$ is evaluated on binary masks $s \in \{0,1\}^n$. To encourage sparsity of the resulting mask, one can still apply ℓ_1-regularization on s, giving rise to the following optimization problem:

$$\min_{\theta \in [0,1]^k} \quad \mathbb{E}_{s \sim \text{Ber}(\theta)} \left[D(x, s, \mathcal{V}_s, \Phi) + \lambda \|s\|_1 \right]. \qquad (\mathcal{P}_2)$$

Optimizing the parameter θ requires a continuous relaxation to apply SGD. This can be done using the concrete distribution [17], which samples s from a continuous relaxation of the Bernoulli distribution.

3.2.3 Matching Pursuit

As an alternative, one can also perform matching pursuit [18]. Here, the non-zero entries of $s \in \{0,1\}^n$ are determined sequentially in a greedy fashion to minimize the resulting distortion in each step. More precisely, we start with a zero mask $s^0 = 0$ and gradually build up the mask by updating s^t at step t by the rule given by

$$s^{t+1} = s^t + \arg\min_{e_j : s_j^t = 0} D(x, s^t + e_j, \mathcal{V}_s, \Phi).$$

Here, the minimization is taken over all standard basis vectors $e_j \in \mathbb{R}^k$ with $s_j^t = 0$. The algorithm terminates when reaching some desired error tolerance or after a prefixed number of iterations. While this means that in each iteration we have to test every entry of s, it is applicable when k is small or when we are only interested in very sparse masks.

4 Experiments

With our experiments, we demonstrate the broad applicability of the generalized RDE framework. Moreover, our experiments illustrate how different choices of obfuscation strategies, optimization procedures, measures of distortion, and input signal representations, discussed in Sect. 3.1, can be leveraged in practice. We explain model decisions on various challenging data modalities and tailor the input signal representation and measure of distortion to the domain and interpretation query. In Sect. 4.1, we focus on image classification, a common baseline task in the interpretability literature. In Sects. 4.2 and 4.3, we consider two other data modalities that are often unexplored. Section 4.2 focuses on audio data, where the underlying task is to classify acoustic instruments based on a short audio sample of distinct notes, while in Sect. 4.3, the underlying task is a regression with data in the form of physical simulations in urban environments. We also believe our explanation framework sustains applications beyond interpretability tasks. An example is given in Sect. 4.3.2, where we add an RDE inspired regularizer to the training objective of a radio map estimation model.

4.1 Images

We begin with the most ordinary domain in the interpretability literature: image classification tasks. The authors of [16] applied RDE to image data before by considering pixel-wise perturbations. We refer to this method as *Pixel RDE*. Other explanation methods [1–3,20], have also previously exclusively operated in the pixel domain. In [11], we challenged this customary practice by successfully applying RDE in a wavelet basis, where sparsity translates into piece-wise smooth images (also called cartoon-like images). The novel explanation method was coined *CartoonX* [11] and extracts the relevant piece-wise smooth part of an image. First, we review the Pixel RDE method and present experiments on the ImageNet dataset [4], which is commonly considered a challenging classification task. Finally, we present CartoonX and discuss its advantages. For all the ImageNet experiments, we use the pre-trained *MobileNetV3-Small* [10], which achieved a top-1 accuracy of 67.668% and a top-5 accuracy of 87.402%, as the classifier.

4.1.1 Pixel RDE

Consider the following pixel-wise representation of an RGB image $x \in \mathbb{R}^{3 \times n}$: $f : \prod_{i=1}^{n} \mathbb{R}^3 \to \mathbb{R}^{n \times 3}$, $x = f(h_1, ..., h_n)$, where $h_i \in \mathbb{R}^3$ represents the three color channel values of the i-th pixel in the image x, i.e. $(x_{i,j})_{j=1,...,3} = h_i$. In pixel RDE a sparse mask $s \in [0, 1]^n$ with n entries—one for each pixel—is optimized to achieve low expected distortion $D(x, s, \mathcal{V}_s, \Phi)$. The obfuscation of an image x with the pixel mask s and a distribution $v \sim \mathcal{V}_s$ on $\prod_{i=1}^{n} \mathbb{R}^3$ is defined as $f(s \odot h + (1 - s) \odot v)$. In our experiments, we initialize the mask with ones, i.e., $s_i = 1$ for every $i \in \{1, \ldots, n\}$, and consider Gaussian noise perturbations $\mathcal{V}_s = \mathcal{N}(\mu, \Sigma)$. We set the noise mean $\mu \in \mathbb{R}^{3 \times n}$ as the pixel value mean of the original image x and the covariance matrix $\Sigma := \sigma^2 \operatorname{Id} \in \mathbb{R}^{3n \times 3n}$ as

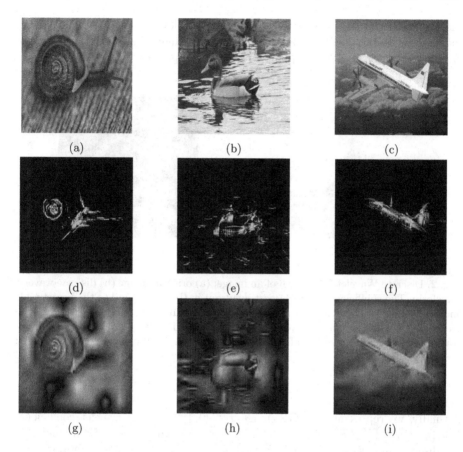

Fig. 1. Top row: original images correctly classified as (a) snail, (b) male duck, and (c) airplane. Middle row: Pixel RDEs. Bottom row: CartoonX. Notably, CartoonX is roughly piece-wise smooth and overall more interpretable than the jittery Pixel RDEs.

a diagonal matrix with $\sigma > 0$ defined as the pixel value standard deviation of the original image x. We then optimize the pixel mask s for 2000 gradient descent steps on the ℓ_1-relaxation of the RDE objective (see Sect. 3.2.1). We computed the distortion $d(\Phi(x), \Phi(y))$ in $D(x, s, \mathcal{V}_s, \Phi)$ in the post-softmax activation of the predicted label multiplied by a constant $C = 100$, i.e., $d(\Phi(x), \Phi(y)) := C(\Phi_{j^*}(x) - \Phi_{j^*}(y))^2$.

The expected distortion $D(x, s, \mathcal{V}_s, \Phi)$ was approximated as a simple Monte-Carlo estimate after sampling 64 noise perturbations. For the sparsity level, we set the Lagrange multiplier to $\lambda = 0.6$. All images were resized to 256×256 pixels. The mask was optimized for 2000 steps using the Adam optimizer with step size 0.003. In the middle row of Fig. 1, we show three example explanations with Pixel RDE for an image of a snail, a male duck, and an airplane, all from the ImageNet dataset. Pixel RDE highlights as relevant both the snail's inner shell and part of its head, the lower segment of the male duck along with various lines in the water, and the airplane's fuselage and part of its rudder.

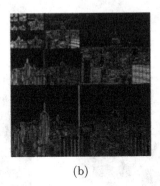

(a)   (b)

Fig. 2. Discrete Wavelet Transform of an image: (a) original image (b) discrete wavelet transform. The coefficients of the largest quadrant in (b) correspond to the lowest scale and coefficients of smaller quadrants gradually build up to the highest scales, which are located in the four smallest quadrants. Three nested L-shaped quadrants represent horizontal, vertical and diagonal edges at a resolution determined by the associated scale.

4.1.2   CartoonX

Formally, we represent an RGB image $x \in [0,1]^{3 \times n}$ in its wavelet coefficients $h = \{h_i\}_{i=1}^n \in \prod_{i=1}^n \mathbb{R}^3$ with $J \in \{1, \ldots, \lfloor \log_2 n \rfloor\}$ scales as $x = f(h)$, where f is the discrete inverse wavelet transform. Each $h_i = (h_{i,c})_{c=1}^3 \subseteq \mathbb{R}^3$ contains three wavelet coefficients of the image, one for each color channel and is associated with a scale $k_i \in \{1, \ldots, J\}$ and a position in the image. Low scales describe high frequencies and high scales describe low frequencies at the respective image position. We briefly illustrate the wavelet coefficients in Fig. 2, which visualizes the discrete wavelet transform of an image. CartoonX [11] is a special case of the generalized RDE framework, particularly a special case of Example 2, and optimizes a sparse mask $s \in [0,1]^n$ on the wavelet coefficients (see Fig. 3c) so that the expected distortion $D(x, s, \mathcal{V}_s, \Phi)$ remains small. The obfuscation of an image x with a wavelet mask s and a distribution $v \sim \mathcal{V}_s$ on the wavelet coefficients is $f(s \odot h + (1 - s) \odot v)$. In our experiments, we used Gaussian noise perturbations and chose the standard deviation and mean adaptively for each scale: the standard deviation and mean for wavelet coefficients of scale $j \in \{1, \ldots, J\}$ were chosen as the standard deviation and mean of the wavelet coefficients of scale $j \in \{1, \ldots, J\}$ of the original image. Figure 3d shows the obfuscation $f(s \odot h + (1 - s) \odot v)$ with the final wavelet mask s after the RDE optimization procedure. In Pixel RDE, the mask itself is the explanation as it lies in pixel space (see middle row in Fig. 1), whereas the CartoonX mask lies in the wavelet domain. To go back to the natural image domain, we multiply the wavelet

Fig. 3. CartoonX machinery: (a) image classified as park-bench, (b) discrete wavelet transform of the image, (c) final mask on the wavelet coefficients after the RDE optimization procedure, (d) obfuscation with final wavelet mask and noise, (e) final CartoonX, (f) Pixel RDE for comparison.

mask element-wise with the wavelet coefficients of the original greyscale image and invert this product back to pixel space with the discrete inverse wavelet transform. The inversion is finally clipped into $[0, 1]$ as are obfuscations during the RDE optimization to avoid overflow (we assume here the pixel values in x are normalized into $[0, 1]$). The clipped inversion in pixel space is the final CartoonX explanation (see Fig. 3e).

The following points should be kept in mind when interpreting the final CartoonX explanation, i.e., the inversion of the wavelet coefficient mask: (1) CartoonX provides the relevant pice-wise smooth part of the image. (2) The inversion of the wavelet coefficient mask was not optimized to be sparse in pixel space but in the wavelet basis. (3) A region that is black in the inversion could nevertheless be relevant if it was already black in the original image. This is due to the multiplication of the mask with the wavelet coefficients of the greyscale image before taking the discrete inverse wavelet transform. (4) Bright high resolution regions are relevant in high resolution and bright low resolution regions are relevant in low resolution. (5) It is inexpensive for CartoonX to mark large regions in low resolution as relevant. (6) It is expensive for CartoonX to mark large regions in high resolution as relevant.

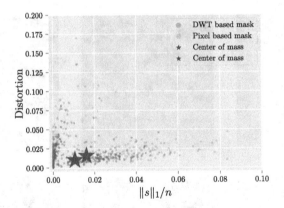

Fig. 4. Scatter plot of rate-distortion in pixel basis and wavelet basis. Each point is an explanation of a distinct image in the ImageNet dataset with distortion and normalized ℓ_1-norm measured for the final mask. The wavelet mask achieves lower distortion than the pixel mask, while using less coefficients.

In Fig. 1, we compare CartoonX to Pixel RDE. The piece-wise smooth wavelet explanations are more interpretable than the jittery Pixel RDEs. In particular, CartoonX asserts that the snail's shell without the head suffices for the classification, unlike Pixel RDE, which insinuated that both the inner shell and part of the head are relevant. Moreover, CartoonX shows that the water gives the classifier context for the classification of the duck, which one could have only guessed from the Pixel RDE. Both Pixel RDE and CartoonX state that the head of the duck is not relevant. Lastly, CartoonX, like Pixel RDE, confirms that the wings play a subordinate role in the classification of the airplane.

4.1.3   Why Explain in the Wavelet Basis?

Wavelets provide optimal representation for piece-wise smooth 1D functions [5], and represent 2D piece-wise smooth images, also called *cartoon-like images* [12], efficiently as well [21]. Indeed, sparse vectors in the wavelet coefficient space encode cartoon-like images reasonably well [27], certainly better than sparse pixel representations. Moreover, the optimization process underlying CartoonX produces sparse vectors in the wavelet coefficient space. Hence CartoonX typically generates cartoon-like images as explanations. This is the fundamental difference to Pixel RDE, which produces rough, jittery, and pixel-sparse explanations. Cartoon-like images are more interpretable and provide a natural model of simplified images. Since the goal of the RDE explanation is to generate an easy to interpret simplified version of the input signal, we argue that CartoonX explanations are more appropriate for image classification than Pixel RDEs. Our experiments confirm that the CartoonX explanations are roughly piece-wise smooth explanations and are overall more interpretable than Pixel RDEs (see Fig. 1).

4.1.4 CartoonX Implementation

Throughout our CartoonX experiments we chose the Daubechies 3 wavelet system, $J = 5$ levels of scales and zero padding for the discrete wavelet transform. For the implementation of the discrete wavelet transform, we used the Pytorch Wavelets package, which supports gradient computation in Pytorch. Distortion was computed as in the Pixel RDE experiments. The perturbations $v \sim \mathcal{V}_s$ on the wavelet coefficients were chosen as Gaussian noise with standard deviation and mean computed adaptively per scale. As in the Pixel RDE experiments, the wavelet mask was optimized for 2000 steps with the Adam optimizer to minimize the ℓ_1-relaxation of the RDE objective. We used $\lambda = 3$ for CartoonX.

4.1.5 Efficiency of CartoonX

Finally, we compare Pixel RDE to CartoonX quantitatively by analyzing the distortion and sparsity associated with the final explanation mask. Intuitively, we expect the CartoonX method to have an efficiency advantage, since the discrete wavelet transform already encodes natural images sparsely, and hence less wavelet coefficients are required to represent images than pixel coefficients. Our experiments confirmed this intuition, as can be seen in the scatter plot in Fig. 4.

4.2 Audio

We consider the NSynth dataset [6], a library of short audio samples of distinct notes played on a variety of instruments. We pre-process the data by computing the power-normalized magnitude spectrum and phase information using the discrete Fourier transform on a logarithmic scale from 20 to 8000 Hertz. Each data instance is then represented by the magnitude and the phase of its Fourier coefficients as well as the discrete inverse Fourier transform (see Example 3).

4.2.1 Explaining the Classifier

Our model Φ is a network trained to classify acoustic instruments. We compute the distortion with respect to the pre-softmax scores, i.e., deploy d_1 in Example 7 as the measure of distortion. We follow the obfuscation strategy described in Example 5 and train an inpainter G to generate the obfuscation $G(h, s, z)$. Here, h corresponds to the representation of a signal, s is a binary mask and z is a normally distributed seed to the generator.

We use a residual CNN architecture for G with added noise in the input and deep features. More details can be found in Sect. 4.2.3. We train G until the outputs are found to be satisfactory, exemplified by the outputs in Fig. 5.

To compute the explanation maps, we numerically solve $(\mathcal{P}_2)$ as discussed in Subsect. 3.2. In particular, s is a binary mask indicating whether the phase and magnitude information of a certain frequency should be dropped and is specified as a Bernoulli variable $s \sim \mathrm{Ber}(\theta)$. We chose a regularization parameter of $\lambda = 50$ and minimized the corresponding objective using the Adam optimizer with a step size of 10^{-5} in 10^6 iterations. For the concrete distribution, we used a temperature of 0.1. Two examples resulting from this process can be seen in Fig. 6.

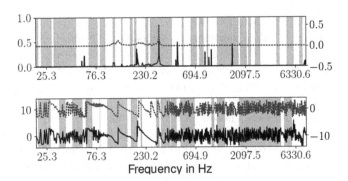

Fig. 5. Inpainted Bass: Example inpainting from G. The bottom plot depicts phase versus frequency and the top plot depicts magnitude versus frequency. The random binary mask is represented by the green parts. The axes for the inpainted signal (black) and the original signal (blue dashed) are offset to improve visibility. Note how the inpainter generates plausible peaks in the magnitude and phase spectra, especially with regard to rapid ($\geq$600 Hz) versus smooth (<270 Hz) changes in phase. (Color figure online)

Notice here that the method actually shows a strong reliance of the classifier on low frequencies (30 Hz–60 Hz) to classify the top sample in Fig. 6 as a guitar, as only the guitar samples have this low frequency slope in the spectrum. We can also see in contrast that classifying the bass sample relies more on the continuous signal 100 Hz and 230 Hz.

4.2.2 Magnitude vs Phase

In the above experiment, we have represented the signals by the magnitude and phase information at each frequency, hence the mask s acts on each frequency. Now we consider the *interpretation query* of whether the entire magnitude spectrum or the entire phase spectrum is more relevant for the prediction. Accordingly, we consider the representation discussed in Example 4 and apply the mask s to turn off or on the whole magnitude spectrum or the phase information. Furthermore, we can optimize s not only for one datum but for all samples from a class. This extracts the information whether magnitude or phase is more important for predicting samples from a specific class.

For this, we again minimized ($\mathcal{P}_2$) (meaned over all samples of a class) with θ as the Bernoulli parameter using the Adam optimizer for 2×10^5 iterations with a step size of 10^{-4} and the regularization parameter $\lambda = 30$. Again, a temperature of $t = 0.1$ was used for the concrete distribution.

From the results of these computations, which can be seen in Table 2, we can observe that there is a clear difference on what the classifier bases its decision on across instruments. The classification of most instruments is largely based on phase information. For the mallet, the values are low for magnitude and phase, which means that the expected distortion is very low compared to the ℓ_1-norm of the mask, even when the signal is completely inpainted. This underlines that the

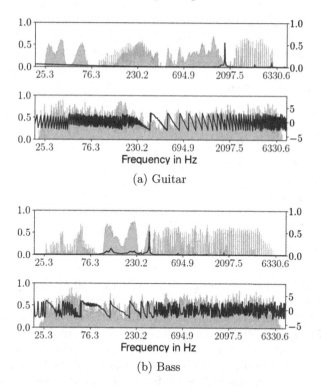

Fig. 6. Interpreting NSynth Model: The optimized importance parameter θ (green) overlayed on top of the DFT (blue). For each of guitar and bass, the top graph shows the power-normalized magnitude and the bottom the phase. Notice the solid peaks 30 Hz and 60 Hz for guitar and 100 Hz and 230 Hz for bass. These occur because the model is relying on those parts of the spectra, for the classification. Notice also how many parts of the spectrum are important even when the magnitude is near zero. This indicates that the model pays attention to whether those frequencies are missing. (Color figure online)

regularization parameter λ may have to be adjusted for different data instances, especially when measuring distortion in the pre-softmax scores.

4.2.3   Architecture of the Inpainting Network G

Here, we briefly describe the architecture of the inpainting network G that was used to generate obfuscations to the target signals. In particular, Fig. 7 shows the diagram of the network G and Table 3 shows information about its layers.

4.3   Radio Maps

In this subsection, we assume a set of transmitting devices (Tx) broadcasting a signal within a city. The received strength varies with location and depends

Table 2. Magnitude importance versus phase importance.

Instrument	Magnitude importance	Phase importance
Organ	0.829	1.0
Guitar	0.0	0.999
Flute	0.092	1.0
Bass	1.0	1.0
Reed	0.136	1.0
Vocal	1.0	1.0
Mallet	0.005	0.217
Brass	0.999	1.0
Keyboard	0.003	1.0
String	1.0	0.0

on physical factors such as line of sight, reflection, and diffraction. We consider the regression problem of estimating a function that assigns the proper signal strength to each location in the city. Our dataset $\mathcal{D}$ is RadioMapSeer [14] containing 700 maps, 80 Tx per map, and a corresponding grayscale label encoding the signal strength at every location. Our model Φ receives as input $x = [x^{(0)}, x^{(1)}, x^{(2)}]$, where $x^{(0)}$ is a binary map of the Tx locations, $x^{(1)}$ is a noisy binary map of the city (where a few buildings are missing), and $x^{(2)}$ is a grayscale image representing a number of ground truth measurements of the strength of the signal at the measured locations and zero elsewhere. We apply the UNet [13,14,22] architecture and train Φ to output the estimation of the signal strength throughout the city that interpolates the input measurements.

Apart from the model Φ, we also have a simpler model Φ_0, which only receives the city map and the Tx locations as inputs and is trained with unperturbed input city maps. This second model Φ_0 will be deployed to inpaint measurements to input to Φ. See Fig. 8a, 8b, and 8c for examples of a ground truth map and estimations for Φ and Φ_0, respectively.

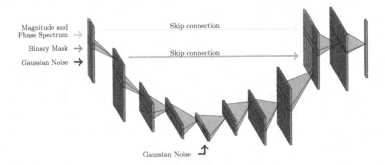

Fig. 7. Diagram of the inpainting network for NSynth.

Table 3. Layer table of the Inpainting model for the NSynth task.

Layer	Filter size	Output shape	# Params
Conv1d-1	21	$[-1, 32, 1024]$	4,736
ReLU-2		$[-1, 32, 1024]$	0
Conv1d-3	21	$[-1, 64, 502]$	43,072
ReLU-4		$[-1, 64, 502]$	0
BatchNorm1d-5		$[-1, 64, 502]$	128
Conv1d-6	21	$[-1, 128, 241]$	172,160
ReLU-7		$[-1, 128, 241]$	0
BatchNorm1d-8		$[-1, 128, 241]$	256
Conv1d-9	21	$[-1, 16, 112]$	43,024
ReLU-10		$[-1, 16, 112]$	0
BatchNorm1d-11		$[-1, 16, 112]$	32
ConvTranspose1d-12	21	$[-1, 64, 243]$	43,072
ReLU-13		$[-1, 64, 243]$	0
BatchNorm1d-14		$[-1, 64, 243]$	128
ConvTranspose1d-15	21	$[-1, 128, 505]$	172,160
ReLU-16		$[-1, 128, 505]$	0
BatchNorm1d-17		$[-1, 128, 505]$	256
ConvTranspose1d-18	20	$[-1, 64, 1024]$	163,904
ReLU-19		$[-1, 64, 1024]$	0
BatchNorm1d-20		$[-1, 64, 1024]$	128
Skip Connection		$[-1, 103, 1024]$	0
Conv1d-21	7	$[-1, 128, 1024]$	92,416
ReLU-22		$[-1, 128, 1024]$	0
Conv1d-23	7	$[-1, 2, 1024]$	1,794
ReLU-24		$[-1, 2, 1024]$	0
Total number of parameters			737,266

4.3.1   Explaining Radio Map Φ

Observe that in Fig. 8a there is a missing building in the input (the black one) and in Fig. 8b, Φ in-fills this building with a shadow. As a black box method, it is unclear why it made this decision. Did it rely on signal measurements or on building patterns? To address this, we consider each building as a cluster of pixels and each measurement as potential targets for our mask $s = [s^{(1)}, s^{(2)}]$, where $s^{(1)}$ acts on buildings and $s^{(2)}$ acts on measurements. We then apply matching pursuit (see Subsect. 3.2.3) to find a minimal mask s of critical components (buildings and measurements).

To be precise, suppose we are given a target input signal $x = [x^{(0)}, x^{(1)}, x^{(2)}]$. Let k_1 denote the number of buildings in $x^{(1)}$ and k_2 denote the number of

(a) Ground Truth   (b) Φ Estimation   (c) Φ_0 Estimation

Fig. 8. Radio map estimations: The radio map (gray), input buildings (blue), and input measurements (red). (Color figure online)

measurements in $x^{(2)}$. Consider the function f_1 that takes as inputs vectors in $\{0,1\}^{k_1}$, which indicate the existence of buildings in $x^{(1)}$, and maps them to the corresponding city map in the original city map format. Analogously, consider the function f_2 that takes as input the measurements in $\mathbb{R}^{k_2}$ and maps them to the corresponding grayscale image of the original measurements format. Then, f_1 and f_2 encode the locations of the buildings and measurements in the target signal $x = [x^{(0)}, f_1(h^{(1)}), f_2(h^{(2)})]$, where $h^{(1)}$ and $h^{(2)}$ denotes the building and measurement representation of x in f_1 and f_2. When $s^{(1)}$ has a zero entry, i.e., a building in $h^{(1)}$ was not selected, we replace the value in the obfuscation with zero (this corresponds to a constant perturbation equal to zero). Then, the obfuscation of the target signal x with a mask $s = [s^{(1)}, s^{(2)}]$ and perturbations $v = [v^{(1)}, v^{(2)}] := [0, v^{(2)}]$ becomes:

$$y := [x^{(0)}, f_1(s^{(1)} \odot h^{(1)}), f_2(s^{(2)} \odot h^{(2)} + (1 - s^{(2)}) \odot v^{(2)})].$$

While it is natural to model masking out a building by simply zeroing out the corresponding cluster of pixels by choosing $v^{(1)} = 0$, we need to also properly choose $v^{(2)}$ for the entries, where the mask $s^{(2)}$ takes value 0, in order to obtain appropriate obfuscations. For this, we can deploy the second model Φ_0 as an inpainter. We consider the following two extreme obfuscation strategies. The first is to set also $v^{(2)}$ to zero, i.e., simply remove the unchosen measurements from the input, with the underlying assumption being that any subset of measurements is valid for a city map. In the other extreme case, we inpaint all unchosen measurements by sampling at their locations the estimated radio map obtained by Φ_0 based on the buildings selected by $s^{(1)}$.

The two extreme measurement completion methods correspond to two extremes of the interpretation query. Filling-in the missing measurements by Φ_0 tends to overestimate the strength of the signal because there are fewer buildings to obstruct the transmissions. The empty mask will complete all measurements to the maximal possible signal strength – the free space radio map. The overestimation in signal strength is reduced when more measurements and buildings are chosen, resulting in darker estimated radio maps. Thus, this strategy is related to the query of which measurements and buildings are important to darken the free space radio map, turning it to the radio map produced by Φ. In the other extreme, adding more measurements to the mask with a fixed set of buildings typically brightens the resulting radio map. This allows us to answer which measurements are most important for brightening the radio map.

Between these two extreme strategies lies a continuum of completion methods where a random subset of the unchosen measurements is sampled from Φ_0, while the rest are set to zero. Examples of explanations of a prediction $\Phi(x)$ according to these methods are presented in Fig. 9. Since we only care about specific small patches exemplified by the green boxes, the distortion here is measured with respect to the ℓ_2 distance between the output images restricted to the corresponding region (see also Example 8).

(a) Estimated map.

(b) Explanation: Inpaint all unchosen measurements.

(c) Explanation: Inpaint 2.5% of unchosen measurements.

Fig. 9. Radio map queries and explanations: The radio map (gray), input buildings (blue), input measurements (red), and area of interest (green box). Middle represents the query "How to fill in the image with shadows", while right is the query "How to fill in the image both with shadows and bright spots?". We inpaint with Φ_0. (Color figure online)

When the query is how to darken the free space radio map (Fig. 9), the optimized mask s suggests that samples in the shadow of the missing building are the most influential in the prediction. These dark measurements are supposed to be in line-of-sight of a Tx, which indicates that the network deduced that there is a missing building. When the query is how to fill in the image both with shadows and bright spots (Fig. 9c), both samples in the shadow of the missing building and samples right before the building are influential. This indicates that the network used the bright measurements in line-of-sight and avoided predicting an overly large building. To understand the chosen buildings, note that Φ is based on a composition of UNets and is thus interpreted as a procedure of extracting high level and global information from the inputs to synthesize the output. The locations of the chosen buildings in Fig. 9 reflect this global nature.

4.3.2   Interpretation-Driven Training

We now discuss an example application of the explanation obtained by the RDE approach described above, called *interpretation driven training* [23,24,28]. When a missing building is in line-of-sight of a Tx, we would like Φ to reconstruct this building relying on samples in the shadow of the building rather than patterns in the city. To reduce the reliance of Φ on the city information in this situation, one can add a regularization term in the training loss which promotes explanations relying on measurements. Suppose $x = [x^{(0)}, x^{(1)}, x^{(2)}]$ contains a missing input building in line-of-sight of the Tx location and denote the subset of pixels of the missing building in the city map as J_x. Denote the prediction by Φ restricted to the subset J_x as Φ_{J_x}. Moreover, define $\tilde{x} := [x^{(0)}, 0, x^{(2)}]$ to be the modification of x with all input buildings masked out. We then define the *interpretation loss* for x as

$$\ell_{\text{int}}(\Phi, x) := \|\Phi_{J_x}(x) - \Phi_{J_x}(\tilde{x})\|_2^2.$$

 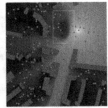  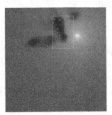

(a) Vanilla Φ estimation   (b) Interpretation-driven Φ_{int} estimation   (c) Vanilla Φ explanation   (d) Interpretation-driven Φ_{int} explanation

Fig. 10. Radio map estimations, interpretation driven training vs vanilla training: The radio map (gray), input buildings (blue), input measurements (red), and domain of the missing building (green box). (Color figure online)

The interpretation driven training objective then regularizes Φ during training by adding the interpretation loss for all inputs x that contain a missing input building in line-of-sight of the Tx location. An example comparison between explanations of the vanilla RadioUNet Φ and the interpretation driven network Φ_{int} is given in Fig. 10.

5   Conclusion

In this chapter, we presented the *Rate-Distortion Explanation* (RDE) framework in a revised and comprehensive manner. Our framework is flexible enough to answer various interpretation queries by considering suitable data representations tailored to the underlying domain and query. We demonstrate the latter and the overall efficacy of the RDE framework on an image classification task, on an audio signal classification task, and on a radio map estimation task, a seldomly explored regression task.

Acknowledgements. G.K. acknowledges partial support by the ONE Munich Strategy Forum (LMU Munich, TU Munich, and the Bavarian Ministery for Science and Art), the German Research Foundation under Grants DFG-SPP-2298, KU 1446/31-1 and KU 1446/32-1, and the BMBF under Grant MaGriDo. R.L. acknowledges support by the DFG SPP 1798, KU 1446/21-2 "Compressed Sensing in Information Processin" through Project Massive MIMO-II.

References

1. Bach, S., Binder, A., Montavon, G., Klauschen, F., Müller, K.R., Samek, W.: On pixel-wise explanations for non-linear classifier decisions by layer-wise relevance propagation. PLoS ONE **10**(7), e0130140 (2015)
2. Chang, C., Creager, E., Goldenberg, A., Duvenaud, D.: Explaining image classifiers by counterfactual generation. In: Proceedings of the 7th International Conference on Learning Representations, ICLR (2019)
3. Dabkowski, P., Gal, Y.: Real time image saliency for black box classifiers. In: Proceedings of the 31st International Conference on Neural Information Processing Systems, NeurIPS, pp. 6970–6979 (2017)
4. Deng, J., Dong, W., Socher, R., Li, L.J., Li, K., Fei-Fei, L.: ImageNet: a large-scale hierarchical image database. In: Proceedings of the 2009 IEEE Conference on Computer Vision and Pattern Recognition, CVPR, pp. 248–255 (2009)
5. DeVore, R.A.: Nonlinear approximation. Acta Numer. **7**, 51–150 (1998)
6. Engel, J., et al.: Neural audio synthesis of musical notes with wavenet autoencoders. In: Proceedings of the 34th International Conference on Machine Learning, ICML, vol. 70, pp. 1068–1077 (2017)
7. Fong, R.C., Vedaldi, A.: Interpretable explanations of black boxes by meaningful perturbation. In: Proceedings of 2017 IEEE International Conference on Computer Vision (ICCV), pp. 3449–3457 (2017)
8. Goodfellow, I.J., et al.: Generative adversarial nets. In: Proceedings of the 27th International Conference on Neural Information Processing Systems, NeurIPS, pp. 2672–2680 (2014)
9. Heiß, C., Levie, R., Resnick, C., Kutyniok, G., Bruna, J.: In-distribution interpretability for challenging modalities. ICML, Interpret. Sci. Discov. (2020)
10. Howard, A., et al.: Searching for MobileNetV3. In: Proceedings of the 2019 IEEE/CVF International Conference on Computer Vision (ICCV), pp. 1314–1324 (2019)
11. Kolek, S., Nguyen, D.A., Levie, R., Bruna, J., Kutyniok, G.: Cartoon explanations of image classifiers. Preprint arXiv:2110.03485 (2021)
12. Kutyniok, G., Lim, W.-Q.: Compactly supported shearlets are optimally sparse. J. Approx. Theory **163**(11), 1564–1589 (2011). https://doi.org/10.1016/j.jat.2011.06.005
13. Levie, R., Yapar, C., Kutyniok, G., Caire, G.: Pathloss prediction using deep learning with applications to cellular optimization and efficient D2D link scheduling. In: ICASSP 2020–2020 IEEE International Conference on Acoustics, Speech and Signal Processing (ICASSP), pp. 8678–8682 (2020). https://doi.org/10.1109/ICASSP40776.2020.9053347
14. Levie, R., Yapar, C., Kutyniok, G., Caire, G.: RadioUNet: fast radio map estimation with convolutional neural networks. IEEE Trans. Wirel. Commun. **20**(6), 4001–4015 (2021)

15. Lundberg, S.M., Lee, S.: A unified approach to interpreting model predictions. In: Proceedings of the 31st International Conference on Neural Information Processing Systems, NeurIPS, pp. 4768–4777 (2017)
16. Macdonald, J., Wäldchen, S., Hauch, S., Kutyniok, G.: A rate-distortion framework for explaining neural network decisions. Preprint arXiv:1905.11092 (2019)
17. Maddison, C.J., Mnih, A., Teh, Y.W.: The concrete distribution: a continuous relaxation of discrete random variables. Preprint arXiv:1611.00712 (2016)
18. Mallat, S., Zhang, Z.: Matching pursuits with time-frequency dictionaries. IEEE Trans. Signal Process. 41(12), 3397–3415 (1993)
19. Narasimha, M., Peterson, A.: On the computation of the discrete cosine transform. IEEE Trans. Commun. 26(6), 934–936 (1978)
20. Ribeiro, M.T., Singh, S., Guestrin, C.: "Why should I trust you?": Explaining the predictions of any classifier. In: Proceedings of the 22nd International Conference on Knowledge Discovery and Data Mining, ACM SIGKDD, pp. 1135–1144. Association for Computing Machinery (2016)
21. Romberg, J.K., Wakin, M.B., Baraniuk, R.G.: Wavelet-domain approximation and compression of piecewise smooth images. IEEE Trans. Image Process. 15, 1071–1087 (2006)
22. Ronneberger, O., Fischer, P., Brox, T.: U-Net: convolutional networks for biomedical image segmentation. In: Navab, N., Hornegger, J., Wells, W.M., Frangi, A.F. (eds.) MICCAI 2015. LNCS, vol. 9351, pp. 234–241. Springer, Cham (2015). https://doi.org/10.1007/978-3-319-24574-4_28
23. Ross, A.S., Hughes, M.C., Doshi-Velez, F.: Right for the right reasons: training differentiable models by constraining their explanations. In: Proceedings of the Twenty-Sixth International Joint Conference on Artificial Intelligence, IJCAI-17, pp. 2662–2670 (2017)
24. Schramowski, P., et al.: Making deep neural networks right for the right scientific reasons by interacting with their explanations. Nat. Mach. Intell. 2, 476–486 (2020)
25. Shrikumar, A., Greenside, P., Kundaje, A.: Learning important features through propagating activation differences. In: Proceedings of the 34th International Conference on Machine Learning, ICML, vol. 70, pp. 3145–3153 (2017)
26. Smilkov, D., Thorat, N., Kim, B., Viégas, F., Wattenberg, M.: SmoothGrad: removing noise by adding noise. In: Workshop on Visualization for Deep Learning, ICML (2017)
27. Stéphane, M.: Chapter 11.3. In: Stéphane, M. (ed.) A Wavelet Tour of Signal Processing, Third Edition, pp. 535–610. Academic Press, Boston (2009)
28. Sun, J., Lapuschkin, S., Samek, W., Binder, A.: Explain and improve: LRP-inference fine-tuning for image captioning models. Inf. Fusion 77, 233–246 (2022)
29. Sundararajan, M., Taly, A., Yan, Q.: Axiomatic attribution for deep networks. In: Proceedings of the 34th International Conference on Machine Learning, ICML, vol. 70, pp. 3319–3328 (2017)
30. Teneggi, J., Luster, A., Sulam, J.: Fast hierarchical games for image explanations. Preprint arXiv:2104.06164 (2021)
31. Wäldchen, S., Macdonald, J., Hauch, S., Kutyniok, G.: The computational complexity of understanding network decisions. J. Artif. Intell. Res. 70 (2019)
32. Yu, J., Lin, Z., Yang, J., Shen, X., Lu, X., Huang, T.S.: Generative image inpainting with contextual attention. In: Proceedings of the 2018 IEEE/CVF Conference on Computer Vision and Pattern Recognition, CVPR, pp. 5505–5514 (2018)

Explaining the Predictions
of Unsupervised Learning Models

Grégoire Montavon[1,2](✉)[iD], Jacob Kauffmann[1][iD], Wojciech Samek[2,3][iD],
and Klaus-Robert Müller[1,2,4,5][iD]

[1] ML Group, Department of Electrical Engineering and Computer Science,
Technische Universität Berlin, Berlin, Germany
gregoire.montavon@tu-berlin.de
[2] BIFOLD – Berlin Institute for Foundations of Learning and Data,
Berlin, Germany
[3] Department of Artificial Intelligence, Fraunhofer Heinrich Hertz Institute,
Berlin, Germany
[4] Department of Artificial Intelligence, Korea University, Seoul, Korea
[5] Max Planck Institut für Informatik, Saarbrücken, Germany

Abstract. Unsupervised learning is a subfield of machine learning that
focuses on learning the structure of data without making use of labels.
This implies a different set of learning algorithms than those used for
supervised learning, and consequently, also prevents a direct transposi-
tion of Explainable AI (XAI) methods from the supervised to the less
studied unsupervised setting. In this chapter, we review our recently proposed 'neuralization-propagation' (NEON) approach for bringing XAI
to workhorses of unsupervised learning such as kernel density estimation and k-means clustering. NEON first converts (without retraining)
the unsupervised model into a functionally equivalent neural network so
that, in a second step, supervised XAI techniques such as layer-wise relevance propagation (LRP) can be used. The approach is showcased on
two application examples: (1) analysis of spending behavior in wholesale
customer data and (2) analysis of visual features in industrial and scene
images.

Keywords: Explainable AI · Unsupervised learning · Neural networks

1 Introduction

Supervised learning has been in the spotlight of machine learning research and
applications for the last decade, with deep neural networks achieving record-
breaking classification accuracy and enabling new machine learning applications
[5,15,23]. The success of deep neural networks can be attributed to their ability
to implement with their multiple layers, complex nonlinear functions in a compact manner [32]. Recently, a significant amount of work has been dedicated to
make deep neural network models more transparent [13,24,40,41], for example,

© The Author(s) 2022
A. Holzinger et al. (Eds.): xxAI 2020, LNAI 13200, pp. 117–138, 2022.
https://doi.org/10.1007/978-3-031-04083-2_7

by proposing algorithms that identify which input features are responsible for a given classification outcome. Methods such as layer-wise relevance propagation (LRP) [3], guided backprop [47], and Grad-CAM [42], have been shown capable of quickly and robustly computing these explanations.

Unsupervised learning is substantially different from supervised learning in that there is no ground-truth supervised signal to match. Consequently, non-neural network models such as kernel density estimation or k-means clustering, where the user controls the scale and the level of abstraction through a particular choice of kernel or feature representation, have remained highly popular. Despite the predominance of unsupervised machine learning in a variety of applications (e.g. [9,22]), research on explaining unsupervised models has remained relatively sparse [18,19,25,28,30] compared to their supervised counterparts. Paradoxically, it might in fact be unsupervised models that most strongly require interpretability. Unsupervised models are indeed notoriously hard to quantitatively validate [51], and the main purpose of applying these models is often to better understand the data in the first place [9,17].

In this chapter, we review the 'neuralization-propagation' (NEON) approach we have developed in the papers [18–20] to make the predictions of unsupervised models, e.g. cluster membership or anomaly score, explainable. NEON proceeds in two steps: (1) the decision function of the unsupervised model is reformulated (without retraining) as a functionally equivalent neural network (i.e. it is 'neuralized'); (2) the extracted neural network structure is then leveraged by the LRP method to produce an explanation of the model prediction. We review the application of NEON to kernel density estimation for outlier detection and k-means clustering, as presented originally in [18–20]. We also extend the reviewed work with a new contribution: explanation of *inlier* detection, and we use the framework of random features [36] for that purpose.

The NEON approach is showcased on several practical examples, in particular, the analysis of wholesale customer data, image-based industrial inspection, and analysis of scene images. The first scenario covers the application of the method directly to the raw input features, whereas the second scenario illustrates how the framework can be applied to unsupervised models built on some intermediate layer of representation of a neural network.

2   A Brief Review of Explainable AI

The field of Explainable AI (XAI) has produced a wealth of explanation techniques and types of explanation. They address the heterogeneity of ML models found in applications and the heterogeneity of questions the user may formulate about the model and its predictions. An explanation may take the form of a simple decision tree (or other intrinsically interpretable model) that approximates the model's input-output relation [10,29]. Alternatively, an explanation may be a prototype for the concept represented at the output of the model, specifically, an input example to which the model reacts most strongly [34,45]. Lastly, an explanation may highlight what input features are the most important for the model's predictions [3,4,7].

In the following, we focus on a well-studied problem of XAI, which is how to attribute the prediction of an individual data point, to the input features [3,4,29,37,45,48,50]. Let us denote by $\mathcal{X} = \mathcal{I}_1 \times \cdots \times \mathcal{I}_d$ the input space formed by the concatenation of d input features (e.g. words, pixels, or sensor measurements). We assume a learned model $f : \mathcal{X} \to \mathbb{R}$ (supervised or unsupervised), mapping each data point in $\mathcal{X}$ to a real-valued score measuring the evidence for a class or some other predicted quantity. The problem of attribution can be abstracted as producing for the given function f a mapping $\mathcal{E}_f : \mathcal{X} \to \mathbb{R}^d$ that associates to each input example a vector of scores representing the (positive or negative) contribution of each feature. Often, one requires attribution techniques to implement a *conservation* (or completeness) property, where for all $x \in \mathcal{X}$ we have $\mathbf{1}^\top \mathcal{E}_f(x) = f(x)$ i.e. for every data point the sum of explanation scores over the input features should match the function value.

2.1   Approaches to Attribution

A first approach, *occlusion-based*, consists of testing the function to explain against various occlusions of the input features [53,54]. An important method of this family (and which was originally developed in the context of game theory) is the Shapley value [29,43,48]. The Shapley value identifies a unique attribution that satisfies some predefined set of axioms of an explanation, including the conservation property stated above. While the approach has strong theoretical underpinnings, computing the explanation however requires an exponential number of function evaluations (an evaluation for every subset of input features). This makes the Shapley value in its basic form intractable for any problem with more than a few input dimensions.

Another approach, *gradient-based*, leverages the gradient of the function, so that a mapping of the function value onto the multiple input dimensions is readily obtained [45,50]. The method of integrated gradients [50], in particular, attributes the prediction to input features by integrating the gradient along a path connecting some reference point (e.g. the origin) to the data point. The method requires somewhere between ten and a hundred function evaluations, and satisfies the aforementioned conservation property. The main advantage of gradient-based methods is that, by leveraging the gradient information in addition to the function value, one no longer has to perturb each input feature individually to produce an explanation.

A further approach, *surrogate-based*, consists of learning a simple local surrogate model of the function which is as accurate as possible, and whose structure makes explanation fast and unambiguous [29,37]. For example, when approximating the function locally with a linear model, e.g. $g(x) = \sum_{i=1}^d x_i w_i$, the output of that linear model can be easily decomposed to the input features by taking the individual summands. While explanation itself is fast to compute, training the surrogate model incurs a significant additional cost, and further care must be taken to ensure that the surrogate model implements the same decision strategy as the original model, in particular, that it uses the same input features.

A last approach, *propagation-based*, assumes that the prediction has been produced by a neural network, and leverages the neural network structure by casting the problem of explanation as performing a backward pass in the network [3,42,47]. The propagation approach is embodied by the Layer-wise Relevance Propagation (LRP) method [3,31]. The backward pass implemented by LRP consists of a sequence of conservative propagation steps where each step is implemented by a propagation rule. Let j and k be indices for neurons at layer l and $l + 1$ respectively, and assume that the function output $f(\boldsymbol{x})$ has been propagated from the top-layer to layer $l + 1$. We denote the resulting attribution onto these neurons as the vector of 'relevance scores' $(R_k)_k$. LRP then defines 'messages' $R_{j \leftarrow k}$ that redistribute the relevance R_k to neurons in the layer below. These messages typically have the structure $R_{j \leftarrow k} = [z_{jk} / \sum_j z_{jk}] \cdot R_k$, where z_{jk} models the contribution of neuron j to activating neuron k. The overall relevance of neuron j is then obtained by computing $R_j = \sum_k R_{j \leftarrow k}$. It is easy to show that application of LRP from one layer to the layer below is conservative. Consequently, the explanation formed by iterating the LRP propagation from the top layer to the input layer is therefore also conservative, i.e. $\sum_i R_i = \cdots = \sum_j R_j = \sum_k R_k = \cdots = f(\boldsymbol{x})$. As a result, explanations satisfying the conservation property can be obtained within a single forward/backward pass, instead of multiple function evaluations, as it was the case for the approaches described above. The runtime advantage of LRP facilitates explanation of large models and datasets (e.g. GPU implementations of LRP can achieve hundreds of image classification explanations per second [1,40]).

2.2   Neuralization-Propagation

Propagation-based explanation techniques such as LRP have a computational advantage over approaches based on multiple function evaluations. However, they assume a preexisting neural network structure associated to the prediction function. Unsupervised learning models such as kernel density estimation or k-means, are a priori not neural networks. However, the fact that these models are not given as neural networks does not preclude the existence of a neural network that implements the same function. If such a network exists (neural network equivalents of some unsupervised models will be presented in Sects. 3 and 4), we can quickly and robustly compute explanations by applying the following two steps:

Step 1: The unsupervised model is *'neuralized'*, that is, rewritten (without retraining) as a functionally equivalent neural network.

Step 2: The LRP method is applied to the resulting neural network, in order to produce an explanation of the prediction of the original model.

These two steps are illustrated in Fig. 1. In practice, for the second step to work well, some restrictions must be imposed on the type of neurons composing the network. In particular neurons should have a clear directionality in their input space to ensure that meaningful propagation to the lower layer can be

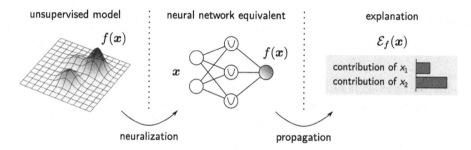

Fig. 1. Overview of the neuralization-propagation (NEON) approach to explain the predictions of an unsupervised model. As a first step, the unsupervised model is transformed without retraining into a functionally equivalent neural network. As a second step, the LRP procedure is applied to identify, with help of the neural network structure, by what amount each input feature has contributed to a given prediction.

achieved. (We will see in Sects. 3 and 4, that this requirement does not always hold.) Hence, the 'neuralized model' must be designed under the double constraint of (1) replicating the decision function of the unsupervised model exactly, and (2) being composed of neurons that enable a meaningful redistribution from the output to the input features.

3 Kernel Density Estimation

Kernel density estimation (KDE) [35] is one of the most common methods for unsupervised learning. The KDE model (or variations of it) has been used, in particular, for anomaly detection [21,26,38]. It assumes an unlabeled dataset $\mathcal{D} = (\boldsymbol{u}_1, \ldots, \boldsymbol{u}_N)$, and a kernel, typically the Gaussian kernel $\mathbb{K}(\boldsymbol{x}, \boldsymbol{x}') = \exp(-\gamma \|\boldsymbol{x} - \boldsymbol{x}'\|^2)$. The KDE model predicts a new data point $\boldsymbol{x}$ by computing:

$$\tilde{p}(\boldsymbol{x}) = \frac{1}{N} \sum_{k=1}^{N} \exp(-\gamma \|\boldsymbol{x} - \boldsymbol{u}_k\|^2). \tag{1}$$

The function $\tilde{p}(\boldsymbol{x})$ can be interpreted as an (unnormalized) probability density function. From this score, one can predict inlierness or outlierness of a data point. For example, one can say that $\boldsymbol{x}$ is more anomalous than $\boldsymbol{x}'$ if the inequality $\tilde{p}(\boldsymbol{x}) < \tilde{p}(\boldsymbol{x}')$ holds. In the following, we consider the task of neuralizing the KDE model so that its inlier/outlier predictions can be explained.

3.1 Explaining Outlierness

A first question to ask is why a particular example $\boldsymbol{x}$ is predicted by KDE to be an *outlier*, more specifically, what features of this example contribute to outlierness. As a first step, we consider what is a suitable measure of outlierness. The function $\tilde{p}(\boldsymbol{x})$ produced by KDE decreases with outlierness, and also saturates to zero even though outlierness continues to grow. A better measure of outlierness is given by [19]:

$$o(\boldsymbol{x}) \triangleq -\frac{1}{\gamma} \log \tilde{p}(\boldsymbol{x}),$$

Unlike the function $\tilde{p}(\boldsymbol{x})$, the function $o(\boldsymbol{x})$ increases as the probability decreases. It also does not saturate as $\boldsymbol{x}$ becomes more distant from the dataset. We now focus on neuralizing the outlier score $o(\boldsymbol{x})$. We find that $o(\boldsymbol{x})$ can be expressed as the two-layer neural network:

$$h_k = \|\boldsymbol{x} - \boldsymbol{u}_k\|^2 \qquad \text{(layer 1)}$$
$$o(\boldsymbol{x}) = \text{LME}_k^{-\gamma}\{h_k\} \qquad \text{(layer 2)}$$

where $\text{LME}_k^{\alpha}\{h_k\} = \frac{1}{\alpha} \log\left(\frac{1}{N}\sum_{k=1}^{N} \exp(\alpha\, h_k)\right)$ is a generalized log-mean-exp pooling. The first layer computes the square distance of the new example from each point in the dataset. The second layer can be interpreted as a soft min-pooling. The structure of the outlier computation is shown for a one-dimensional toy example in Fig. 2.

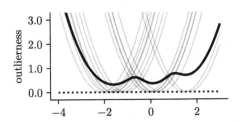

Fig. 2. Neuralized view of kernel density estimation for outlier prediction. The outlier function can be represented as a soft min-pooling over square distances. These distances also provide directionality in input space.

This structure is particularly amenable to explanation. In particular, redistribution of $o(\boldsymbol{x})$ in the intermediate layer can be achieved by a soft argmin operation, e.g.

$$R_k = \frac{\exp(-\beta h_k)}{\sum_k \exp(-\beta h_k)} \cdot o(\boldsymbol{x}),$$

where β is a hyperparameter to be selected. Then, propagation on the input features can leverage the geometry of the distance function, by computing

$$R_i = \sum_k \frac{[\boldsymbol{x} - \boldsymbol{u}_k]_i^2}{\epsilon + \|\boldsymbol{x} - \boldsymbol{u}_k\|^2} R_k.$$

The hyperparameter ϵ in the denominator is a stabilization term that 'dissipates' some of the relevance when $\boldsymbol{x}$ and $\boldsymbol{u}_k$ coincide.

Referring back to Sect. 2.1 we want to stress that computing the relevance of input features with LRP has the same computational complexity as a single forward pass, and does not require to train an explainable surrogate model.

3.2 Explaining Inlierness: Direct Approach

In Sect. 3.1, we have focused on explaining what makes a given example an outlier. An equally important question to ask is why a given example x is predicted by the KDE model to be an *inlier*. Inlierness is naturally modeled by the KDE output $\tilde{p}(x)$. Hence we can define the measure of inlierness as $\mathrm{i}(x) \triangleq \tilde{p}(x)$. An inspection of Eq. (1) suggests the following two-layer neural network:

$$
\begin{aligned}
h_k &= \exp(-\gamma \left\| x - u_k \right\|^2) &\quad \text{(layer 1)} \\
\mathrm{i}(x) &= \tfrac{1}{N} \sum_{k=1}^{N} h_k &\quad \text{(layer 2)}
\end{aligned}
$$

The first layer performs a mapping on Gaussian functions at different locations, and the second layer performs an average pooling. We now consider the task of propagation. A natural way of redistributing in the top layer is in proportion to the activations. This gives us the scores

$$
R_k = \frac{h_k}{\sum_k h_k} \mathrm{i}(x).
$$

A decomposition of R_k on the input features is however difficult. Because the relevance R_k can be rewritten as a product:

$$
R_k = \frac{1}{N} \prod_{i=1}^{d} \exp(-\gamma \left(x_i - u_{ik} \right)^2)
$$

and observing that the contribution R_k can be made nearly zero by perturbing any of the input features significantly, we can conclude that every input feature contributes equally to R_k and should therefore be attributed an equal share of it. Application of this strategy for every neuron k would result in an uniform redistribution of the score $\mathrm{i}(x)$ to the input features. The explanation would therefore be qualitatively always the same, regardless of the data point x and the overall shape of the inlier function $\mathrm{i}(x)$. While uniform attribution may be a good baseline, we usually strive for a more informative explanation.

3.3 Explaining Inlierness: Random Features Approach

To overcome the limitations of the approach above, we explore a second approach to explaining inlierness, where the neuralization is based on a feature map representation of the KDE model. For this, we first recall that any kernel-based model also admits a formulation in terms of the feature map $\Phi(x)$ associated to the kernel, i.e. $\mathbb{K}(\boldsymbol{x}, \boldsymbol{x}') = \langle \varPhi(\boldsymbol{x}), \varPhi(\boldsymbol{x}') \rangle$. In particular Eq. (1) can be equivalently rewritten as:

$$\tilde{p}(\boldsymbol{x}) = \Big\langle \varPhi(\boldsymbol{x}), \frac{1}{N} \sum_{k=1}^{N} \varPhi(\boldsymbol{u}_k) \Big\rangle, \tag{2}$$

i.e. the product in feature space of the current example and the dataset mean. Here, we first recall that there is no explicit finite-dimensional feature map associated to the Gaussian kernel. However, such feature map can be approximated using the framework of random features [36]. In particular, for a Gaussian kernel, features can be sampled as

$$\widehat{\varPhi}(\boldsymbol{x}) = \frac{\sqrt{2}}{H} \big(\cos(\boldsymbol{\omega}_j^\top \boldsymbol{x} + b_j) \big)_{j=1}^{H}, \tag{3}$$

with $\boldsymbol{\omega}_j \sim \mathcal{N}(\boldsymbol{\mu}, \sigma^2 I)$ and $b_j \sim \mathcal{U}(0, 2\pi)$, and where the mean and scale parameters of the Gaussian are $\boldsymbol{\mu} = \mathbf{0}$ and $\sigma = \sqrt{2\gamma}$. The dot product $\langle \widehat{\varPhi}(\boldsymbol{x}), \widehat{\varPhi}(\boldsymbol{x}') \rangle$ converges to the Gaussian kernel as more and more features are being drawn. In practice, we settle for a fixed number H of features. Injecting the random features in Eq. (2) yields the two-layer architecture:

$$h_j = \sqrt{2} \cos \big(\boldsymbol{\omega}_j^\top \boldsymbol{x} + b_j \big) \cdot \mu_j \qquad \text{(layer 1)}$$

$$\widehat{\mathbf{i}}(\boldsymbol{x}) = \tfrac{1}{H} \sum_{j=1}^{H} h_j \qquad \text{(layer 2)}$$

where $\mu_j = \frac{1}{N} \sum_{k=1}^{N} \sqrt{2} \cos(\boldsymbol{\omega}_j^\top \boldsymbol{u}_k + b_j)$ and with $(\boldsymbol{\omega}_j, b_j)_j$ drawn from the distribution given above. This architecture produces at its output an approximation of the true inlierness score $\mathbf{i}(\boldsymbol{x})$ which becomes increasingly accurate as H becomes large. Here, the first layer is a detection layer with a cosine nonlinearity, and the second layer performs average pooling. The structure of the neural network computation is illustrated on our one-dimensional example in Fig. 3.

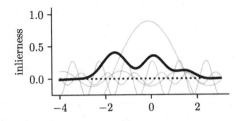

Fig. 3. Kernel density estimation approximated with random features (four of them are depicted in the figure). Unlike the Gaussian kernel, random features have a clear directionality in input space, thereby enabling a feature-wise explanation.

This structure of the inlierness computation is more amenable to explanation. In the top layer, the pooling operation can be attributed based on the summands. In order words, we can apply

$$R_j = \frac{h_j}{\sum_j h_j} \widehat{\mathrm{i}}(\boldsymbol{x})$$

for the first step of redistribution of $\widehat{\mathrm{i}}(\boldsymbol{x})$. More importantly, in the first layer, the random features have now a clear directionality (given by the vectors $(\boldsymbol{\omega}_j)_j$), which we can use for attribution on the input features. In particular, we can apply the propagation rule:

$$R_i = \sum_j \frac{[\boldsymbol{\omega}_j]_i^2}{\|\boldsymbol{\omega}_j\|^2} \cdot R_j.$$

Compared to the direct approach of Sect. 3.2, the explanation produced here assigns different scores for each input feature. Moreover, while the estimate of inlierness $\widehat{\mathrm{i}}(\boldsymbol{x})$ converges to the true KDE inlierness score $\mathrm{i}(\boldsymbol{x})$ as more random features are being drawn, we observe similar convergence for the *explanation* associated to the inlier prediction.

4   K-Means Clustering

Another important class of unsupervised models is clustering. K-means is a popular algorithm for identifying clusters in the data. The k-means model represents each cluster c with a centroid $\boldsymbol{\mu}_c \in \mathbb{R}^d$ corresponding to the mean of the cluster members. It assigns data onto clusters by first computing the distance between the data point and each cluster, e.g.

$$d_c(\boldsymbol{x}) = \|\boldsymbol{x} - \boldsymbol{\mu}_c\| \tag{4}$$

and chooses the cluster with the lowest distance $d_c(\boldsymbol{x})$. Once the data has been clustered, it is often the case that we would like to gain understanding of why a given data point has been assigned to a particular cluster, either for validating a given clustering model or for getting novel insights on the cluster structure of the data.

4.1   Explaining Cluster Assignments

As a starting point for applying our explanation framework, we need to identify a function $f_c(\boldsymbol{x})$ that represents well the assignment onto a particular cluster c, e.g. a function that is larger than zero when the data point is assigned to a given cluster, and less than zero otherwise.

The distance function $d_c(\boldsymbol{x})$ on which the clustering algorithm is based is however not directly suitable for the purpose of explanation. Indeed, $d_c(\boldsymbol{x})$ tends to be inversely related to cluster membership, and it also does not take into account how far the data point is from other clusters. In [18], it is proposed to contrast the assigned cluster with the competing clusters. In particular, k-means cluster membership can be modeled as the difference of (squared) distances between the nearest competing cluster and the assigned cluster c:

$$f_c(\boldsymbol{x}) = \min_{k \neq c} \left\{ d_k^2(\boldsymbol{x}) \right\} - d_c^2(\boldsymbol{x}) \tag{5}$$

The paper [18] shows that this contrastive strategy results in a two-layer neural network. In particular, Eq. (5) can be rewritten as the two-layer neural network:

$$h_k = \boldsymbol{w}_k^\top \boldsymbol{x} + b_k \qquad \text{(layer 1)}$$

$$f_c(\boldsymbol{x}) = \min_{k \neq c} \{h_k\} \qquad \text{(layer 2)}$$

where $\boldsymbol{w}_k = 2(\boldsymbol{\mu}_c - \boldsymbol{\mu}_k)$ and $b_k = \|\boldsymbol{\mu}_k\|^2 - \|\boldsymbol{\mu}_c\|^2$. The first layer is a linear layer that depends on the centroid locations and provides a clear directionality in input space. The second layer is a hard min-pooling. Once the neural network structure of cluster membership has been extracted, we can proceed with explanation techniques such as LRP by first reverse-propagating cluster evidence in the top layer (contrasting the given cluster with all cluster competitors) and then further propagating in the layer below. In particular, we first apply the soft argmin redistribution

$$R_k = \frac{\exp(-\beta h_k)}{\sum_{k \neq c} \exp(-\beta h_k)} \cdot f_c(\boldsymbol{x})$$

where β is a hyperparameter to be selected. An advantage of the soft argmin over its hard counterpart is that this does not create an abrupt transition between nearest competing clusters, which would in turn cause nearly identical data points with the same cluster decision to result in a substantially different explanation. Finally, the last step of redistribution on the input features can be achieved by leveraging the orientation of linear functions in the first layer, and applying the redistribution rule:

$$R_i = \sum_{k \neq c} \frac{[\boldsymbol{w}_k]_i^2}{\|\boldsymbol{w}_k\|^2} R_k.$$

Overall, these two redistribution steps provide us with a way of meaningfully attributing the cluster evidence onto the input features.

5   Experiments

We showcase the neuralization approaches presented above on two examples with two types of data: standard vector data representing wholesale customer spending behavior, and image data, more specifically, industrial inspection and scene images.

5.1   Wholesale Customer Analysis

Our first use case is the analysis of a wholesale customer dataset [11]. The dataset consists of 440 instances representing different customers, and for each instance, the annual consumption of the customer in monetary units (m.u.) for the categories 'fresh', 'milk', 'grocery', 'frozen', 'detergents/paper', 'delicatessen' is given. Two additional geographic features are also part of this dataset, however we do not include them in our experiment. We will place our focus on two particular data points with feature values shown in the table below:

Table 1. Excerpt of the Wholesale Customer Dataset [11] where we show feature values, expressed in monetary units (m.u.), for two instances as well as the average values over the whole dataset.

Index	Fresh	Milk	Grocery	Frozen	Detergents/ Paper	Delicatessen
338	9351 m.u	1347 m.u	2611 m.u	8170 m.u	442 m.u	868 m.u.
339	3 m.u	333 m.u	7201 m.u	15601 m.u	15 m.u	550 m.u.
AVG	12000 m.u.	5796 m.u.	7951 m.u.	3072 m.u.	2881 m.u.	1525 m.u.

Instance 338 has rather typical levels of spending across categories, in general slightly lower than average, but with high spending on frozen products. Instance 339 has more extreme spending with almost no spending on fresh products and detergents and very high spending on frozen products.

To get further insights into the data, we construct a KDE model on the whole data and apply our analysis to the selected instances. Each input feature is first mapped to the logarithm and standardized (mean 0 and variance 1). We choose the kernel parameter $\gamma = 1$. We use a leave-one-out approach where the data used to build the KDE model is the whole data except the instance to be predicted and analyzed. The number of random features is set to $H = 2500$ such that the computational complexity of the inlier model stays within one order of magnitude to the original kernel model. Predictions on the whole dataset and analysis for the selected instances is shown in Fig. 4.

Instance 338 is predicted to be an inlier, which is consistent with our initial observation that the levels of spending across categories are on the lower end but remain usual. We can characterize this instance as a typical small customer. We also note that the feature 'frozen' contributes less to inlierness according to our analysis, probably due to the spending on that category being unusually high for a typical small customer.

Instance 339 has an inlierness score almost zero, which is consistent with the observation in Table 1 that spending behavior is extremal for multiple product categories. The decomposition of an inlierness score of almost zero on the different categories is rather uninformative, hence, for this customer, we look at what explains outlierness (bottom of Fig. 4). We observe as expected that categories where spending behavior diverges for this instance are indeed strongly represented in the explanation of outlierness, with 'fresh', 'milk', 'frozen' and 'detergents/paper' contributing almost all evidence for outlierness. Surprisingly, we observe that extremely low spending on 'fresh' is underrepresented in the outlierness score, compared to other categories such as 'milk' or 'frozen' where

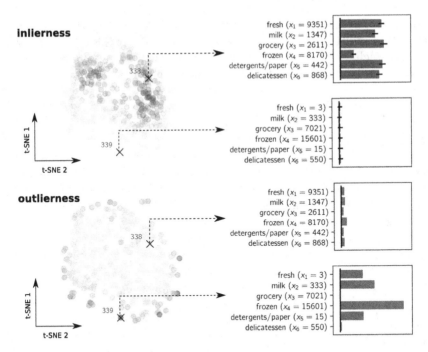

Fig. 4. Explanation of different predictions on the Wholesale Customers Dataset. The dataset is represented on the left as a t-SNE plot (perplexity 100) and each data point is color-coded according to its predicted inlierness and outlierness. On the right, explanation of inlierness and outlierness in terms of input features for two selected instances. Large bars in the plot correspond to strongly contributing features. For explanation of inlierness, error bars are computed over 100 trials of newly drawn random features. (Color figure online)

spending is less extreme. This apparent contradiction will be resolved by a cluster analysis.

Using the same logarithmic mapping and standardization step as for the KDE model, we now train a k-means model on the data and set the number of clusters to 6. Training is repeated 10 times with different centroid initializations, and we retain the model that has reached the lowest k-means objective. The outcome of the clustering is shown in Fig. 5 (left).

We observe that Instance 338 falls somewhere at the border between the green and red clusters, whereas Instance 339 is well into the yellow cluster at the bottom. The decomposition of cluster evidence for these two instances is shown on the right. Because Instance 338 is at the border between two clusters, there is no evidence of membership to one or another cluster, and the decomposition of such (lack of) evidence results in an explanation that is zero for all categories. The decomposition of the cluster evidence for Instance 339, however, reveals that its cluster membership is mainly due to a singular spending pattern on the category 'fresh'. To shed further light into this decision, we look at the cluster

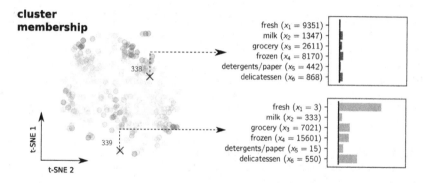

Fig. 5. On the left, a t-SNE representation of the Wholesale Customers Dataset, color-coded by cluster membership according to our k-means model, and where opacity represents evidence for the assigned cluster, i.e. how deep into its cluster the data point is. On the right, explanation of cluster assignments for two selected instances. (Color figure online)

to which this instance has been assigned, in particular, the average spending of cluster members on each category. This information is shown in Table 2.

Table 2. Average spending per category in the cluster to which Instance 339 has been assigned.

Cluster	Fresh	Milk	Grocery	Frozen	Detergents/Paper	Delicatessen
Yellow	616 m.u.	3176 m.u.	6965 m.u.	1523 m.u.	1414 m.u.	135 m.u.

We observe that this cluster is characterized by low spending on fresh products and delicatessen. It may be a cluster of small retailers that, unlike supermarkets, do not have substantial refrigeration capacity. Hence, the very low level of spending of Instance 339 on 'fresh' products puts it well into that cluster, and it also explains why the outlierness of Instance 339 is not attributed to 'fresh' but to other features (cf. Fig. 4). In particular, what distinguishes Instance 339 from its cluster is a very high level of spending on frozen products, and this is also the category that contributes the most to outlierness of this instance according to our analysis of the KDE model.

Traditionally, cluster membership has been characterized by more basic approaches such as population statistics of individual features (e.g. [8]). Figure 6 shows such analysis for Instances 338 and 339 of the Wholesale Customer Dataset. Although similar observations to the ones above can be made from this simple statistical analysis, e.g. the feature 'frozen' appears to contradict

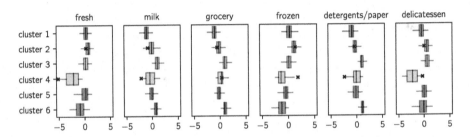

Fig. 6. Population statistics of individual features for the 6 clusters. The black cross in Cluster 2 is Instance 338, the black cross in Cluster 4 is Instance 339. Features are mapped to the logarithm and standardized.

the membership of Instance 339 to Cluster 4, it is not clear from this simple analysis what makes Instance 339 a member of Cluster 4 in the first place. For example, while the feature 'grocery' of Instance 339 is within the inter quartile range (IQR) of Cluster 4 and can therefore be considered typical of that cluster, other clusters have similar IQRs for that feature. Moreover, Instance 339 falls significantly outside Cluster 4's IQR for other features. In comparison, our LRP approach more directly and reliably explains the cluster membership and outlierness of the considered instances. Furthermore, population statistics of individual features may be misleading on non-linear models (such as kernel clustering) and does not scale to high-dimensional data, such as image data.

Overall, our analysis allows to identify on a single-instance basis features that contribute to various properties relating this instance to the rest of the data, such as inlierness/outlierness and cluster membership. As our analysis has revealed, the insights that are obtained go well beyond a traditional data analysis based on looking at population statistics for individual features, or a simple inspection of unsupervised learning outcomes.

5.2   Image Analysis

Our next experiment looks at explanation of inlierness, outlierness, and cluster membership for image data. Unlike the example above, relevant image statistics are better expressed at a more abstract level than directly on the pixels. A popular approach consists of using a pretrained neural model (e.g. the VGG-16 network [46]), and use the activations produced at a certain layer as input.

We first consider the problem of anomaly detection for industrial inspection and use for this an image of the MVTec AD dataset [6], specifically, an image of wood where an anomalous horizontal scratch can be observed. The image is shown in Fig. 7 (left). We feed that image to a pretrained VGG-16 network and collect the activations at the output of Block 5 (i.e. at the output of the feature extractor). We consider each spatial location at the output of that block as a data point and build a KDE model (with $\gamma = 0.05$) on the resulting dataset.

We then apply our analysis to attribute the predicted inlierness/outlierness to the activations of Block 5. In practice, we need to consider the fact that any attribution on a deactivated neuron cannot be redistributed further to input pixels as there is no pattern in pixel space to attach to. Hence, the propagation procedure must be carefully implemented to address this constraint, possibly by only redistributing a limited share of the model output. The details are given in Appendix A. As a last step, we take relevance scores computed at the output of Block 5 and pursue the relevance propagation procedure in the VGG-16 network using standard LRP rules until the pixels are reached. Explanations obtained for inlierness and outlierness of the wood image of interest are shown in Fig. 7.

input image	inlierness	outlierness

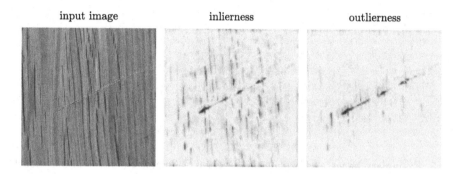

Fig. 7. Exemplary image from the MVTec AD dataset along with the explanation of an inlier/outlier prediction of a KDE model built at the output of the VGG-16 feature extractor. Red color indicates positively contributing pixels, blue color indicates negatively contributing pixels, and gray indicates irrelevant pixels. (Color figure online)

It can be observed that pixels associated to regular wood stripes are the main contributors to inlierness. Instead, the horizontal scratch on the wood panel is a contributing factor for outlierness. Hence, with our explanation method, we can precisely identify, on a pixel-wise basis what are the factors that contribute for/against predicted inlierness and outlierness.

We now consider some image of the SUN 2010 database [52], an indoor scene containing different pieces of furniture and home appliances. We consider the same VGG-16 network as in the experiment above and build a dataset by collecting activations at each spatial location of the output of Block 5. We then apply the k-means algorithm on this dataset with the number of clusters hardcoded to 5. Once the clustering model has been built, we rescale each cluster centroid to fixed norm. We then apply our analysis attribute the cluster membership scores to the activations at the output of Block 5. As for the industrial inspection example above, we must adjust the LRP rules so that deactivated neurons are not attributed relevance. The details of the LRP procedure are given in Appendix A. Obtained relevance scores are then propagated further to the input pixels using standard LRP rules. Resulting explanations are shown in Fig. 8.

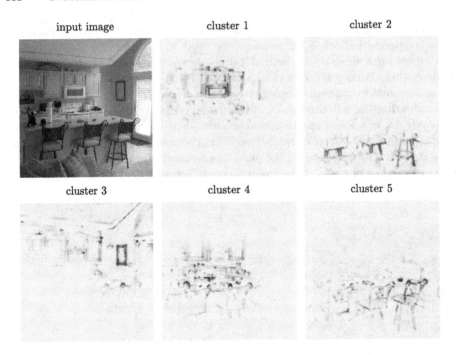

input image cluster 1 cluster 2

cluster 3 cluster 4 cluster 5

Fig. 8. Exemplary image and explanation of cluster assignments of a k-means model built at the output of the VGG-16 feature extractor. Red, blue and gray indicate positively contributing, negatively contributing, and irrelevant pixels respectively. (Color figure online)

We observe that different clusters identify distinct concepts. For example, one cluster focuses on the microwave oven and the surrounding cupboards, a second cluster represents the bottom part of the bar chairs, a third cluster captures the kitchen's background with a particular focus on a painting on the wall, the fourth cluster captures various objects on the table and in the background, and a last cluster focuses on the top-part of the chairs. While the clustering representation extracts distinct human-recognizable image features, it also shows some limits of the given representation, for example, the concept 'bar chair' is split in two distinct concepts (the bottom and top part of the chair respectively), whereas the clutter attached to Cluster 4 is not fully disentangled from the surrounding chairs and cupboards.

Overall, our experiments on image data demonstrate that neuralization of unsupervised learning models can be naturally integrated with existing procedures for explaining deep neural networks. This enables an application of our method to a broad range of practical problems where unsupervised modeling is better tackled at a certain level of abstraction and not directly in input space.

6   Conclusion and Outlook

In this paper, we have considered the problem of explaining the predictions of unsupervised models, in particular, we have reviewed and extended the neuralization/propagation approach of [18,19] which consists of rewriting, without retraining, the unsupervised model as a functionally equivalent neural network, and applying LRP in a second step. On two models of interest, kernel density estimation and k-means, we have highlighted a variety of techniques that can be used for neuralization. This includes the identification of log-mean-exp pooling structures, the use of random features, and the transformation of a difference of (squared) distances into a linear layer. The capacity of our approach to deliver meaningful explanations was highlighted on two examples covering simple tabular data and images including their mapping on some layer of a convolutional network.

While our approach delivers good quality explanations at low computational cost, there are however still a number of open questions that remain to be addressed to further solidify the neuralization-propagation approach, and the explanation of unsupervised models in general.

A first question concerns the applicability of our method to a broader range of practical scenarios. We have highlighted how neuralized models can be built not only in input space but also on some layer of a deep neural network, thereby bringing explanations to much more complex unsupervised models. However, there is a higher diversity of unsupervised learning algorithms that are encountered in practice, including energy-based models [16], spectral methods [33,44], linkage clustering [12], non-Euclidean methods [27], or prototype-based anomaly detection [14]. An important future work will therefore be to extend the proposed framework to handle this heterogeneity of unsupervised machine learning approaches.

Another question is that of validation. There are many possible LRP propagation rules that one can define in practice, as well as potentially multiple neural network reformulations of the same unsupervised model. This creates a need for reliable techniques to evaluate the quality of different explanation methods. While techniques to evaluate explanation quality have been proposed and successfully applied in the context of supervised learning (e.g. based on feature removal [39]), further care needs to be taken in the unsupervised scenario, in particular, to avoid that the outcome of the evaluation is spuriously affected by such feature removals. As an example, removing some feature responsible for some predicted anomaly may unintentionally cause some new artefact to be created in the data. That would in turn increase the anomaly score instead of lowering it as it was originally intended [19].

In addition to further extending and validating the neuralization-propagation approach, one needs to ask how to develop these explanation techniques beyond their usage as a simple visualization or data exploration tool. For example, it remains to demonstrate whether these explanation techniques, in combination with user feedback, can be used to systematically verify and improve the unsupervised model at hand (e.g. as recently demonstrated for supervised models [2,49]). Some initial steps have already been taken in this direction [20,38].

Acknowledgements. This work was supported by the German Ministry for Education and Research under Grant 01IS14013A-E, Grant 01GQ1115, Grant 01GQ0850, as BIFOLD (ref. 01IS18025A and ref. 01IS18037A) and Patho234 (ref. 031LO207), the European Union's Horizon 2020 programme (grant no. 965221), and the German Research Foundation (DFG) as Math+: Berlin Mathematics Research Center (EXC 2046/1, project-ID: 390685689). This work was supported in part by the Institute of Information & Communications Technology Planning & Evaluation (IITP) grants funded by the Korea Government under Grant 2017-0-00451 (Development of BCI Based Brain and Cognitive Computing Technology for Recognizing User's Intentions using Deep Learning) and Grant 2019-0-00079 (Artificial Intelligence Graduate School Program, Korea University).

A   Attribution on CNN Activations

Propagation rules mentioned in Sects. 3 and 4 are not suited for identifying relevant neurons at some layer of a neural network when the goal is to propagate the relevance further down the layers of the neural network, e.g. to obtain a pixelwise explanation. What we need to ensure in such scenario is that all relevant information is expressed in terms of *activated* neurons as they are the only ones for which the associated relevance can be grounded to a specific pattern in the pixel space. One possible approach is to decompose the relevance propagation into a propagating term and a non-propagating (or 'dissipating') one, which leads to a partial (although still useful) explanation. In the following, we describe the approaches we have taken to achieve our extension of explanations to deep models.

A.1   Attributing Outlierness

The activations in the first layer of the neuralized outlier model is

$$h_k = \|\boldsymbol{a} - \boldsymbol{u}_k\|^2$$

and the relevance that arrives on the corresponding neuron is given by $R_k = p_k \mathrm{LME}_{k'}^{-\gamma}\{h_{k'}\}$ with $p_k = \frac{\exp(-\beta h_k)}{\sum_{k'} \exp(-\beta h_{k'})}$. Relevance associated to neuron k can be expressed as:

$$R_k = \underbrace{p_k \cdot \boldsymbol{a}^\top (\boldsymbol{a} - \boldsymbol{u}_k)}_{R_k^{\mathrm{dot}}} + \underbrace{p_k \cdot (\boldsymbol{u}_k^\top (\boldsymbol{u}_k - \boldsymbol{a}) + \mathrm{LME}_{k'}^{-\gamma}\{h_{k'} - h_k\})}_{R_k^{\mathrm{res}}}$$

where we have used the commutativity of the LME function and the distributivity of the squared norm to decompose the relevance in two terms, one that can be meaningfully redistributed on the activations, and one that cannot be redistributed. Redistribution in the first layer can then proceed as:

$$R_i = \sum_k \frac{a_i \cdot (a_i - u_{ik})}{\sum_i a_i \cdot (a_i - u_{ik})} R_k^{\text{dot}}$$

It is easy to demonstrate from this equation that any neuron with $a_i = 0$ (i.e. deactivated) will not be attributed any relevance.

A.2   Attributing Inlierness

Neurons in the first layer of the inlierness model based on random features, have activations given by:

$$h_j = \sqrt{2}\cos(\boldsymbol{\omega}_j^\top \boldsymbol{a} + b_j) \cdot \mu_j$$

and relevance scores $R_j = h_j/H$. Using a simple trigonometric identity, we can rewrite the relevance scores in terms of unphased sine and cosine functions as:

$$R_j = \underbrace{\left(- \sin(\boldsymbol{\omega}_j^\top \boldsymbol{a}) \sin(b_j) \cdot c_j \right)}_{R_j^{\text{sin}}} + \underbrace{\cos(\boldsymbol{\omega}_j^\top \boldsymbol{a}) \cos(b_j) \cdot c_j}_{R_j^{\text{cos}}}$$

where $c_j = \frac{1}{H}\sqrt{2}\mu_j$. We propose the redistribution rule:

$$R_i = \sum_j \frac{a_i \omega_{ij}}{\sum_i a_i \omega_{ij}} R_j^{\text{sin}} + \sum_j \frac{a_i \omega_{ij}}{\epsilon_j + \sum_i a_i \omega_{ij}} R_j^{\text{cos}}$$

where ϵ_j is a term set to be of same sign as the denominator, and that addresses the case where a positive R_j^{cos} comes with a near-zero response $\boldsymbol{\omega}_j^\top \boldsymbol{a}$, by 'dissipating' some of the relevance R_j^{cos}.

A.3   Attributing Cluster Membership

The activation in the first layer of the neuralized cluster membership model is:

$$h_k = \boldsymbol{w}_k^\top \boldsymbol{a} + b_k$$

and the relevance score is given by $R_k = p_k \min_{k' \neq c}\{h_{k'}\}$ with $p_k = \frac{\exp(-\beta h_k)}{\sum_{k'} \exp(-\beta h_{k'})}$. Similar to the outlier case, we decompose the relevance score as:

$$R_k = \underbrace{p_k \cdot \boldsymbol{a}^\top \boldsymbol{w}_k}_{R_k^{\text{dot}}} + \underbrace{p_k \cdot (b_k + \min_{k' \neq c}\{h_{k'} - h_k\})}_{R_k^{\text{res}}}$$

and only consider the first term for propagation. Specifically, we apply the propagation rule:

$$R_i = \sum_k \frac{a_i w_{ik}}{\sum_i a_i w_{ik}} R_k^{\text{dot}}$$

where it can again be shown that only activated neurons are attributed relevance.

References

1. Alber, M., et al.: iNNvestigate neural networks! J. Mach. Learn. Res. **20**, 93:1–93:8 (2019)
2. Anders, C.J., Weber, L., Neumann, D., Samek, W., Müller, K.-R., Lapuschkin, S.: Finding and removing Clever Hans: using explanation methods to debug and improve deep models. Inf. Fusion **77**, 261–295 (2022)
3. Bach, S., Binder, A., Montavon, G., Klauschen, F., Müller, K.-R., Samek, W.: On pixel-wise explanations for non-linear classifier decisions by layer-wise relevance propagation. PLoS ONE **10**(7), e0130140 (2015)
4. Baehrens, D., Schroeter, T., Harmeling, S., Kawanabe, M., Hansen, K., Müller, K.-R.: How to explain individual classification decisions. J. Mach. Learn. Res. **11**, 1803–1831 (2010)
5. Bahdanau, D., Cho, K., Bengio, Y.: Neural machine translation by jointly learning to align and translate. In: ICLR (2015)
6. Bergmann, P., Batzner, K., Fauser, M., Sattlegger, D., Steger, C.: The Mvtec anomaly detection dataset: a comprehensive real-world dataset for unsupervised anomaly detection. Int. J. Comput. Vis. **129**(4), 1038–1059 (2021). https://doi.org/10.1007/s11263-020-01400-4
7. Blum, A., Langley, P.: Selection of relevant features and examples in machine learning. Artif. Intell. **97**(1–2), 245–271 (1997)
8. Chapfuwa, P., Li, C., Mehta, N., Carin, L., Henao, R.: Survival cluster analysis. In: Ghassemi, M. (ed.) ACM Conference on Health, Inference, and Learning, pp. 60–68. ACM (2020)
9. Ciriello, G., Miller, M.L., Aksoy, B.A., Senbabaoglu, Y., Schultz, N., Sander, C.: Emerging landscape of oncogenic signatures across human cancers. Nat. Genet. **45**(10), 1127–1133 (2013)
10. Craven, M.V., Shavlik, J.W.: Extracting tree-structured representations of trained networks. In: NIPS, pp. 24–30. MIT Press (1995)
11. de Abreu, N.G.C.F.M.: Análise do perfil do cliente recheio e desenvolvimento de um sistema promocional. Master's thesis, Instituto Universitário de Lisboa (2011)
12. Gower, J.C., Ross, G.J.S.: Minimum spanning trees and single linkage cluster analysis. Appl. Stat. **18**(1), 54 (1969)
13. Guidotti, R., Monreale, A., Ruggieri, S., Turini, F., Giannotti, F., Pedreschi, D.: A survey of methods for explaining black box models. ACM Comput. Surv. **51**(5), 93:1–93:42 (2019)
14. Harmeling, S., Dornhege, G., Tax, D., Meinecke, F., Müller, K.-R.: From outliers to prototypes: ordering data. Neurocomputing **69**(13–15), 1608–1618 (2006)
15. He, K., Zhang, X., Ren, S., Sun, J.: Deep residual learning for image recognition. In: CVPR, pp. 770–778. IEEE Computer Society (2016)
16. Hinton, G.E.: Training products of experts by minimizing contrastive divergence. Neural Comput. **14**(8), 1771–1800 (2002)
17. Kau, A.K., Tang, Y.E., Ghose, S.: Typology of online shoppers. J. Consum. Mark. **20**(2), 139–156 (2003)
18. Kauffmann, J.R., Esders, M., Montavon, G., Samek, W., Müller, K.-R.: From clustering to cluster explanations via neural networks. CoRR, abs/1906.07633 (2019)
19. Kauffmann, J.R., Müller, K.-R., Montavon, G.: Towards explaining anomalies: a deep Taylor decomposition of one-class models. Pattern Recognit. **101**, 107198 (2020)
20. Kauffmann, J.R., Ruff, L., Montavon, G., Müller, K.-R.: The Clever Hans effect in anomaly detection. CoRR, abs/2006.10609 (2020)

21. Kim, J., Scott, C.D.: Robust kernel density estimation. J. Mach. Learn. Res. **13**, 2529–2565 (2012)
22. Koren, Y., Bell, R.M., Volinsky, C.: Matrix factorization techniques for recommender systems. Computer **42**(8), 30–37 (2009)
23. Krizhevsky, A., Sutskever, I., Hinton, G.E.: ImageNet classification with deep convolutional neural networks. In: NIPS, pp. 1106–1114 (2012)
24. Lapuschkin, S., Wäldchen, S., Binder, A., Montavon, G., Samek, W., Müller, K.-R.: Unmasking Clever Hans predictors and assessing what machines really learn. Nat. Commun. **10**(1096), 1–8 (2019)
25. Laskov, P., Rieck, K., Schäfer, C., Müller, K.-R.: Visualization of anomaly detection using prediction sensitivity. In: Sicherheit, volume P-62 of LNI, pp. 197–208. GI (2005)
26. Latecki, L.J., Lazarevic, A., Pokrajac, D.: Outlier detection with kernel density functions. In: Perner, P. (ed.) MLDM 2007. LNCS (LNAI), vol. 4571, pp. 61–75. Springer, Heidelberg (2007). https://doi.org/10.1007/978-3-540-73499-4_6
27. Liu, F.T., Ting, K.M., Zhou, Z.: Isolation forest. In: Proceedings of the 8th IEEE International Conference on Data Mining, pp. 413–422. IEEE Computer Society (2008)
28. Liu, N., Shin, D., Hu, X.: Contextual outlier interpretation. In: IJCAI, pp. 2461–2467. ijcai.org (2018)
29. Lundberg, S.M., Lee, S.: A unified approach to interpreting model predictions. In: Advances in Neural Information Processing Systems, vol. 30, pp. 4765–4774 (2017)
30. Micenková, B., Ng, R.T., Dang, X., Assent, I.: Explaining outliers by subspace separability. In: ICDM, pp. 518–527. IEEE Computer Society (2013)
31. Montavon, G., Binder, A., Lapuschkin, S., Samek, W., Müller, K.-R.: Layer-wise relevance propagation: an overview. In: Samek, W., Montavon, G., Vedaldi, A., Hansen, L.K., Müller, K.-R. (eds.) Explainable AI: Interpreting, Explaining and Visualizing Deep Learning. LNCS (LNAI), vol. 11700, pp. 193–209. Springer, Cham (2019). https://doi.org/10.1007/978-3-030-28954-6_10
32. Montúfar, G.F., Pascanu, R., Cho, K., Bengio, Y.: On the number of linear regions of deep neural networks. In: NIPS, pp. 2924–2932 (2014)
33. Ng, A.Y., Jordan, M.I., Weiss, Y.: On spectral clustering: analysis and an algorithm. In: NIPS, pp. 849–856. MIT Press (2001)
34. Nguyen, A., Dosovitskiy, A., Yosinski, J., Brox, T., Clune, J.: Synthesizing the preferred inputs for neurons in neural networks via deep generator networks. In: NIPS, pp. 3387–3395 (2016)
35. Parzen, E.: On estimation of a probability density function and mode. Ann. Math. Stat. **33**(3), 1065–1076 (1962)
36. Rahimi, A., Recht, B.: Random features for large-scale kernel machines. In: NIPS, pp. 1177–1184 (2007)
37. Ribeiro, M.T., Singh, S., Guestrin, C.: "Why should I trust you?": Explaining the predictions of any classifier. In: KDD, pp. 1135–1144. ACM (2016)
38. Ruff, L., et al.: A unifying review of deep and shallow anomaly detection. Proc. IEEE **109**(5), 756–795 (2021)
39. Samek, W., Binder, A., Montavon, G., Lapuschkin, S., Müller, K.-R.: Evaluating the visualization of what a deep neural network has learned. IEEE Trans. Neural Netw. Learn. Syst. **28**(11), 2660–2673 (2017)
40. Samek, W., Montavon, G., Lapuschkin, S., Anders, C.J., Müller, K.-R.: Explaining deep neural networks and beyond: a review of methods and applications. Proc. IEEE **109**(3), 247–278 (2021)

41. Samek, W., Montavon, G., Vedaldi, A., Hansen, L.K., Müller, K.-R. (eds.): Explainable AI: Interpreting, Explaining and Visualizing Deep Learning. LNCS (LNAI), vol. 11700. Springer, Cham (2019). https://doi.org/10.1007/978-3-030-28954-6

42. Selvaraju, R.R., Cogswell, M., Das, A., Vedantam, R., Parikh, D., Batra, D.: Gradcam: visual explanations from deep networks via gradient-based localization. Int. J. Comput. Vis. **128**(2), 336–359 (2020)

43. Shapley, L.S.: 17. A value for n-person games. In: Contributions to the Theory of Games (AM-28), vol. II. Princeton University Press (1953)

44. Shi, J., Malik, J.: Normalized cuts and image segmentation. IEEE Trans. Pattern Anal. Mach. Intell. **22**(8), 888–905 (2000)

45. Simonyan, K., Vedaldi, A., Zisserman, A.: Deep inside convolutional networks: visualising image classification models and saliency maps. In: ICLR (Workshop Poster) (2014)

46. Simonyan, K., Zisserman, A.: Very deep convolutional networks for large-scale image recognition. In: ICLR (2015)

47. Springenberg, J.T., Dosovitskiy, A., Brox, T., Riedmiller, M.: Striving for simplicity: the all convolutional net. In: ICLR (Workshop) (2015)

48. Strumbelj, E., Kononenko, I.: An efficient explanation of individual classifications using game theory. J. Mach. Learn. Res. **11**, 1–18 (2010)

49. Sun, J., Lapuschkin, S., Samek, W., Binder, A.: Explain and improve: LRP-inference fine tuning for image captioning models. Inf. Fusion **77**, 233–246 (2022)

50. Sundararajan, M., Taly, A., Yan, Q.: Axiomatic attribution for deep networks. In: ICML, Proceedings of Machine Learning Research, vol. 70, pp. 3319–3328. PMLR (2017)

51. von Luxburg, U., Williamson, R.C., Guyon, I.: Clustering: science or art? In: ICML Unsupervised and Transfer Learning, JMLR Proceedings, vol. 27, pp. 65–80. JMLR.org (2012)

52. Xiao, J., Ehinger, K.A., Hays, J., Torralba, A., Oliva, A.: SUN database: exploring a large collection of scene categories. Int. J. Comput. Vis. **119**(1), 3–22 (2016)

53. Zeiler, M.D., Fergus, R.: Visualizing and understanding convolutional networks. In: Fleet, D., Pajdla, T., Schiele, B., Tuytelaars, T. (eds.) ECCV 2014. LNCS, vol. 8689, pp. 818–833. Springer, Cham (2014). https://doi.org/10.1007/978-3-319-10590-1_53

54. Zintgraf, L.M., Cohen, T.S., Adel, T., Welling, M.: Visualizing deep neural network decisions: prediction difference analysis. In: ICLR (Poster). OpenReview.net (2017)

Towards Causal Algorithmic Recourse

Amir-Hossein Karimi[1,2]([✉]), Julius von Kügelgen[1,3], Bernhard Schölkopf[1,2], and Isabel Valera[1,4]

[1] Max Planck Institute for Intelligent Systems, Tübingen, Germany
`{amir,jvk,bs}@tue.mpg.de`
[2] Max Planck ETH Center for Learning Systems, Zürich, Switzerland
[3] Department of Engineering, University of Cambridge, Cambridge, UK
[4] Department of Computer Science, Saarland University, Saarbrücken, Germany
`ivalera@cs.uni-saarland.de`

Abstract. Algorithmic recourse is concerned with aiding individuals who are unfavorably treated by automated decision-making systems to overcome their hardship, by offering recommendations that would result in a more favorable prediction when acted upon. Such recourse actions are typically obtained through solving an optimization problem that minimizes changes to the individual's feature vector, subject to various plausibility, diversity, and sparsity constraints. Whereas previous works offer solutions to the optimization problem in a variety of settings, they critically overlook real-world considerations pertaining to the environment in which recourse actions are performed.

The present work emphasizes that changes to a subset of the individual's attributes may have consequential down-stream effects on other attributes, thus making recourse a fundamcausal problem. Here, we model such considerations using the framework of structural causal models, and highlight pitfalls of not considering causal relations through examples and theory. Such insights allow us to reformulate the optimization problem to directly optimize for minimally-costly recourse over a space of feasible actions (in the form of causal interventions) rather than optimizing for minimally-distant "counterfactual explanations". We offer both the optimization formulations and solutions to deterministic and probabilistic recourse, on an individualized and sub-population level, overcoming the steep assumptive requirements of offering recourse in general settings. Finally, using synthetic and semi-synthetic experiments based on the German Credit dataset, we demonstrate how such methods can be applied in practice under minimal causal assumptions.

A.-H. Karimi and J. von Kügelgen—Equal contribution.
This chapter is mostly based on the following two works:
1. Karimi, A. H., Schölkopf, B., & Valera, I. Algorithmic recourse: from counterfactual explanations to interventions. In: Proceedings of the 4th Conference on Fairness, Accountability, and Transparency (FAccT 2021). pp. 353–362 (2021).
2. Karimi, A. H., von Kügelgen, J., Schölkopf, B., & Valera, I. Algorithmic recourse under imperfect causal knowledge: a probabilistic approach. In: Advances in Neural Information Processing Systems 33 (NeurIPS 2020). pp. 265–277 (2020).

A. Holzinger et al. (Eds.): xxAI 2020, LNAI 13200, pp. 139–166, 2022.
https://doi.org/10.1007/978-3-031-04083-2_8

1   Introduction

Predictive models are being increasingly used to support consequential decision-making in a number of contexts, e.g., denying a loan, rejecting a job applicant, or prescribing life-altering medication. As a result, there is mounting social and legal pressure [64,72] to provide explanations that help the affected individuals to understand "why a prediction was output", as well as "how to act" to obtain a desired outcome. Answering these questions, for the different stakeholders involved, is one of the main goals of explainable machine learning [15,19,32,37,42,53,54].

In this context, several works have proposed to explain a model's predictions of an affected individual using *counterfactual explanations*, which are defined as statements of "how the world would have (had) to be different for a desirable outcome to occur" [76]. Of specific importance are *nearest counterfactual explanations*, presented as the most similar *instances* to the feature vector describing the individual, that result in the desired prediction from the model [25,35]. A closely related term is *algorithmic recourse*—the actions required for, or "the systematic process of reversing unfavorable decisions by algorithms and bureaucracies across a range of counterfactual scenarios"—which is argued as the underwriting factor for temporally extended agency and trust [70].

Counterfactual explanations have shown promise for practitioners and regulators to validate a model on metrics such as fairness and robustness [25,58,69]. However, in their raw form, such explanations do not seem to fulfill one of the primary objectives of "explanations as a means to help a data-subject *act* rather than merely *understand*" [76].

The translation of counterfactual explanations to recourse actions, i.e., to a recommendable set of actions to help an individual achieve a favorable outcome, was first explored in [69], where additional *feasibility* constraints were imposed to support the concept of actionable features (e.g., to prevent asking the individual to reduce their age or change their race). While a step in the right direction, this work and others that followed [25,41,49,58] implicitly assume that the set of actions resulting in the desired output would directly follow from the counterfactual explanation. This arises from the assumption that "what would *have had to be* in the past" (retrodiction) not only translates to "what *should be* in the future" (prediction) but also to "what *should be done* in the future" (recommendation) [63]. We challenge this assumption and attribute the shortcoming of existing approaches to their lack of consideration for real-world properties, specifically the *causal relationships* governing the physical world in which actions are performed.

1.1   Motivating Examples

Example 1. Consider, for example, the setting in Fig. 1 where an individual has been denied a loan and seeks an explanation and recommendation on how to proceed. This individual has an annual salary (X_1) of $\$75,000$ and an account balance (X_2) of $\$25,000$ and the predictor grants a loan based on the binary

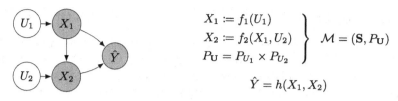

$$X_1 := f_1(U_1)$$
$$X_2 := f_2(X_1, U_2) \left.\right\} \; \mathcal{M} = (\mathbf{S}, P_\mathbf{U})$$
$$P_\mathbf{U} = P_{U_1} \times P_{U_2}$$

$$\hat{Y} = h(X_1, X_2)$$

Fig. 1. Illustration of an example bivariate causal generative process, showing both the graphical model $\mathcal{G}$ (left), and the corresponding structural causal model (SCM) $\mathcal{M}$ (right) [45]. In this example, X_1 represents an individual's annual salary, X_2 represents their bank balance, and $\hat{Y}$ denotes the output of a fixed deterministic predictor h, predicting an individual's eligibility to receive a loan. U_1 and U_2 denote unobserved (exogenous) random variables.

output of $h(X_1, X_2) = \text{sgn}(X_1 + 5 \cdot X_2 - \$225,000)$. Existing approaches may identify nearest counterfactual explanations as another individual with an annual salary of $\$100,000$ $(+33\%)$ or a bank balance of $\$30,000$ $(+20\%)$, therefore encouraging the individual to reapply when either of these conditions are met. On the other hand, assuming actions take place in a world where home-seekers save 30% of their salary, up to external fluctuations in circumstance, (i.e., $X_2 := 0.3X_1 + U_2$), a salary increase of *only* $+14\%$ to $\$85,000$ would automatically result in $\$3,000$ additional savings, with a net positive effect on the loan-granting algorithm's decision.

Example 2. Consider now another instance of the setting of Fig. 1 in which an agricultural team wishes to increase the yield of their rice paddy. While many factors influence yield (temperature, solar radiation, water supply, seed quality, ...), assume that the primary actionable capacity of the team is their choice of paddy location. Importantly, the altitude (X_1) at which the paddy sits has an effect on other variables. For example, the laws of physics may imply that a $100m$ increase in elevation results in an average decrease of $1°C$ in temperature (X_2). Therefore, it is conceivable that a counterfactual explanation suggesting an increase in elevation for optimal yield, without consideration for downstream effects of the elevation increase on other variables (e.g., a decrease in temperature), may actually result in the prediction *not* changing.

These two examples illustrate the pitfalls of generating recourse actions directly from counterfactual explanations without consideration for the (causal) structure of the world in which the actions will be performed. Actions derived directly from counterfactual explanations may ask too much effort from the individual (Example 1) or may not even result in the desired output (Example 2).

We also remark that merely accounting for correlations between features (instead of modeling their causal relationships) would be insufficient as this would not align with the asymmetrical nature of causal interventions: for Example 1, increasing bank balance (X_2) would *not* lead to a higher salary (X_1), and for Example 2, increasing temperature (X_2) would *not* affect altitude (X_1), contrary to what would be predicted by a purely correlation-based approach.

1.2   Summary of Contributions and Structure of This Chapter

In the present work, we remedy this situation via a *fundamental reformulation of the recourse problem*: we rely on *causal reasoning* (Sect. 2.2) to incorporate knowledge of causal dependencies between features into the process of recommending recourse actions that, if acted upon, would result in a counterfactual instance that favorably changes the output of the predictive model (Sect. 2.1).

First, we illuminate the intrinsic limitations of an approach in which recourse actions are directly derived from counterfactual explanations (Sect. 3.1). We show that actions derived from pre-computed (nearest) counterfactual explanations may prove sub-optimal in the sense of higher-than-necessary cost, or, even worse, ineffective in the sense of not actually achieving recourse. To address these limitations, we emphasize that, from a causal perspective, actions correspond to interventions which not only model changes to the intervened-upon variables, but also downstream effects on the remaining (non-intervened-upon) variables. This insight leads us to propose a new framework of *recourse through minimal interventions* in an underlying structural causal model (SCM) (Sect. 3.2). We complement this formulation with a negative result showing that recourse guarantees are generally only possible if the true SCM is known (Sect. 3.3).

Second, since real-world SCMs are rarely known we focus on the problem of *algorithmic recourse under imperfect causal knowledge* (Sect. 4). We propose two probabilistic approaches which allow to relax the strong assumption of a fully-specified SCM. In the first (Sect. 4.1), we assume that the true SCM, while unknown, is an additive Gaussian noise model [23, 47]. We then use Gaussian processes (GPs) [79] to average predictions over a whole family of SCMs to obtain a distribution over *counterfactual* outcomes which forms the basis for *individualised* algorithmic recourse. In the second (Sect. 4.2), we consider a different *subpopulation-based* (i.e., interventional rather than counterfactual) notion of recourse which allows us to further relax our assumptions by removing any assumptions on the form of the structural equations. This approach proceeds by estimating the effect of interventions on individuals similar to the one for which we aim to achieve recourse (i.e., the conditional average treatment effect [1]), and relies on conditional variational autoencoders [62] to estimate the interventional distribution. In both cases, we assume that the causal graph is known or can be postulated from expert knowledge, as without such an assumption causal reasoning from observational data is not possible [48, Prop. 4.1]. To find minimum cost interventions that achieve recourse with a given probability, we propose a gradient-based approach to solve the resulting optimisation problems (Sect. 4.3).

Our experiments (Sect. 5) on synthetic and semi-synthetic loan approval data, show the need for probabilistic approaches to achieve algorithmic recourse in practice, as point estimates of the underlying true SCM often propose invalid recommendations or achieve recourse only at higher cost. Importantly, our results also suggest that subpopulation-based recourse is the right approach to adopt when assumptions such as additive noise do not hold. A user-friendly implementation of all methods that only requires specification of the causal graph and a training set is available at https://github.com/amirhk/recourse.

2   Preliminaries

In this work, we consider algorithmic recourse through the lens of causality. We begin by reviewing the main concepts.

2.1   XAI: Counterfactual Explanations and Algorithmic Recourse

Let $\mathbf{X} = (X_1, ..., X_d)$ denote a tuple of random variables, or features, taking values $\mathbf{x} = (x_1, ..., x_d) \in \mathcal{X} = \mathcal{X}_1 \times ... \times \mathcal{X}_d$. Assume that we are given a binary probabilistic classifier $h : \mathcal{X} \to [0, 1]$ trained to make decisions about i.i.d. samples from the data distribution $P_{\mathbf{X}}$.[1]

For ease of illustration, we adopt the setting of loan approval as a running example, i.e., $h(\mathbf{x}) \geq 0.5$ denotes that a loan is granted and $h(\mathbf{x}) < 0.5$ that it is denied. For a given ("factual") individual $\mathbf{x}^{\mathrm{F}}$ that was denied a loan, $h(\mathbf{x}^{\mathrm{F}}) < 0.5$, we aim to answer the following questions: "Why did individual $\mathbf{x}^{\mathrm{F}}$ not get the loan?" and "What would they have to change, preferably with minimal effort, to increase their chances for a future application?".

A popular approach to this task is to find so-called (nearest) *counterfactual explanations* [76], where the term "counterfactual" is meant in the sense of the closest possible world with a different outcome [36]. Translating this idea to our setting, a nearest counterfactual explanation $\mathbf{x}^{\mathrm{CFE}}$ for an individual $\mathbf{x}^{\mathrm{F}}$ is given by a solution to the following optimisation problem:

$$\mathbf{x}^{\mathrm{CFE}} \in \underset{\mathbf{x} \in \mathcal{X}}{\arg\min} \quad \mathrm{dist}(\mathbf{x}, \mathbf{x}^{\mathrm{F}}) \quad \text{subject to} \quad h(\mathbf{x}) \geq 0.5, \tag{1}$$

where $\mathrm{dist}(\cdot, \cdot)$ is a distance on $\mathcal{X} \times \mathcal{X}$, and additional constraints may be added to reflect plausibility, feasibility, or diversity of the obtained counterfactual explanations [22,24,25,39,41,49,58]. Most existing approaches have focused on providing solutions to (1) by exploring semantically meaningful choices of $\mathrm{dist}(\cdot, \cdot)$ for measuring similarity between individuals (e.g., $\ell_0, \ell_1, \ell_\infty$, percentile-shift), accommodating different predictive models h (e.g., random forest, multilayer perceptron), and realistic plausibility constraints $\mathcal{P} \subseteq \mathcal{X}$.[2]

Although nearest counterfactual explanations provide an *understanding* of the most similar set of features that result in the desired prediction, they stop short of giving explicit *recommendations* on how to act to realize this set of features. The lack of specification of the actions required to realize $\mathbf{x}^{\mathrm{CFE}}$ from $\mathbf{x}^{\mathrm{F}}$ leads to uncertainty and limited agency for the individual seeking recourse. To

---

[1] Following the related literature, we consider a binary classification task by convention; most of our considerations extend to multi-class classification or regression settings as well though.

[2] In particular, [14,41,76] solve (1) using gradient-based optimization; [55,69] employ mixed-integer linear program solvers to support mixed numeric/binary data; [49] use graph-based shortest path algorithms; [35] use a heuristic search procedure by growing spheres around the factual instance; [18,58] build on genetic algorithms for model-agnostic behavior; and [25] solve (1) using satisfiability solvers with closeness guarantees. For a more complete exposition, see the recent surveys [26,71].

shift the focus from explaining a decision to providing recommendable actions to achieve recourse, Ustun et al. [69] reformulated (1) as:

$$\boldsymbol{\delta}^* \in \underset{\boldsymbol{\delta} \in \mathcal{F}}{\arg \min} \quad \text{cost}^{\text{F}}(\boldsymbol{\delta}) \quad \text{subject to} \quad h(\mathbf{x}^{\text{F}} + \boldsymbol{\delta}) \geq 0.5, \quad \mathbf{x}^{\text{F}} + \boldsymbol{\delta} \in \mathcal{P}, \quad (2)$$

where $\text{cost}^{\text{F}}(\cdot)$ is a user-specified cost function that encodes preferences between feasible actions from $\mathbf{x}^{\text{F}}$, and $\mathcal{F}$ and $\mathcal{P}$ are optional sets of feasibility and plausibility constraints,[3] restricting the actions and the resulting counterfactual explanation, respectively. The feasibility constraints in (2), as introduced in [69], aim at restricting the set of features that the individual may act upon. For instance, recommendations should not ask individuals to change their gender or reduce their age. Henceforth, we refer to the optimization problem in (2) as *CFE-based recourse* problem, where the emphasis is shifted from minimising a distance as in (1) to optimising a personalised cost function $\text{cost}^{\text{F}}(\cdot)$ over a set of actions $\boldsymbol{\delta}$ which individual $\mathbf{x}^{\text{F}}$ can perform.

The seemingly innocent reformulation of the counterfactual explanation problem in (1) as a recourse problem in (2) is founded on two key assumptions.

Assumption 1. The feature-wise difference between factual and nearest counterfactual instances, $\mathbf{x}^{\text{CFE}} - \mathbf{x}^{\text{F}}$, directly translates to minimal action sets $\boldsymbol{\delta}^*$, such that performing the actions in $\boldsymbol{\delta}^*$ starting from $\mathbf{x}^{\text{F}}$ will result in $\mathbf{x}^{\text{CFE}}$.

Assumption 2. There is a 1-1 mapping between $\text{dist}(\cdot, \mathbf{x}^{\text{F}})$ and $\text{cost}^{\text{F}}(\cdot)$, whereby more effortful actions incur larger distance and higher cost.

Unfortunately, these assumptions only hold in restrictive settings, rendering solutions of (2) *sub-optimal* or *ineffective* in many real-world scenarios. Specifically, Assumption 1 implies that features X_i for which $\delta_i^* = 0$ are unaffected. However, this generally holds only if (i) the individual applies effort in a world where changing a variable does not have downstream effects on other variables (i.e., features are independent of each other); or (ii) the individual changes the value of a subset of variables while simultaneously enforcing that the values of all other variables remain unchanged (i.e., breaking dependencies between features). Beyond the *sub-optimality* that arises from assuming/reducing to an independent world in (i), and disregarding the *feasibility* of non-altering actions in (ii), non-altering actions may naturally incur a cost which is not captured in the current definition of cost, and hence Assumption 2 does not hold either. Therefore, except in trivial cases where the model designer actively inputs pairwise independent features (independently manipulable inputs) to the classifier h (see Fig. 2a), generating recommendations from counterfactual explanations in this manner, i.e., ignoring the potentially rich causal structure over $\mathbf{X}$ and the resulting downstream effects that changes to some features may have on others (see Fig. 2b), warrants reconsideration. A number of authors have argued for the need to consider causal relations between variables when generating counterfactual explanations [25,39,41,69,76], however, this has not yet been formalized.

---

[3] Here, "feasible" means *possible to do*, whereas "plausible" means *possibly true, believable or realistic*. Optimization terminology refers to both as *feasibility* sets.

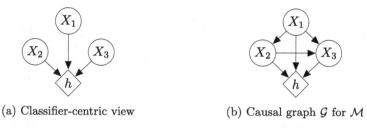

(a) Classifier-centric view   (b) Causal graph $\mathcal{G}$ for $\mathcal{M}$

Fig. 2. A view commonly adopted for counterfactual explanations (a) treats features as independently manipulable inputs to a given fixed and deterministic classifier h. In the causal approach to algorithmic recourse taken in this work, we instead view variables as causally related to each other by a structural causal model (SCM) $\mathcal{M}$ with associated causal graph $\mathcal{G}$ (b).

2.2 Causality: Structural Causal Models, Interventions, and Counterfactuals

To reason formally about causal relations between features $\mathbf{X} = (X_1, ..., X_d)$, we adopt the *structural causal model* (SCM) framework [45].[4] Specifically, we assume that the data-generating process of $\mathbf{X}$ is described by an (unknown) underlying SCM $\mathcal{M}$ of the general form

$$\mathcal{M} = (\mathbf{S}, P_\mathbf{U}), \quad \mathbf{S} = \{X_r := f_r(\mathbf{X}_{\mathrm{pa}(r)}, U_r)\}_{r=1}^d, \quad P_\mathbf{U} = P_{U_1} \times \ldots \times P_{U_d}, \tag{3}$$

where the structural equations $\mathbf{S}$ are a set of assignments generating each observed variable X_r as a deterministic function f_r of its causal parents $\mathbf{X}_{\mathrm{pa}(r)} \subseteq \mathbf{X} \setminus X_r$ and an unobserved noise variable U_r. The assumption of mutually independent noises (i.e., a fully factorised $P_\mathbf{U}$) entails that there is no hidden confounding and is referred to as *causal sufficiency*. An SCM is often illustrated by its associated causal graph $\mathcal{G}$, which is obtained by drawing a directed edge from each node in $\mathbf{X}_{\mathrm{pa}(r)}$ to X_r for $r \in [d] := \{1, \ldots, d\}$, see Fig. 1 and Fig. 2b for examples. We assume throughout that $\mathcal{G}$ is acyclic. In this case, $\mathcal{M}$ implies a unique observational distribution $P_\mathbf{X}$, which factorises over $\mathcal{G}$, defined as the push-forward of $P_\mathbf{U}$ via $\mathbf{S}$.[5]

Importantly, the SCM framework also entails *interventional distributions* describing a situation in which some variables are manipulated externally. E.g., using the *do*-operator, an intervention which fixes $\mathbf{X}_\mathcal{I}$ to $\boldsymbol{\theta}$ (where $\mathcal{I} \subseteq [d]$) is denoted by $do(\mathbf{X}_\mathcal{I} = \boldsymbol{\theta})$. The corresponding distribution of the remaining variables $\mathbf{X}_{-\mathcal{I}}$ can be computed by replacing the structural equations for $\mathbf{X}_\mathcal{I}$ in $\mathbf{S}$ to obtain the new set of equations $\mathbf{S}^{do(\mathbf{X}_\mathcal{I} = \boldsymbol{\theta})}$. The interventional distribution $P_{\mathbf{X}_{-\mathcal{I}}|do(\mathbf{X}_\mathcal{I} = \boldsymbol{\theta})}$ is then given by the observational distribution implied by the manipulated SCM $\left(\mathbf{S}^{do(\mathbf{X}_\mathcal{I} = \boldsymbol{\theta})}, P_\mathbf{U}\right)$.

---

[4] Also known as non-parametric structural equation model with independent errors.

[5] I.e., for $r \in [d]$, $P_{X_r|\mathbf{X}_{\mathrm{pa}(r)}}(X_r|\mathbf{X}_{\mathrm{pa}(r)}) := P_{U_r}(f_r^{-1}(X_r|\mathbf{X}_{\mathrm{pa}(r)}))$, where $f_r^{-1}(X_r|\mathbf{X}_{\mathrm{pa}(r)})$ denotes the pre-image of X_r given $\mathbf{X}_{\mathrm{pa}(r)}$ under f_r, i.e., $f_r^{-1}(X_r|\mathbf{X}_{\mathrm{pa}(r)}) := \{u \in \mathcal{U}_r : f_r(\mathbf{X}_{\mathrm{pa}(r)}, u) = X_r\}$.

Similarly, an SCM also implies distributions over *counterfactuals*—statements about a world in which a hypothetical intervention was performed *all else being equal*. For example, *given* observation $\mathbf{x}^F$ we can ask what would have happened if $\mathbf{X}_\mathcal{I}$ had instead taken the value $\boldsymbol{\theta}$. We denote the counterfactual variable by $\mathbf{X}(do(\mathbf{X}_\mathcal{I} = \boldsymbol{\theta}))|\mathbf{x}^F$, whose distribution can be computed in three steps [45]:

1. **Abduction**: compute the posterior distribution $P_{\mathbf{U}|\mathbf{x}^F}$ of the exogenous variables $\mathbf{U}$ given the factual observation $\mathbf{x}^F$;
2. **Action**: perform the intervention $do(\mathbf{X}_\mathcal{I} = \boldsymbol{\theta})$ by replacing the structural equations for $\mathbf{X}_\mathcal{I}$ by $\mathbf{X}_\mathcal{I} := \boldsymbol{\theta}$ to obtain the new structural equations $\mathbf{S}^{do(\mathbf{X}_\mathcal{I} = \boldsymbol{\theta})}$;
3. **Prediction**: the counterfactual distribution $P_{\mathbf{X}(do(\mathbf{X}_\mathcal{I} = \boldsymbol{\theta}))|\mathbf{x}^F}$ is the distribution induced by the resulting SCM $\left(\mathbf{S}^{do(\mathbf{X}_\mathcal{I} = \boldsymbol{\theta})}, P_{\mathbf{U}|\mathbf{x}^F}\right)$.

For instance, the counterfactual variable for individual $\mathbf{x}^F$ had action $a = do(\mathbf{X}_\mathcal{I} = \boldsymbol{\theta}) \in \mathcal{F}$ been performed would be $\mathbf{X}^{\text{SCF}}(a) := \mathbf{X}(a)|\mathbf{x}^F$. For a worked-out example of computing counterfactuals in SCMs, we refer to Sect. 3.2.

3   Causal Recourse Formulation

3.1   Limitations of CFE-Based Recourse

Here, we use causal reasoning to formalize the limitations of the CFE-based recourse approach in (2). To this end, we first reinterpret the actions resulting from solving the CFE-based recourse problem, i.e., $\boldsymbol{\delta}^*$, as structural interventions by defining the set of indices $\mathcal{I}$ of observed variables that are intervened upon.

Definition 1 (CFE-based actions). *Given an individual $\mathbf{x}^F$ in world $\mathcal{M}$ and a solution $\boldsymbol{\delta}^*$ of (2), denote by $\mathcal{I} = \{i \mid \delta_i^* \neq 0\}$ the set of indices of observed variables that are acted upon. A CFE-based action then refers to a set of structural interventions of the form $a^{\text{CFE}}(\boldsymbol{\delta}^*, \mathbf{x}^F) := \text{do}(\{X_i := x_i^F + \delta_i^*\}_{i\in\mathcal{I}})$.*

Using Definition 1, we can derive the following key results that provide necessary and sufficient conditions for CFE-based actions to guarantee recourse.

Proposition 1. *A CFE-based action $a^{\text{CFE}}(\boldsymbol{\delta}^*, \mathbf{x}^F)$ in general (i.e., for arbitrary underlying causal models) results in the structural counterfactual $\mathbf{x}^{\text{SCF}} = \mathbf{x}^{\text{CFE}} := \mathbf{x}^F + \boldsymbol{\delta}^*$ and thus guarantees recourse (i.e., $h(\mathbf{x}^{\text{SCF}}) \neq h(\mathbf{x}^F)$) if and only if the set of descendants of the acted upon variables determined by $\mathcal{I}$ is the empty set.*

Corollary 1. *If all features in the true world $\mathcal{M}$ are mutually independent, (i.e., if they are all root-nodes in the causal graph), then CFE-based actions always guarantee recourse.*

While the above results are formally proven in Appendix A of [28], we provide a sketch of the proof below. If the intervened-upon variables do not have descendants, then by definition $\mathbf{x}^{\text{SCF}} = \mathbf{x}^{\text{CFE}}$. Otherwise, the value of the descendants will depend on the counterfactual value of their parents, leading to a structural counterfactual that does not resemble the nearest counterfactual explanation, $\mathbf{x}^{\text{SCF}} \neq \mathbf{x}^{\text{CFE}}$, and thus may not result in recourse. Moreover, in an independent world the set of descendants of all the variables is by definition the empty set.

Unfortunately, the independent world assumption is not realistic, as it requires all the features selected to train the predictive model h to be independent of each other. Moreover, limiting changes to only those variables without descendants may unnecessarily limit the agency of the individual, e.g., in Example 1, restricting the individual to only changing bank balance without e.g., pursuing a new/side job to increase their income would be limiting. Thus, for a given non-independent $\mathcal{M}$ capturing the true causal dependencies between features, CFE-based actions require the individual seeking recourse to enforce (at least partially) an independent post-intervention model $\mathcal{M}^{a^{\text{CFE}}}$ (so that Assumption 1 holds), by intervening on all the observed variables for which $\delta_i \neq 0$ as well as on their descendants (even if their $\delta_i = 0$). However, such requirement suffers from two main issues. First, it conflicts with Assumption 2, since holding the value of variables may still imply potentially *infeasible* and costly interventions in $\mathcal{M}$ to sever all the incoming edges to such variables, and even then it may be ineffective and not change the prediction (see Example 2). Second, as will be proven in the next section (see also, Example 1), CFE-based actions may still be *suboptimal*, as they do not benefit from the causal effect of actions towards changing the prediction. Thus, even when equipped with knowledge of causal dependencies, recommending actions directly from counterfactual explanations in the manner of existing approaches is not satisfactory.

3.2   Recourse Through Minimal Interventions

We have demonstrated that actions which immediately follow from counterfactual explanations may require unrealistic assumptions, or alternatively, result in sub-optimal or even infeasible recommendations. To solve such limitations we rewrite the recourse problem so that instead of finding the minimal (independent) shift of features as in (2), we seek the minimal cost set of actions (in the form of structural interventions) that results in a counterfactual instance yielding the favorable output from h. For simplicity, we present the formulation for the case of an invertible SCM (i.e., one with invertible structural equations $\mathbf{S}$) such that the ground-truth counterfactual $\mathbf{x}^{\text{SCF}} = \mathbf{S}^a(\mathbf{S}^{-1}(\mathbf{x}^{\text{F}}))$ is a unique point. The resulting optimisation formulation is as follows:

$$a^* \in \arg\min_{a \in \mathcal{F}} \quad \text{cost}^{\text{F}}(a) \quad \text{subject to} \quad h(\mathbf{x}^{\text{SCF}}(a)) \geq 0.5,$$
$$\mathbf{x}^{\text{SCF}}(a) = \mathbf{x}(a)|\mathbf{x}^{\text{F}} \in \mathcal{P}, \tag{4}$$

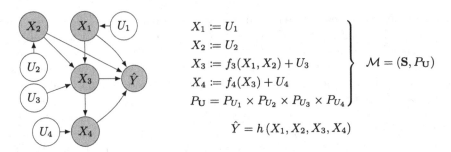

$$X_1 := U_1$$
$$X_2 := U_2$$
$$X_3 := f_3(X_1, X_2) + U_3$$
$$X_4 := f_4(X_3) + U_4$$
$$P_U = P_{U_1} \times P_{U_2} \times P_{U_3} \times P_{U_4}$$
$$\mathcal{M} = (\mathbf{S}, P_U)$$

$$\hat{Y} = h(X_1, X_2, X_3, X_4)$$

Fig. 3. The structural causal model (graph and equations) for the working example and demonstration in Sect. 3.2.

where $a^* \in \mathcal{F}$ directly specifies the set of feasible actions to be performed for minimally costly recourse, with $\text{cost}^F(\cdot)$.[6]

Importantly, using the formulation in (4) it is now straightforward to show the suboptimality of CFE-based actions (proof in Appendix A of [28]):

Proposition 2. *Given an individual* $\mathbf{x}^F$ *observed in world* $\mathcal{M}$, *a set of feasible actions* $\mathcal{F}$, *and a solution* $a^* \in \mathcal{F}$ *of (4), assume that there exists a CFE-based action* $a^{CFE}(\delta^*, \mathbf{x}^F) \in \mathcal{F}$ *(see Definition 1) that achieves recourse, i.e.,* $h(\mathbf{x}^F) \neq h(\mathbf{x}^{CFE})$. *Then,* $\text{cost}^F(a^*) \leq \text{cost}^F(a^{CFE})$.

Thus, for a *known* causal model capturing the dependencies among observed variables, and a family of feasible interventions, the optimization problem in (4) yields *Recourse through Minimal Interventions* (MINT). Generating minimal interventions through solving (4) requires that we be able to compute the structural counterfactual, $\mathbf{x}^{SCF}$, of the individual $\mathbf{x}^F$ in world $\mathcal{M}$, given *any* feasible action $a \in \mathcal{F}$. To this end, and for the purpose of demonstration, we consider a class of invertible SCMs, specifically, additive noise models (ANM) [23], where the structural equations $\mathbf{S}$ are of the form

$$\mathbf{S} = \{X_r := f_r(\mathbf{X}_{\text{pa}(r)}) + U_r\}_{r=1}^d \implies u_r^F = x_r^F - f_r(\mathbf{x}_{\text{pa}(r)}^F), \quad r \in [d], \quad (5)$$

and propose to use the three steps of structural counterfactuals in [45] to assign a single counterfactual $\mathbf{x}^{SCF}(a) := \mathbf{x}(a)|\mathbf{x}^F$ to each action $a = do(\mathbf{X}_\mathcal{I} = \boldsymbol{\theta}) \in \mathcal{F}$ as below.

Working Example. Consider the model in Fig. 3, where $\{U_i\}_{i=1}^4$ are mutually independent exogenous variables, and $\{f_i\}_{i=1}^4$ are deterministic (linear or

---

[6] We note that, although $\mathbf{x}^{*SCF} := \mathbf{x}(a^*)|\mathbf{x}^F = \mathbf{S}^{a^*}(\mathbf{S}^{-1}(\mathbf{x}^F))$ is a counterfactual instance, it does not need to correspond to the nearest counterfactual explanation, $\mathbf{x}^{*CFE} := \mathbf{x}^F + \delta^*$, resulting from (2) (see, e.g., Example 1). This further emphasizes that minimal interventions are not necessarily obtainable via pre-computed nearest counterfactual instances, and recourse actions should be obtained by solving (4) rather than indirectly through the solution of (2).

nonlinear) functions. Let $\mathbf{x}^{\mathrm{F}} = (x_1^{\mathrm{F}}, x_2^{\mathrm{F}}, x_3^{\mathrm{F}}, x_4^{\mathrm{F}})^{\top}$ be the observed features belonging to the (factual) individual seeking recourse. Also, let $\mathcal{I}$ denote the set of indices corresponding to the subset of endogenous variables that are intervened upon according to the action set a. Then, we obtain a structural counterfactual, $\mathbf{x}^{\mathrm{SCF}}(a) := \mathbf{x}(a)|\mathbf{x}^{\mathrm{F}} = \mathbf{S}^a(\mathbf{S}^{-1}(\mathbf{x}^{\mathrm{F}}))$, by applying the Abduction-Action-Prediction steps [46] as follows:

Step 1. Abduction uniquely determines the value of all exogenous variables $\mathbf{U}$ given the observed evidence $\mathbf{X} = \mathbf{x}^{\mathrm{F}}$:

$$
\begin{aligned}
u_1 &= x_1^{\mathrm{F}}, \\
u_2 &= x_2^{\mathrm{F}}, \\
u_3 &= x_3^{\mathrm{F}} - f_3(x_1^{\mathrm{F}}, x_2^{\mathrm{F}}), \\
u_4 &= x_4^{\mathrm{F}} - f_4(x_3^{\mathrm{F}}).
\end{aligned}
\tag{6}
$$

Step 2. Action modifies the SCM according to the hypothetical interventions, $\mathrm{do}(\{X_i := a_i\}_{i \in \mathcal{I}})$ (where $a_i = x_i^F + \delta_i$), yielding $\mathbf{S}^a$:

$$
\begin{aligned}
X_1 &:= [1 \in \mathcal{I}] \cdot a_1 + [1 \notin \mathcal{I}] \cdot U_1, \\
X_2 &:= [2 \in \mathcal{I}] \cdot a_2 + [2 \notin \mathcal{I}] \cdot U_2, \\
X_3 &:= [3 \in \mathcal{I}] \cdot a_3 + [3 \notin \mathcal{I}] \cdot \big(f_3(X_1, X_2) + U_3\big), \\
X_4 &:= [4 \in \mathcal{I}] \cdot a_4 + [4 \notin \mathcal{I}] \cdot \big(f_4(X_3) + U_4\big),
\end{aligned}
\tag{7}
$$

where $[\cdot]$ denotes the Iverson bracket.

Step 3. Prediction recursively determines the values of all endogenous variables based on the computed exogenous variables $\{u_i\}_{i=1}^4$ from Step 1 and $\mathbf{S}^a$ from Step 2, as:

$$
\begin{aligned}
x_1^{\mathrm{SCF}} &:= [1 \in \mathcal{I}] \cdot a_1 + [1 \notin \mathcal{I}] \cdot \big(u_1\big), \\
x_2^{\mathrm{SCF}} &:= [2 \in \mathcal{I}] \cdot a_2 + [2 \notin \mathcal{I}] \cdot \big(u_2\big), \\
x_3^{\mathrm{SCF}} &:= [3 \in \mathcal{I}] \cdot a_3 + [3 \notin \mathcal{I}] \cdot \big(f_3(x_1^{\mathrm{SCF}}, x_2^{\mathrm{SCF}}) + u_3\big), \\
x_4^{\mathrm{SCF}} &:= [4 \in \mathcal{I}] \cdot a_4 + [4 \notin \mathcal{I}] \cdot \big(f_4(x_3^{\mathrm{SCF}}) + u_4\big).
\end{aligned}
\tag{8}
$$

General Assignment Formulation for ANMs. As we have not made any restricting assumptions about the structural equations (only that we operate with additive noise models[7] where noise variables are pairwise independent), the solution for the working example naturally generalizes to SCMs corresponding to other DAGs with more variables. The assignment of structural counterfactual values can generally be written as:

---

[7] We remark that the presented formulation also holds for more general SCMs (for example where the exogenous variable contribution is not additive) as long as the sequence of structural equations $\mathbf{S}$ is invertible, i.e., there exists a sequence of equations $\mathbf{S}^{-1}$ such that $\mathbf{x} = \mathbf{S}(\mathbf{S}^{-1}(\mathbf{x}))$ (in other words, the exogenous variables are uniquely identifiable via the abduction step).

$$x_i^{\text{SCF}} = [i \in \mathcal{I}] \cdot (x_i^{\text{F}} + \delta_i) + [i \notin \mathcal{I}] \cdot \left(x_i^{\text{F}} + f_i(\mathbf{pa}_i^{\text{SCF}}) - f_i(\mathbf{pa}_i^{\text{F}}) \right). \qquad (9)$$

In words, the counterfactual value of the i-th feature, x_i^{SCF}, takes the value $x_i^{\text{F}} + \delta_i$ if such feature is intervened upon (i.e., $i \in \mathcal{I}$). Otherwise, x_i^{SCF} is computed as a function of both the factual and counterfactual values of its parents, denoted respectively by $f_i(\mathbf{pa}_i^{\text{F}})$ and $f_i(\mathbf{pa}_i^{\text{SCF}})$. The closed-form expression in (9) can replace the counterfactual constraint in (4), i.e.,

$$\mathbf{x}^{\text{SCF}}(a) := \mathbf{x}(a)|\mathbf{x}^{\text{F}} = \mathbf{S}^a(\mathbf{S}^{-1}(\mathbf{x}^{\text{F}})),$$

after which the optimization problem may be solved by building on existing frameworks for generating nearest counterfactual explanations, including gradient-based, evolutionary-based, heuristics-based, or verification-based approaches as referenced in Sect. 2.1. It is important to note that unlike CFE-based actions where the precise value of all covariates post-intervention are specified, MINT-based actions require that the user focus only on the features upon which interventions are to be performed, which may better align with factors under the users control (e.g., some features may be non-actionable but mutable through changes to other features; see also [6]).

3.3 Negative Result: No Recourse Guarantees for Unknown Structural Equations

In practice, the structural counterfactual $\mathbf{x}^{\text{SCF}}(a)$ can only be computed using an approximate (and likely imperfect) SCM $\mathcal{M} = (\mathbf{S}, P_{\mathbf{U}})$, which is estimated from data assuming a particular form of the structural equation as in (5). However, assumptions on the form of the true structural equations $\mathbf{S}_\star$ are generally untestable—not even with a randomized experiment—since there exist multiple SCMs which imply the same observational and interventional distributions, but entail different structural counterfactuals.

Example 3 (adapted from 6.19 in [48]). Consider the following two SCMs $\mathcal{M}_A$ and $\mathcal{M}_B$ which arise from the general form in Fig. 1 by choosing $U_1, U_2 \sim$ Bernoulli(0.5) and $U_3 \sim$ Uniform($\{0, \ldots, K\}$) independently in both $\mathcal{M}_A$ and $\mathcal{M}_B$, with structural equations

$$
\begin{aligned}
X_1 &:= U_1, & \text{in} \quad &\{\mathcal{M}_A, \mathcal{M}_B\}, \\
X_2 &:= X_1(1 - U_2), & \text{in} \quad &\{\mathcal{M}_A, \mathcal{M}_B\}, \\
X_3 &:= \mathbb{I}_{X_1 \neq X_2}(\mathbb{I}_{U_3 > 0} X_1 + \mathbb{I}_{U_3 = 0} X_2) + \mathbb{I}_{X_1 = X_2} U_3, & \text{in} \quad &\mathcal{M}_A, \\
X_3 &:= \mathbb{I}_{X_1 \neq X_2}(\mathbb{I}_{U_3 > 0} X_1 + \mathbb{I}_{U_3 = 0} X_2) + \mathbb{I}_{X_1 = X_2}(K - U_3), & \text{in} \quad &\mathcal{M}_B.
\end{aligned}
$$

Then $\mathcal{M}_A$ and $\mathcal{M}_B$ both imply exactly the same observational and interventional distributions, and thus are indistinguishable from empirical data. However, having observed $\mathbf{x}^{\text{F}} = (1, 0, 0)$, they predict different counterfactuals had X_1 been 0, i.e., $\mathbf{x}^{\text{SCF}}(X_1 = 0) = (0, 0, 0)$ and $(0, 0, K)$, respectively.[8]

---

[8] This follows from abduction on $\mathbf{x}^{\text{F}} = (1, 0, 0)$ which for both $\mathcal{M}_A$ and $\mathcal{M}_B$ implies $U_3 = 0$.

Confirming or refuting an assumed form of $\mathbf{S}_\star$ would thus require counterfactual data which is, by definition, never available. Thus, Example 3 proves the following proposition by contradiction.

Proposition 3 (Lack of Recourse Guarantees). *If the set of descendants of intervened-upon variables is non-empty, algorithmic recourse can be guaranteed in general (i.e., without further restrictions on the underlying causal model) only if the true structural equations are known, irrespective of the amount and type of available data.*

Remark 1. The converse of Proposition 3 does not hold. E.g., given $\mathbf{x}^F = (1, 0, 1)$ in Example 3, abduction in either model yields $U_3 > 0$, so the counterfactual of X_3 cannot be predicted exactly.

Building on the framework of [28], we next present two novel approaches for causal algorithmic recourse under unknown structural equations. The first approach in Sect. 4.1 aims to estimate the counterfactual distribution under the assumption of ANMs (5) with Gaussian noise for the structural equations. The second approach in Sect. 4.2 makes no assumptions about the structural equations, and instead of approximating the structural equations, it considers the effect of interventions on a sub-population similar to $\mathbf{x}^F$. We recall that the causal graph is assumed to be known throughout.

4 Recourse Under Imperfect Causal Knowledge

4.1 Probabilistic Individualised Recourse

Since the true SCM $\mathcal{M}_\star$ is unknown, one approach to solving (4) is to learn an approximate SCM $\mathcal{M}$ within a given model class from training data $\{\mathbf{x}^i\}_{i=1}^n$. For example, for an ANM (5) with zero-mean noise, the functions f_r can be learned via linear or kernel (ridge) regression of X_r given $\mathbf{X}_{\mathrm{pa}(r)}$ as input. We refer to these approaches as $\mathcal{M}_{\mathrm{LIN}}$ and $\mathcal{M}_{\mathrm{KR}}$, respectively. $\mathcal{M}$ can then be used in place of $\mathcal{M}_\star$ to infer the noise values as in (5), and subsequently to predict a *single-point counterfactual* $\mathbf{x}^{\mathrm{SCF}}(a)$ to be used in (4). However, the learned causal model $\mathcal{M}$ may be imperfect, and thus lead to wrong counterfactuals due to, e.g., the finite sample of the observed data, or more importantly, due to model misspecification (i.e., assuming a wrong parametric form for the structural equations).

To solve such limitation, we adopt a Bayesian approach to account for the uncertainty in the estimation of the structural equations. Specifically, we assume additive Gaussian noise and rely on probabilistic regression using a Gaussian process (GP) prior over the functions f_r; for an overview of regression with GPs, we refer to [79, § 2].

Definition 2 (GP-SCM). *A Gaussian process SCM (GP-SCM) over $\mathbf{X}$ refers to the model*

$$X_r := f_r(\mathbf{X}_{pa(r)}) + U_r, \qquad f_r \sim \mathcal{GP}(0, k_r), \qquad U_r \sim \mathcal{N}(0, \sigma_r^2), \qquad r \in [d],$$
$$\tag{10}$$

with covariance functions $k_r : \mathcal{X}_{pa(r)} \times \mathcal{X}_{pa(r)} \to \mathbb{R}$, e.g., RBF kernels for continuous $X_{pa(r)}$.

While GPs have previously been studied in a causal context for structure learning [16,73], estimating treatment effects [2,56], or learning SCMs with latent variables and measurement error [61], our goal here is to account for the uncertainty over f_r in the computation of the posterior over U_r, and thus to obtain a *counterfactual distribution*, as summarised in the following propositions.

Proposition 4 (GP-SCM Noise Posterior). *Let $\{x^i\}_{i=1}^n$ be an observational sample from (10). For each $r \in [d]$ with non empty parent set $|pa(r)| > 0$, the posterior distribution of the noise vector $\mathbf{u}_r = (u_r^1, ..., u_r^n)$, conditioned on $\mathbf{x}_r = (x_r^1, ..., x_r^n)$ and $\mathbf{X}_{pa(r)} = (\mathbf{x}_{pa(r)}^1, ..., \mathbf{x}_{pa(r)}^n)$, is given by*

$$\mathbf{u}_r | \mathbf{X}_{pa(r)}, \mathbf{x}_r \sim \mathcal{N}\left(\sigma_r^2 (\mathbf{K} + \sigma_r^2 \mathbf{I})^{-1} \mathbf{x}_r, \sigma_r^2 \left(\mathbf{I} - \sigma_r^2 (\mathbf{K} + \sigma_r^2 \mathbf{I})^{-1}\right)\right), \qquad (11)$$

where $\mathbf{K} := \left(k_r(\mathbf{x}_{pa(r)}^i, \mathbf{x}_{pa(r)}^j)\right)_{ij}$ denotes the Gram matrix.

Next, in order to compute counterfactual distributions, we rely on ancestral sampling (according to the causal graph) of the descendants of the intervention targets $\mathbf{X}_{\mathcal{I}}$ using the noise posterior of (11). The counterfactual distribution of each descendant X_r is given by the following proposition.

Proposition 5 (GP-SCM Counterfactual Distribution). *Let $\{x^i\}_{i=1}^n$ be an observational sample from (10). Then, for $r \in [d]$ with $|pa(r)| > 0$, the counterfactual distribution over X_r had $\mathbf{X}_{pa(r)}$ been $\tilde{\mathbf{x}}_{pa(r)}$ (instead of $\mathbf{x}_{pa(r)}^F$) for individual $\mathbf{x}^F \in \{x^i\}_{i=1}^n$ is given by*

$$\begin{aligned} X_r(\mathbf{X}_{pa(r)} &= \tilde{\mathbf{x}}_{pa(r)}) | \mathbf{x}^F, \{x^i\}_{i=1}^n \\ &\sim \mathcal{N}\left(\mu_r^F + \tilde{\mathbf{k}}^T (\mathbf{K} + \sigma_r^2 \mathbf{I})^{-1} \mathbf{x}_r, s_r^F + \tilde{k} - \tilde{\mathbf{k}}^T (\mathbf{K} + \sigma_r^2 \mathbf{I})^{-1} \tilde{\mathbf{k}}\right), \end{aligned} \qquad (12)$$

where $\tilde{k} := k_r(\tilde{\mathbf{x}}_{pa(r)}, \tilde{\mathbf{x}}_{pa(r)})$, $\tilde{\mathbf{k}} := \left(k_r(\tilde{\mathbf{x}}_{pa(r)}, \mathbf{x}_{pa(r)}^1), ..., k_r(\tilde{\mathbf{x}}_{pa(r)}, \mathbf{x}_{pa(r)}^n)\right)$, $\mathbf{x}_r$ and $\mathbf{K}$ as defined in Proposition 4, and μ_r^F and s_r^F are the posterior mean and variance of u_r^F given by (11).

All proofs can be found in Appendix A of [27]. We can now generalise the recourse problem (4) to our probabilistic setting by replacing the single-point counterfactual $\mathbf{x}^{SCF}(a)$ with the counterfactual random variable $\mathbf{X}^{SCF}(a) := \mathbf{X}(a)|\mathbf{x}^F$. As a consequence, it no longer makes sense to consider a hard constraint of the form $h(\mathbf{x}^{SCF}(a)) > 0.5$, i.e., that the prediction needs to change. Instead, we can reason about the expected classifier output under the counterfactual distribution, leading to the following *probabilistic version of the individualised recourse optimisation problem*:

$$\begin{aligned} &\min_{a = do(\mathbf{X}_{\mathcal{I}} = \theta) \in \mathcal{F}} \quad \text{cost}^F(a) \\ &\text{subject to} \quad \mathbb{E}_{\mathbf{X}^{SCF}(a)} \left[h\left(\mathbf{X}^{SCF}(a)\right)\right] \geq \text{thresh}(a). \end{aligned} \qquad (13)$$

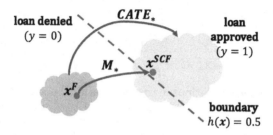

Fig. 4. Illustration of point- and subpopulation-based recourse approaches.

Note that the threshold `thresh(a)` is allowed to depend on a. For example, an intuitive choice is

$$\texttt{thresh}(a) = 0.5 + \gamma_{\mathrm{LCB}} \sqrt{\mathrm{Var}_{\mathbf{X}^{\mathrm{SCF}}(a)} \left[h \left(\mathbf{X}^{\mathrm{SCF}}(a) \right) \right]} \tag{14}$$

which has the interpretation of the lower-confidence bound crossing the decision boundary of 0.5. Note that larger values of the hyperparameter γ_{LCB} lead to a more conservative approach to recourse, while for $\gamma_{\mathrm{LCB}} = 0$ merely crossing the decision boundary with $\geq 50\%$ chance suffices.

4.2   Probabilistic Subpopulation-Based Recourse

The GP-SCM approach in Sect. 4.1 allows us to average over an infinite number of (non-)linear structural equations, under the assumption of additive Gaussian noise. However, this assumption may still not hold under the true SCM, leading to sub-optimal or inefficient solutions to the recourse problem. Next, we remove any assumptions about the structural equations, and propose a second approach that does not aim to approximate an individualized counterfactual distribution, but instead considers the effect of interventions on a subpopulation defined by certain shared characteristics with the given (factual) individual $\mathbf{x}^{\mathrm{F}}$. The key idea behind this approach resembles the notion of conditional average treatment effects (CATE) [1] (illustrated in Fig. 4) and is based on the fact that any intervention $do(\mathbf{X}_{\mathcal{I}} = \boldsymbol{\theta})$ only influences the descendants $\mathrm{d}(\mathcal{I})$ of the intervened-upon variables, while the non-descendants $\mathrm{nd}(\mathcal{I})$ remain unaffected. Thus, when evaluating an intervention, we can condition on $\mathbf{X}_{\mathrm{nd}(\mathcal{I})} = \mathbf{x}^{\mathrm{F}}_{\mathrm{nd}(\mathcal{I})}$, thus selecting a subpopulation of individuals similar to the factual subject.

Specifically, we propose to solve the following *subpopulation-based recourse optimization problem*

$$\min_{a = do(\mathbf{X}_{\mathcal{I}} = \boldsymbol{\theta}) \in \mathcal{F}} \mathrm{cost}^{\mathrm{F}}(a)$$
$$\text{subject to} \quad \mathbb{E}_{\mathbf{X}_{\mathrm{d}(\mathcal{I})} | do(\mathbf{X}_{\mathcal{I}} = \boldsymbol{\theta}), \mathbf{x}^{\mathrm{F}}_{\mathrm{nd}(\mathcal{I})}} \left[h \left(\mathbf{x}^{\mathrm{F}}_{\mathrm{nd}(\mathcal{I})}, \boldsymbol{\theta}, \mathbf{X}_{\mathrm{d}(\mathcal{I})} \right) \right] \geq \texttt{thresh}(a), \tag{15}$$

where, in contrast to (13), the expectation is taken over the corresponding interventional distribution.

In general, this interventional distribution does not match the conditional distribution, i.e.,

$$P_{\mathbf{X}_{d(\mathcal{I})}|do(\mathbf{X}_{\mathcal{I}} = \boldsymbol{\theta}),\mathbf{x}^{F}_{nd(\mathcal{I})}} \neq P_{\mathbf{X}_{d(\mathcal{I})}|\mathbf{X}_{\mathcal{I}} = \boldsymbol{\theta},\mathbf{x}^{F}_{nd(\mathcal{I})}},$$

because some spurious correlations in the observational distribution do not transfer to the interventional setting. For example, in Fig. 2b we have that

$$P_{X_2|do(X_1 = x_1, X_3 = x_3)} = P_{X_2|X_1 = x_1} \neq P_{X_2|X_1 = x_1, X_3 = x_3}.$$

Fortunately, the interventional distribution can still be identified from the observational one, as stated in the following proposition.

Proposition 6. *Subject to causal sufficiency,* $P_{\mathbf{X}_{d(\mathcal{I})}|do(\mathbf{X}_{\mathcal{I}} = \boldsymbol{\theta}),\mathbf{x}^{F}_{nd(\mathcal{I})}}$ *is observationally identifiable (i.e., computable from the observational distribution) via:*

$$p\big(\mathbf{X}_{d(\mathcal{I})}|do(\mathbf{X}_{\mathcal{I}} = \boldsymbol{\theta}), \mathbf{x}^{F}_{nd(\mathcal{I})}\big) = \prod_{r \in d(\mathcal{I})} p\big(X_r|\mathbf{X}_{pa(r)}\big)\Bigg|_{\mathbf{X}_{\mathcal{I}} = \boldsymbol{\theta},\mathbf{X}_{nd(\mathcal{I})} = \mathbf{x}^{F}_{nd(\mathcal{I})}}.$$
$$(16)$$

As evident from Proposotion 6, tackling the optimization problem in (15) in the general case (i.e., for arbitrary graphs and intervention sets $\mathcal{I}$) requires estimating the stable conditionals $P_{X_r|\mathbf{X}_{pa(r)}}$ (a.k.a. causal Markov kernels) in order to compute the interventional expectation via (16). For convenience (see Sect. 4.3 for details), here we opt for latent-variable implicit density models, but other conditional density estimation approaches may be also be used [e.g., 7,10,68]. Specifically, we model each conditional $p(x_r|\mathbf{x}_{pa(r)})$ with a conditional variational autoencoder (CVAE) [62] as:

$$p(x_r|\mathbf{x}_{pa(r)}) \approx p_{\psi_r}(x_r|\mathbf{x}_{pa(r)}) = \int p_{\psi_r}(x_r|\mathbf{x}_{pa(r)}, \mathbf{z}_r)p(\mathbf{z}_r)d\mathbf{z}_r, \quad p(\mathbf{z}_r) := \mathcal{N}(\mathbf{0}, \mathbf{I}).$$
$$(17)$$

To facilitate sampling x_r (and in analogy to the deterministic mechanisms f_r in SCMs), we opt for deterministic decoders in the form of neural nets D_r parametrised by ψ_r, i.e., $p_{\psi_r}(x_r|\mathbf{x}_{pa(r)}, \mathbf{z}_r) = \delta(x_r - D_r(\mathbf{x}_{pa(r)}, \mathbf{z}_r; \psi_r))$, and rely on variational inference [77], amortised with approximate posteriors $q_{\phi_r}(\mathbf{z}_r|x_r, \mathbf{x}_{pa(r)})$ parametrised by encoders in the form of neural nets with parameters ϕ_r. We learn both the encoder and decoder parameters by maximising the evidence lower bound (ELBO) using stochastic gradient descend [11,30,31,50]. For further details, we refer to Appendix D of [27]

Remark 2. The collection of CVAEs can be interpreted as learning an approximate SCM of the form

$$\mathcal{M}_{\text{CVAE}}: \quad \mathbf{S} = \{X_r := D_r(\mathbf{X}_{pa(r)}, \mathbf{z}_r; \psi_r)\}_{r=1}^{d}, \quad \mathbf{z}_r \sim \mathcal{N}(\mathbf{0}, \mathbf{I}) \quad \forall r \in [d] \quad (18)$$

However, this family of SCMs may not allow to identify the true SCM (provided it can be expressed as above) from data without additional assumptions.

Moreover, exact posterior inference over $\mathbf{z}_r$ given $\mathbf{x}^{\mathrm{F}}$ is intractable, and we need to resort to approximations instead. It is thus unclear whether sampling from $q_{\phi_r}(\mathbf{z}_r|x_r^{\mathrm{F}}, \mathbf{x}_{\mathrm{pa}(r)}^{\mathrm{F}})$ instead of from $p(\mathbf{z}_r)$ in (17) can be interpreted as a counterfactual within (18). For further discussion on such "pseudo-counterfactuals" we refer to Appendix C of [27]

4.3   Solving the Probabilistic Recourse Optimization Problem

We now discuss how to solve the resulting optimization problems in (13) and (15). First, note that both problems differ only on the distribution over which the expectation in the constraint is taken: in (13) this is the counterfactual distribution of the descendants given in Proposition 5; and in (15) it is the interventional distribution identified in Proposition 6. In either case, computing the expectation for an arbitrary classifier h is intractable. Here, we approximate these integrals via Monte Carlo by sampling $\mathbf{x}_{\mathrm{d}(\mathcal{I})}^{(m)}$ from the interventional or counterfactual distributions resulting from $a = do(\mathbf{X}_{\mathcal{I}} = \boldsymbol{\theta})$, i.e.,

$$\mathbb{E}_{\mathbf{X}_{\mathrm{d}(\mathcal{I})|\theta}}\left[h\big(\mathbf{x}_{\mathrm{nd}(\mathcal{I})}^{\mathrm{F}}, \boldsymbol{\theta}, \mathbf{X}_{\mathrm{d}(\mathcal{I})}\big)\right] \approx \frac{1}{M}\sum_{m=1}^{M} h\big(\mathbf{x}_{\mathrm{nd}(\mathcal{I})}^{\mathrm{F}}, \boldsymbol{\theta}, \mathbf{x}_{\mathrm{d}(\mathcal{I})}^{(m)}\big).$$

Brute-Force Approach. A way to solve (13) and (15) is to (i) iterate over $a \in \mathcal{F}$, with $\mathcal{F}$ being a finite set of feasible actions (possibly as a result of discretizing in the case of a continuous search space); (ii) approximately evaluate the constraint via Monte Carlo ; and (iii) select a minimum cost action amongst all evaluated candidates satisfying the constraint. However, this may be computationally prohibitive and yield suboptimal interventions due to discretisation.

Gradient-based Approach. Recall that, for actions of the form $a = do(\mathbf{X}_{\mathcal{I}} = \boldsymbol{\theta})$, we need to optimize over both the intervention *targets* $\mathcal{I}$ and the intervention *values* $\boldsymbol{\theta}$. Selecting targets is a hard combinatorial optimization problem, as there are $2^{d'}$ possible choices for $d' \leq d$ actionable features, with a potentially infinite number of intervention values. We therefore consider different choices of targets $\mathcal{I}$ in parallel, and propose a gradient-based approach suitable for differentiable classifiers to efficiently find an optimal $\boldsymbol{\theta}$ for a given intervention set $\mathcal{I}$.[9] In particular, we first rewrite the constrained optimization problem in unconstrained form with Lagrangian [29,33]:

$$\mathcal{L}(\boldsymbol{\theta}, \lambda) := \mathrm{cost}^{\mathrm{F}}(a) + \lambda\big(\mathtt{thresh}(a) - \mathbb{E}_{\mathbf{X}_{\mathrm{d}(\mathcal{I})|\theta}}\big[h\big(\mathbf{x}_{\mathrm{nd}(\mathcal{I})}^{\mathrm{F}}, \boldsymbol{\theta}, \mathbf{X}_{\mathrm{d}(\mathcal{I})}\big)\big]\big). \quad (19)$$

We then solve the saddle point problem $\min_{\theta}\max_{\lambda} \mathcal{L}(\boldsymbol{\theta}, \lambda)$ arising from (19) with stochastic gradient descent [11,30]. Since both the GP-SCM counterfac-

---

[9] For large d when enumerating all $\mathcal{I}$ becomes computationally prohibitive, we can upper-bound the allowed number of variables to be intervened on simultaneously (e.g., $|\mathcal{I}| \leq 3$), or choose a greedy approach to select $\mathcal{I}$.

tual (12) and the CVAE interventional distributions (17) admit a reparametrization trick [31,50], we can differentiate through the constraint:

$$\nabla_\theta \mathbb{E}_{\mathbf{X}_{d(\mathcal{I})}}\left[h\left(\mathbf{x}^{\mathrm{F}}_{\mathrm{nd}(\mathcal{I})}, \boldsymbol{\theta}, \mathbf{X}_{d(\mathcal{I})}\right)\right] = \mathbb{E}_{\mathbf{z}\sim\mathcal{N}(\mathbf{0},\mathbf{I})}\left[\nabla_\theta h\left(\mathbf{x}^{\mathrm{F}}_{\mathrm{nd}(\mathcal{I})}, \boldsymbol{\theta}, \mathbf{x}_{d(\mathcal{I})}(\mathbf{z})\right)\right]. \quad (20)$$

Here, $\mathbf{x}_{d(\mathcal{I})}(\mathbf{z})$ is obtained by iteratively computing all descendants in topological order: either substituting $\mathbf{z}$ together with the other parents into the decoders D_r for the CVAEs, or by using the Gaussian reparametrization $x_r(\mathbf{z}) = \mu + \sigma\mathbf{z}$ with μ and σ given by (12) for the GP-SCM. A similar gradient estimator for the variance which enters $\mathtt{thresh}(a)$ for $\gamma_{\mathrm{LCB}} \neq 0$ is derived in Appendix F of [27].

5   Experiments

In our experiments, we compare different approaches for *causal* algorithmic recourse on synthetic and semi-synthetic data sets. Additional results can be found in Appendix B of [27].

5.1   Compared Methods

We compare the naive point-based recourse approaches $\mathcal{M}_{\mathrm{LIN}}$ and $\mathcal{M}_{\mathrm{KR}}$ mentioned at the beginning of Sect. 4.1 as baselines with the proposed counterfactual GP-SCM $\mathcal{M}_{\mathrm{GP}}$ and the CVAE approach for sub-population-based recourse (CATE$_{\mathrm{CVAE}}$). For completeness, we also consider a CATE$_{\mathrm{GP}}$ approach as a GP can also be seen as modelling each conditional as a Gaussian,[10] and also evaluate the "pseudo-counterfactual" $\mathcal{M}_{\mathrm{CVAE}}$ approach discussed in Remark 2. Finally, we report oracle performance for individualised $\mathcal{M}_\star$ and sub-population-based recourse methods CATE$_\star$ by sampling counterfactuals and interventions from the true underlying SCM. We note that a comparison with non-causal recourse approaches that assume independent features [58,69] or consider causal relations to generate counterfactual explanations but not recourse actions [24,39] is neither natural nor straight-forward, because it is unclear whether descendant variables should be allowed to change, whether keeping their value constant should incur a cost, and, if so, how much, c.f. [28].

5.2   Metrics

We compare recourse actions recommended by the different methods in terms of *cost*, computed as the L2-norm between the intervention $\boldsymbol{\theta}_{\mathcal{I}}$ and the factual value $\mathbf{x}^{\mathrm{F}}_{\mathcal{I}}$, normalised by the range of each feature $r \in \mathcal{I}$ observed in the training data; and *validity*, computed as the percentage of individuals for which the recommended actions result in a favourable prediction under the true (oracle) SCM. For our probabilistic recourse methods, we also report the lower confidence bound LCB := $\mathbb{E}[h] - \gamma_{\mathrm{LCB}}\sqrt{\mathrm{Var}[h]}$ of the selected action under the given method.

---

[10] Sampling from the noise prior instead of the posterior in (11) leads to an interventional distribution in (12).

Table 1. Experimental results for the gradient-based approach on different 3-variable SCMs. We show average performance ± 1 standard deviation for $N_{\text{runs}} = 100$, $N_{\text{MC-samples}} = 100$, and $\gamma_{\text{LCB}} = 2$.

Method	LINEAR SCM			NON-LINEAR ANM			NON-ADDITIVE SCM		
	Valid* (%)	LCB	Cost (%)	Valid* (%)	LCB	Cost (%)	Valid* (%)	LCB	Cost (%)
$\mathcal{M}_*$	100	–	10.9 ± 7.9	100	–	20.1 ± 12.3	100	–	13.2 ± 11.0
$\mathcal{M}_{\text{LIN}}$	100	–	11.0 ± 7.0	54	–	20.6 ± 11.0	98	–	14.0 ± 13.5
$\mathcal{M}_{\text{KR}}$	90	–	10.7 ± 6.5	91	–	20.6 ± 12.5	70	–	13.2 ± 11.6
$\mathcal{M}_{\text{GP}}$	100	$.55 \pm .04$	12.2 ± 8.3	100	$.54 \pm .03$	21.9 ± 12.9	95	$.52 \pm .04$	13.4 ± 12.8
$\mathcal{M}_{\text{CVAE}}$	100	$.55 \pm .07$	11.8 ± 7.7	97	$.54 \pm .05$	22.6 ± 12.3	95	$.51 \pm .01$	13.4 ± 12.2
CATE$_*$	90	$.56 \pm .07$	11.9 ± 9.2	97	$.55 \pm .05$	26.3 ± 21.4	100	$.52 \pm .02$	13.5 ± 13.0
CATE$_{\text{GP}}$	93	$.56 \pm .05$	12.2 ± 8.4	94	$.55 \pm .06$	25.0 ± 14.8	94	$.52 \pm .03$	13.2 ± 13.1
CATE$_{\text{CVAE}}$	89	$.56 \pm .08$	12.1 ± 8.9	98	$.54 \pm .05$	26.0 ± 14.3	100	$.52 \pm .05$	13.6 ± 12.9

5.3   Synthetic 3-Variable SCMs Under Different Assumptions

In our first set of experiments, we consider three classes of SCM s over three variables with the same causal graph as in Fig. 2b. To test robustness of the different methods to assumptions about the form of the true structural equations, we consider a linear SCM, a non-linear ANM, and a more general, multi-modal SCM with non-additive noise. For further details on the exact form we refer to Appendix E of [27].

Results are shown in Table 1 we observe that the point-based recourse approaches perform (relatively) well in terms of both validity and cost, when their underlying assumptions are met (i.e., $\mathcal{M}_{\text{LIN}}$ on the linear SCM and $\mathcal{M}_{\text{KR}}$ on the nonlinear ANM). Otherwise, validity significantly drops as expected (see, e.g., the results of $\mathcal{M}_{\text{LIN}}$ on the non-linear ANM, or of $\mathcal{M}_{\text{KR}}$ on the non-additive SCM). Moreover, we note that the inferior performance of $\mathcal{M}_{\text{KR}}$ compared to $\mathcal{M}_{\text{LIN}}$ on the linear SCM suggests an overfitting problem, which does not occur for its more conservative probabilistic counterpart $\mathcal{M}_{\text{GP}}$. Generally, the individualised approaches $\mathcal{M}_{\text{GP}}$ and $\mathcal{M}_{\text{CVAE}}$ perform very competitively in terms of cost and validity, especially on the linear and nonlinear ANMs. The subpopulation-based CATE approaches on the other hand, perform particularly well on the challenging non-additive SCM (on which the assumptions of GP approaches are violated) where CATE$_{\text{CVAE}}$ achieves perfect validity as the only non-oracle method. As expected, the subpopulation-based approaches generally lead to higher cost than the individualised ones, since the latter only aim to achieve recourse only for a given individual while the former do it for an entire group (see Fig. 4).

5.4   Semi-synthetic 7-Variable SCM for Loan-Approval

We also test our methods on a larger semi-synthetic SCM inspired by the German Credit UCI dataset [43]. We consider the variables age A, gender G, education-level E, loan amount L, duration D, income I, and savings S with causal graph shown in Fig. 5. We model age A, gender G and loan duration D as

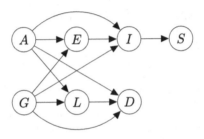

Fig. 5. Assumed causal graph for the semi-synthetic loan approval dataset.

Table 2. Experimental results for the 7-variable SCM for loan-approval. We show average performance ±1 standard deviation for $N_{\text{runs}} = 100$, $N_{\text{MC-samples}} = 100$, and $\gamma_{\text{LCB}} = 2.5$. For linear and non-linear logistic regression as classifiers, we use the gradient-based approach, whereas for the non-differentiable random forest classifier we rely on the brute-force approach (with 10 discretised bins per dimension) to solve the recourse optimisation problems.

Method	LINEAR LOG. REGR.			NON-LIN. LOG. REGR. (MLP)			RANDOM FOREST(BRUTE-FORCE)		
	Valid⋆ (%)	LCB	Cost (%)	Valid⋆ (%)	LCB	Cost (%)	Valid⋆ (%)	LCB	Cost (%)
$\mathcal{M}_\star$	100	–	15.8 ± 7.6	100	–	11.0 ± 7.0	100	–	15.2 ± 7.5
$\mathcal{M}_{\text{LIN}}$	19	–	15.4 ± 7.4	80	–	11.0 ± 6.9	94	–	15.6 ± 7.6
$\mathcal{M}_{\text{KR}}$	41	–	15.6 ± 7.5	87	–	11.1 ± 7.0	92	–	15.1 ± 7.4
$\mathcal{M}_{\text{GP}}$	100	.50 ± .00	18.0 ± 7.7	100	.52 ± .04	11.7 ± 7.3	100	.66 ± .14	16.3 ± 7.4
$\mathcal{M}_{\text{CVAE}}$	100	.50 ± .00	16.6 ± 7.6	99	.51 ± .01	11.3 ± 6.9	100	.66 ± .14	15.9 ± 7.4
CATE⋆	93	.50 ± .01	22.0 ± 9.4	95	.52 ± .05	12.0 ± 7.7	98	.66 ± .15	17.0 ± 7.3
CATE$_{\text{GP}}$	93	.50 ± .02	21.7 ± 9.2	93	.51 ± .06	12.0 ± 7.4	100	.67 ± .15	17.1 ± 7.4
CATE$_{\text{CVAE}}$	94	.49 ± .01	23.7 ± 11.3	95	.51 ± .03	12.0 ± 7.8	100	.68 ± .15	17.9 ± 7.4

non-actionable variables, but consider D to be mutable, i.e., it cannot be manipulated directly but is allowed to change (e.g., as a consequence of an intervention on L). The SCM includes linear and non-linear relationships, as well as different types of variables and noise distributions, and is described in more detail in Appendix B of [27].

The results are summarised in Table 2, where we observe that the insights discussed above similarly apply for data generated from a more complex SCM, and for different classifiers.

Finally, we show the influence of γ_{LCB} on the performance of the proposed probabilistic approaches in Fig. 6. We observe that lower values of γ_{LCB} lead to lower validity (and cost), especially for the CATE approaches. As γ_{LCB} increases validity approaches the corresponding oracles $\mathcal{M}_\star$ and CATE$_\star$, outperforming the point-based recourse approaches. In summary, our probabilistic recourse approaches are not only more robust, but also allow controlling the trade-off between validity and cost using γ_{LCB}.

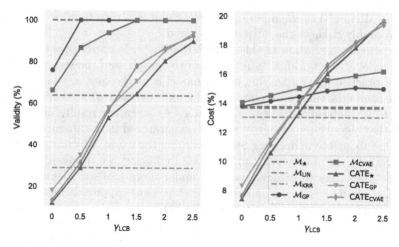

Fig. 6. Trade-off between validity and cost which can be controlled via γ_{LCB} for the probabilistic recourse methods.

6 Discussion

In this paper, we have focused on the problem of algorithmic recourse, i.e., the process by which an individual can change their situation to obtain a desired outcome from a machine learning model. Using the tools from causal reasoning (i.e., structural interventions and counterfactuals), we have shown that in their current form, counterfactual explanations only bring about agency for the individual to achieve recourse in unrealistic settings. In other words, counterfactual explanations imply recourse actions that may neither be optimal nor even result in favorably changing the prediction of h when acted upon. This shortcoming is primarily due to the lack of consideration of causal relations governing the world and thus, the failure to model the downstream effect of actions in the predictions of the machine learning model. In other words, although "counterfactual" is a term from causal language, we observed that existing approaches fall short in terms of taking causal reasoning into account when generating counterfactual explanations and the subsequent recourse actions. Thus, building on the statement by Wachter et al. [76] that counterfactual explanations "do not rely on knowledge of the causal structure of the world," it is perhaps more appropriate to refer to existing approaches as *contrastive*, rather than *counterfactual*, explanations [14, 40]. See [26, §2] for more discussion.

To directly take causal consequences of actions into account, we have proposed a fundamental reformulation of the recourse problem, where actions are performed as interventions and we seek to minimize the cost of performing actions in a world governed by a set of (physical) laws captured in a structural causal model. Our proposed formulation in (4), complemented with several examples and a detailed discussion, allows for *recourse through minimal inter-*

ventions (MINT), that when performed will result in a *structural counterfactual* that favourably changes the output of the model.

The primary limitation of this formulation in (4) is its reliance on the true causal model of the world, subsuming both the graph, and the structural equations. In practice, the underlying causal model is rarely known, which suggests that the counterfactual constraint in (4), i.e., $\mathbf{x}^{\text{SCF}}(a) := \mathbf{x}(a)|\mathbf{x}^{\text{F}} = \mathbf{S}^a(\mathbf{S}^{-1}(\mathbf{x}^{\text{F}}))$, may not be (deterministically) identifiable. As negative result, however, we showed that algorithmic recourse cannot be guaranteed in the absence of perfect knowledge about the underlying SCM governing the world, which unfortunately is not available in practice. To address this limitation, we proposed two probabilistic approaches to achieve recourse under more realistic assumptions. In particular, we derived i) an individual-level recourse approach based on GPs that approximates the counterfactual distribution by averaging over the family of additive Gaussian SCMs; and ii) a subpopulation-based approach, which assumes that only the causal graph is known and makes use of CVAEs to estimate the conditional average treatment effect of an intervention on a subpopulation of individuals similar to the one seeking recourse. Our experiments showed that the proposed probabilistic approaches not only result in more robust recourse interventions than approaches based on point estimates of the SCM, but also allows to trade-off validity and cost.

Assumptions, Limitations, and Extensions. Throughout the present work, we have assumed a known causal graph and causal sufficiency. While this may not hold for all settings, it is the minimal necessary set of assumptions for causal reasoning from observational data alone. Access to instrumental variables or experimental data may help further relax these assumptions [3,13,66]. Moreover, if only a partial graph is available or some relations are known to be confounded, one will need to restrict recourse actions to the subset of interventions that are still identifiable [59,60,67]. An alternative approach could address causal sufficiency violations by relying on latent variable models to estimate confounders from multiple causes [78] or proxy variables [38], or to work with bounds on causal effects instead [5,65,74].

Perhaps more concerningly, our work highlights the implicit causal assumptions made by existing approaches (i.e., that of independence, or feasible and cost-free interventions), which may portray a false sense of recourse guarantees where one does not exists (see Example 2 and all of Sect. 3.1). Our work aims to highlight existing imperfect assumptions, and to offer an alternative formulation, backed with proofs and demonstrations, which would guarantee recourse if assumptions about the causal structure of the world were satisfied. Future research on causal algorithmic recourse may benefit from the rich literature in causality that has developed methods to verify and perform inference under various assumptions [45,48].

This is not to say that counterfactual explanations should be abandoned altogether. On the contrary, we believe that counterfactual explanations hold promise for "guided audit of the data" [76] and evaluating various desirable model properties, such as robustness [21,58] or fairness [20,25,58,69,75]. Besides this, it has been shown that designers of interpretable machine learning systems use counterfactual explanations for predicting model behavior [34] or uncovering inaccuracies in the data profile of individuals [70]. Complementing these offerings of counterfactual explanations, we offer minimal interventions as a way to guarantee algorithmic recourse in general settings, which is not implied by counterfactual explanations.

On the Counterfactual vs Interventional Nature of Recourse. Given that we address two different notions of recourse—counterfactual/individualised (rung 3) vs. interventional/subpopulation-based (rung 2)—one may ask which framing is more appropriate. Since the main difference is whether the background variables $\mathbf{U}$ are assumed fixed (counterfactual) or not (interventional) when reasoning about actions, we believe that this question is best addressed by thinking about the type of environment and interpretation of $\mathbf{U}$: if the environment is static, or if $\mathbf{U}$ (mostly) captures unobserved information about the individual, the counterfactual notion seems to be the right one; if, on the other hand, $\mathbf{U}$ also captures environmental factors which may change, e.g., between consecutive loan applications, then the interventional notion of recourse may be more appropriate. In practice, both notions may be present (for different variables), and the proposed approaches can be combined depending on the available domain knowledge since each parent-child causal relation is treated separately. We emphasise that the subpopulation-based approach is also practically motivated by a reluctance to make (parametric) assumptions about the structural equations which are untestable but necessary for counterfactual reasoning. It may therefore be useful to avoid problems of misspecification, even for counterfactual recourse, as demonstrated experimentally for the non-additive SCM.

7   Conclusion

In this work, we explored one of the main, but often overlooked, objectives of explanations as a means to allow people to act rather than just understand. Using counterexamples and the theory of structural causal models (SCM), we showed that actionable recommendations cannot, in general, be inferred from counterfactual explanations. We show that this shortcoming is due to the lack of consideration of causal relations governing the world and thus, the failure to model the downstream effect of actions in the predictions of the machine learning model. Instead, we proposed a shift of paradigm from *recourse via nearest counterfactual explanations* to *recourse through minimal interventions* (MINT), and presented a new optimization formulation for the common class of additive noise models. Our technical contributions were complemented with an extensive discussion on the *form, feasibility,* and *scope* of interventions in real-world settings. In follow-up work, we further investigated the epistemological differences between counterfactual explanations and consequential recommendations and argued that their technical treatment requires consideration at different levels of the *causal history* [52] of events [26]. Whereas MINT provided exact recourse under strong assumptions (requiring the true SCM), we next explored how to offer recourse under milder and more realistic assumptions (requiring only the causal graph). We present two probabilistic approaches that offer recourse with high probability. The first captures uncertainty over structural equations under additive Gaussian noise, and uses Bayesian model averaging to estimate the counterfactual distribution. The second removes any assumptions on the structural equations by instead computing the average effect of recourse actions on individuals similar to the person who seeks recourse, leading to a novel subpopulation-based interventional notion of recourse. We then derive a gradient-based procedure for selecting optimal recourse actions, and empirically show that the proposed approaches lead to more reliable recommendations under imperfect causal knowledge than non-probabilistic baselines. This contribution is important as it enables recourse recommendations to be generated in more practical settings and under uncertain assumptions.

As a final note, while for simplicity, we have focused in this chapter on credit loan approvals, recourse can have potential applications in other domains such as healthcare [8,9,17,51], justice (e.g., pretrial bail) [4], and other settings (e.g., hiring) [12,44,57] whereby actionable recommendations for individuals are sought.

References

1. Abrevaya, J., Hsu, Y.C., Lieli, R.P.: Estimating conditional average treatment effects. J. Bus. Econ. Stat. **33**(4), 485–505 (2015)
2. Alaa, A.M., van der Schaar, M.: Bayesian inference of individualized treatment effects using multi-task Gaussian processes. In: Advances in Neural Information Processing Systems, pp. 3424–3432 (2017)
3. Angrist, J.D., Imbens, G.W., Rubin, D.B.: Identification of causal effects using instrumental variables. J. Am. Stat. Assoc. **91**(434), 444–455 (1996)
4. Angwin, J., Larson, J., Mattu, S., Kirchner, L.: Machine Bias. ProPublica, New York (2016)
5. Balke, A., Pearl, J.: Counterfactual probabilities: computational methods, bounds and applications. In:: Uncertainty Proceedings 1994, pp. 46–54. Elsevier (1994)
6. Barocas, S., Selbst, A.D., Raghavan, M.: The hidden assumptions behind counterfactual explanations and principal reasons. In: Proceedings of the 2020 Conference on Fairness, Accountability, and Transparency, pp. 80–89 (2020)
7. Bashtannyk, D.M., Hyndman, R.J.: Bandwidth selection for kernel conditional density estimation. Comput. Stat. Data Anal. **36**(3), 279–298 (2001)
8. Bastani, O., Kim, C., Bastani, H.: Interpretability via model extraction. arXiv preprint arXiv:1706.09773 (2017)
9. Begoli, E., Bhattacharya, T., Kusnezov, D.: The need for uncertainty quantification in machine-assisted medical decision making. Nat. Mach. Intell. **1**(1), 20–23 (2019)
10. Bishop, C.M.: Mixture density networks (1994)
11. Bottou, L., Bousquet, O.: The tradeoffs of large scale learning. In: Advances in Neural Information Processing Systems, pp. 161–168 (2008)
12. Cohen, L., Lipton, Z.C., Mansour, Y.: Efficient candidate screening under multiple tests and implications for fairness. arXiv preprint arXiv:1905.11361 (2019)

13. Cooper, G.F., Yoo, C.: Causal discovery from a mixture of experimental and observational data. In: Proceedings of the Fifteenth Conference on Uncertainty in Artificial Intelligence, pp. 116–125 (1999)
14. Dhurandhar, A., et al.: Explanations based on the missing: towards contrastive explanations with pertinent negatives. In: Advances in Neural Information Processing Systems, pp. 592–603 (2018)
15. Doshi-Velez, F., Kim, B.: Towards a rigorous science of interpretable machine learning. arXiv preprint arXiv:1702.08608 (2017)
16. Friedman, N., Nachman, I.: Gaussian process networks. In: Proceedings of the Sixteenth Conference on Uncertainty in Artificial Intelligence, pp. 211–219 (2000)
17. Grote, T., Berens, P.: On the ethics of algorithmic decision-making in healthcare. J. Med. Ethics 46(3), 205–211 (2020)
18. Guidotti, R., Monreale, A., Ruggieri, S., Pedreschi, D., Turini, F., Giannotti, F.: Local rule-based explanations of black box decision systems. arXiv preprint arXiv:1805.10820 (2018)
19. Gunning, D.: DARPA'S explainable artificial intelligence (XAI) program. In: Proceedings of the 24th International Conference on Intelligent User Interfaces, p. ii. ACM (2019)
20. Gupta, V., Nokhiz, P., Roy, C.D., Venkatasubramanian, S.: Equalizing recourse across groups. arXiv preprint arXiv:1909.03166 (2019)
21. Hancox-Li, L.: Robustness in machine learning explanations: does it matter? In: Proceedings of the 2020 Conference on Fairness, Accountability, and Transparency, pp. 640–647 (2020)
22. Holzinger, A., Malle, B., Saranti, A., Pfeifer, B.: Towards multi-modal causability with graph neural networks enabling information fusion for explainable AI. Inf. Fusion 71, 28–37 (2021)
23. Hoyer, P., Janzing, D., Mooij, J.M., Peters, J., Schölkopf, B.: Nonlinear causal discovery with additive noise models. In: Advances in Neural Information Processing Systems, pp. 689–696 (2009)
24. ShJoshi, S., Koyejo, O., Vijitbenjaronk, W., Kim, B., Ghosh, J.: REVISE: towards realistic individual recourse and actionable explanations in black-box decision making systems. arXiv preprint arXiv:1907.09615 (2019)
25. Karimi, A.H., Barthe, G., Balle, B., Valera, I.: Model-agnostic counterfactual explanations for consequential decisions. In: International Conference on Artificial Intelligence and Statistics, pp. 895–905 (2020)
26. Karimi, A.-H., Barthe, G., Schölkopf, B., Valera, I.: A survey of algorithmic recourse: contrastive explanations and consequential recommendations. arXiv preprint arXiv:2010.04050 (2020)
27. Karimi, A.-H., von Kügelgen, J., Schölkopf, B., Valera, I.: Algorithmic recourse under imperfect causal knowledge: a probabilistic approach. In: Advances in Neural Information Processing Systems, pp. 265–277 (2020)
28. Karimi, A.-H., Schölkopf, B., Valera, I.: Algorithmic recourse: from counterfactual explanations to interventions. In: 4th Conference on Fairness, Accountability, and Transparency (FAccT 2021), pp. 353–362 (2021)
29. Karush, W.: Minima of functions of several variables with inequalities as side conditions. Master's thesis, Department of Mathematics, University of Chicago (1939)
30. Kingma, D.P., Ba, J.: Adam: a method for stochastic optimization. In: 3rd International Conference for Learning Representations (2015)
31. Kingma, D.P., Welling, M.: Auto-encoding variational Bayes. In: 2nd International Conference on Learning Representations (2014)

32. Kodratoff, Y.: The comprehensibility manifesto. KDD Nugget Newsl. **94**(9) (1994)
33. Kuhn, H.W., Tucker, A.W.: Nonlinear programming. In: Neyman, J. (ed.) Proceedings of the Second Berkeley Symposium on Mathematical Statistics and Probability. University of California Press, Berkeley (1951)
34. Lage, I.: An evaluation of the human-interpretability of explanation. arXiv preprint arXiv:1902.00006 (2019)
35. Laugel, T., Lesot, M.J., Marsala, C., Renard, X., Detyniecki, M.: Inverse classification for comparison-based interpretability in machine learning. arXiv preprint arXiv:1712.08443 (2017)
36. Lewis, D.K.: Counterfactuals. Harvard University Press, Cambridge (1973)
37. Lipton, Z.C.: The mythos of model interpretability. Queue **16**(3), 31–57 (2018)
38. Louizos, C., Shalit, U., Mooij, J.M., Sontag, D., Zemel, R., Welling, M.: Causal effect inference with deep latent-variable models. In: Advances in Neural Information Processing Systems, pp. 6446–6456 (2017)
39. Mahajan, D., Tan, C., Sharma, A.: Preserving causal constraints in counterfactual explanations for machine learning classifiers. arXiv preprint arXiv:1912.03277 (2019)
40. Miller, T.: Explanation in artificial intelligence: insights from the social sciences. Artif. Intell. **267**, 1–38 (2019)
41. Mothilal, R.K., Sharma, A., Tan, C.: DiCE: explaining machine learning classifiers through diverse counterfactual explanations. arXiv preprint arXiv:1905.07697 (2019)
42. Murdoch, W.J., Singh, C., Kumbier, K., Abbasi-Asl, R., Yu, B.: Definitions, methods, and applications in interpretable machine learning. Proc. Natl. Acad. Sci. **116**(44), 22071–22080 (2019)
43. Murphy, P.M.: UCI repository of machine learning databases. ftp:/pub/machine-learning-databaseonics. uci. edu (1994)
44. Nabi, R., Shpitser, I.: Fair inference on outcomes. In: Proceedings of the... AAAI Conference on Artificial Intelligence. AAAI Conference on Artificial Intelligence, vol. 2018, p. 1931. NIH Public Access (2018)
45. Pearl, J.: Causality. Cambridge University Press, Cambridge (2009)
46. Pearl, J.: Structural counterfactuals: a brief introduction. Cogn. Sci. **37**(6), 977–985 (2013)
47. Peters, J., Bühlmann, P.: Identifiability of Gaussian structural equation models with equal error variances. Biometrika **101**(1), 219–228 (2014)
48. Peters, J., Janzing, D., Schölkopf, B.: Elements of Causal Inference. The MIT Press, Cambridge (2017)
49. Poyiadzi, R., Sokol, K., Santos-Rodriguez, R., De Bie, T., Flach, P.: FACE: feasible and actionable counterfactual explanations. arXiv preprint arXiv:1909.09369 (2019)
50. Rezende, D.J., Mohamed, S., Wierstra, D.: Stochastic backpropagation and approximate inference in deep generative models. In: International Conference on Machine Learning, pp. 1278–1286 (2014)
51. Rieckmann, A., et al.: Causes of outcome learning: a causal inference-inspired machine learning approach to disentangling common combinations of potential causes of a health outcome. medRxiv (2020)
52. Ruben, D.-H.: Explaining Explanation. Routledge, London (2015)
53. Rudin, C.: Stop explaining black box machine learning models for high stakes decisions and use interpretable models instead. Nat. Mach. Intell. **1**(5), 206–215 (2019)

54. Rüping, S.: Learning interpretable models. Ph.D. dissertation, Technical University of Dortmund (2006)
55. Russell, C.: Efficient search for diverse coherent explanations. In: Proceedings of the Conference on Fairness, Accountability, and Transparency, FAT* '19, pp. 20–28. ACM (2019)
56. Schulam, P., Saria, S.: Reliable decision support using counterfactual models. In: Advances in Neural Information Processing Systems, pp. 1697–1708 (2017)
57. Schumann, C., Foster, J.S., Mattei, N., Dickerson, J.P.: We need fairness and explainability in algorithmic hiring. In: Proceedings of the 19th International Conference on Autonomous Agents and MultiAgent Systems, pp. 1716–1720 (2020)
58. Sharma, S., Henderson, J., Ghosh, J.: CERTIFAI: a common framework to provide explanations and analyse the fairness and robustness of black-box models. In: Proceedings of the AAAI/ACM Conference on AI, Ethics, and Society, pp. 166–172 (2020)
59. Shpitser, I., Pearl, J.: Identification of conditional interventional distributions. In: 22nd Conference on Uncertainty in Artificial Intelligence, UAI 2006, pp. 437–444 (2006)
60. Shpitser, I., Pearl, J.: Complete identification methods for the causal hierarchy. J. Mach. Learn. Res. **9**(Sep), 1941–1979 (2008)
61. Silva, R., Gramacy, R.B.: Gaussian process structural equation models with latent variables. In: Proceedings of the Twenty-Sixth Conference on Uncertainty in Artificial Intelligence, pp. 537–545 (2010)
62. Sohn, K., Lee, H., Yan, X.: Learning structured output representation using deep conditional generative models. In: Advances in Neural Information Processing Systems, pp. 3483–3491 (2015)
63. Starr, W.: Counterfactuals. In: Zalta, E.N. (ed.) The Stanford Encyclopedia of Philosophy. Metaphysics Research Lab, Stanford University, fall 2019 edition (2019)
64. Stöger, K., Schneeberger, D., Holzinger, A.: Medical artificial intelligence: the European legal perspective. Commun. ACM **64**(11), 34–36 (2021)
65. Tian, J., Pearl, J.: Probabilities of causation: bounds and identification. Ann. Math. Artif. Intell. **28**(1–4), 287–313 (2000). https://doi.org/10.1023/A:1018912507879
66. Tian, J., Pearl, J.: Causal discovery from changes. In: Proceedings of the Seventeenth Conference on Uncertainty in Artificial Intelligence, pp. 512–521 (2001)
67. Tian, J., Pearl, J.: A general identification condition for causal effects. In: Eighteenth national conference on Artificial intelligence, pp. 567–573 (2002)
68. Trippe, B.L., Turner, R.E.: Conditional density estimation with Bayesian normalising flows. arXiv preprint arXiv:1802.04908 (2018)
69. Ustun, B., Spangher, A., Liu, Y.: Actionable recourse in linear classification. In: Proceedings of the Conference on Fairness, Accountability, and Transparency, pp. 10–19. ACM (2019)
70. Venkatasubramanian, S. , Alfano, M.: The philosophical basis of algorithmic recourse. In: Proceedings of the Conference on Fairness, Accountability, and Transparency. ACM (2020)
71. Verma, S., Dickerson, J., Hines, K.: Counterfactual explanations for machine learning: a review. arXiv preprint arXiv:2010.10596 (2020)
72. Voigt, P., Von dem Bussche, A.: The EU General Data Protection Regulation (GDPR). A Practical Guide, 1st edn. Springer, Cham (2017). https://doi.org/10.1007/978-3-319-57959-7

73. von Kügelgen, J., Rubenstein, P.K., Schölkopf, B., Weller, A.: Optimal experimental design via Bayesian optimization: active causal structure learning for Gaussian process networks. In: NeurIPS Workshop "Do the right thing": machine learning and causal inference for improved decision making (2019)

74. von Kügelgen, J., Agarwal, N., Zeitler, J., Mastouri, A., Schölkopf, B.: Algorithmic recourse in partially and fully confounded settings through bounding counterfactual effects. In: ICML Workshop on Algorithmic Recourse (2021)

75. von Kügelgen, J., Karimi, A.-H., Bhatt, U., Valera, I., Weller, A., Schölkopf, B.: On the fairness of causal algorithmic recourse. In: Proceedings of the 36th AAAI Conference on Artificial Intelligence (2022)

76. Wachter, S., Mittelstadt, B., Russell, C.: Counterfactual explanations without opening the black box: automated decisions and the GDPR. Harvard J. Law Technol. **31**(2) (2017)

77. Wainwright, M.J., Jordan, M.I.: Graphical models, exponential families, and variational inference. Found. Trends® Mach. Learn. **1**(1–2), 1–305 (2008)

78. Wang, Y., Blei, D.M.: The blessings of multiple causes. J. Am. Stat. Assoc. **114**, 1–71 (2019)

79. Williams, C.K.I., Rasmussen, C.E.: Gaussian Processes for Machine Learning, vol. 2. MIT Press, Cambridge (2006)

Interpreting Generative Adversarial Networks for Interactive Image Generation

Bolei Zhou[✉][iD]

Department of Computer Science, University of California, Los Angeles, USA
bolei@cs.ucla.edu

Abstract. Significant progress has been made by the advances in Generative Adversarial Networks (GANs) for image generation. However, there lacks enough understanding of how a realistic image is generated by the deep representations of GANs from a random vector. This chapter gives a summary of recent works on interpreting deep generative models. The methods are categorized into the supervised, the unsupervised, and the embedding-guided approaches. We will see how the human-understandable concepts that emerge in the learned representation can be identified and used for interactive image generation and editing.

Keywords: Explainable machine learning · Generative adversarial networks · Image generation

1 Introduction

Over the years, great progress has been made in image generation by the advances in Generative Adversarial Networks (GANs) [6,12]. As shown in Fig. 1 the generation quality and diversity have been improved substantially from the early DCGAN [16] to the very recent Alias-free GAN [11]. After the adversarial training of the generator and the discriminator, we can have the generator as a pretrained feedforward network for image generation. After feeding a vector sampled from some random distribution, this generator can synthesize a realistic image as the output. However, such an image generation pipeline doesn't allow users to customize the output image, such as changing the lighting condition of the output bedroom image or adding a smile to the output face image. Moreover, it is less understood how a realistic image can be generated from the layer-wise representations of the generator. Therefore, we need to interpret the learned representation of deep generative models for understanding and the practical application of interactive image editing.

This chapter will introduce the recent progress of the explainable machine learning for deep generative models. I will show how we can identify the human-understandable concepts in the generative representation and use them to steer the generator for interactive image generation. Readers might also be interested

A. Holzinger et al. (Eds.): xxAI 2020, LNAI 13200, pp. 167–175, 2022.
https://doi.org/10.1007/978-3-031-04083-2_9

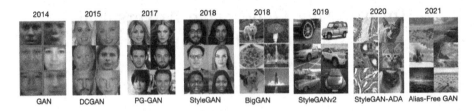

Fig. 1. Progress of image generation made by different GAN models over the years.

in watching a relevant tutorial talk I gave at CVPR'21 Tutorial on Interpretable Machine Learning for Computer Vision[1]. A more detailed survey paper on GAN interpretation and inversion can be found in [21].

This chapter focuses on interpreting the pretrained GAN models, but a similar methodology can be extended to other generative models such as VAE. Recent interpretation methods can be summarized into the following three approaches: the supervised approach, the unsupervised approach, and the embedding-guided approach. The supervised approach uses labels or classifiers to align the meaningful visual concept with the deep generative representation; the unsupervised approach aims to identify the steerable latent factors in the deep generative representation through solving an optimization problem; the embedding-guided approach uses the recent pretrained language-image embedding CLIP [15] to allow a text description to guide the image generation process.

In the following sections, I will select representative methods from each approach and briefly introduce them as primers for this rapidly growing direction.

2   Supervised Approach

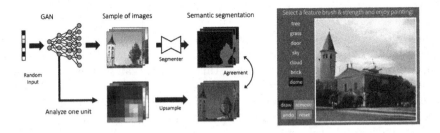

Fig. 2. GAN dissection framework and interactive image editing interface. Images are extracted from [3]. The method aligns the unit activation with the semantic mask of the output image, thus by turning up or down the unit activation we can include or remove the corresponding visual concept in the output image.

---

[1] https://youtu.be/PtRU2B6Iml4.

The supervised approach uses labels or trained classifiers to probe the representation of the generator. One of the earliest interpretation methods is the GAN Dissection [4]. Derived from the previous work Network Dissection [3], GAN Dissection aims to visualize and understand the individual convolutional filters (we term them as units) in the pretrained generator. It uses semantic segmentation networks [24] to segment the output images. It then calculates the agreement between the spatial location of the unit activation map and the semantic mask of the output image. This method can identify a group of interpretable units closely related to object concepts, such as sofa, table, grass, buildings. Those units are then used as switches where we can add or remove some objects such as a tree or lamp by turning up or down the activation of the corresponding units. The framework of GAN Dissection and the image editing interface are shown in Fig. 2. In the interface of GAN Dissection, the user can select the object to be manipulated and brush the output image where it should be removed or added.

Besides steering the filters at the intermediate convolutional layer of the generator as the GAN Dissection does, the latent space where we sample the latent vector as input to the generator is also being explored. The underlying interpretable subspaces aligning with certain attributes of the output image can be identified. Here we denote the pretrained generator as $G(.)$ and the random vector sampled from the latent space as $\mathbf{z}$, and then the output image becomes $I = G(\mathbf{z})$. Under different vectors, the output images become different. Thus the latent space encodes various attributes of images. If we can steer the vector $\mathbf{z}$ through one relevant subspace and preserve its projection to the other subspaces, we can edit one attribute of the output image in a disentangled way.

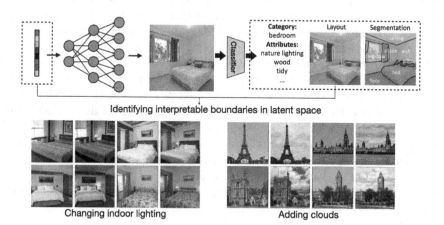

Fig. 3. We can use classifier to predict various attributes from the output image then go back to the latent space to identify the attribute boundaries. Images below show the image editing results achieved by [22].

To align the latent space with the semantic space, we can first apply off-the-shelf classifiers to extract the attributes of the synthesized images and then compute the causality between the occurring attributes in the generated images and the corresponding vectors in the latent space. The HiGAN method proposed in [22] follows such a supervised approach as illustrated in Fig. 3: (1) Thousands of latent vectors are sampled, and the images are generated. (2) Various levels of attributes are predicted from the generated images by applying the off-the-shelf classifiers. (3) For each attribute a, a linear boundary $\mathbf{n}_a$ is trained in the latent space using the predicted labels and the latent vectors. We consider it a binary classification and train a linear SVM to recognize each attribute. The weight of the trained SVM is $\mathbf{n}_a$. (4) a counterfactual verification step is taken to pick up the reliable boundary. Here we follow a linear model to shift the latent code as

$$I' = G(\mathbf{z} + \lambda \mathbf{n}_a), \tag{1}$$

where the normal vector of the trained attribute boundary is denoted as $\mathbf{n}_a$ and I' is the edited image compared to the original image I. Then the difference between predicted attribute scores before and after manipulation becomes,

$$\Delta a = \frac{1}{K} \sum_{k=1}^{K} \max(F(G(\mathbf{z}_k + \mathbf{n}_a)) - F(G(\mathbf{z}_k)), 0), \tag{2}$$

here $F(.)$ is the attribute predictor with the input image, and K is the number of synthesized images. Ranking Δa allows us to identify the reliable attribute boundaries out of the candidate set $\{\mathbf{n}_a\}$, where there are about one hundred attribute boundaries trained from step 3 of the HiGAN method. After that, we can then edit the output image from the generator by adding or removing the normal vector of the target attribute on the original latent code. Some image manipulation results are shown in Fig. 3.

Similar supervised methods have been developed to edit the facial attributes [17,18] and improve the image memorability [5]. Steerability of various attributes in GANs has also been analyzed [9]. Besides, the work of Style-Flow [1] replaces the linear model with a nonlinear invertible flow-based model in the latent space with more precise facial editing. Some recent work uses a differentiable renderer to extract 3D information from the image GANs for more controllable view synthesis [23]. For the supervised approach, many challenges remain for future work, such as expanding the annotation dictionary, achieving more disentangled manipulation, and aligning latent space with image region.

3   Unsupervised Approach

As generative models become more and more popular, people start training them on a wide range of images, such as cats and anime. To steer the generative models trained for cat or anime generation, following the previous supervised approach, we have to define the attributes of the images and annotate many images to train the classifiers. It is a very time-consuming process.

Alternatively, the unsupervised approach aims to identify the controllable dimensions of the generator without using labels/classifiers.

SeFa [19] is an unsupervised approach for discovering the interpretable representation of a generator. It directly decomposes the pre-trained weights. More specifically, in the pre-trained generator of the popular StyleGAN [12] or PGGAN [10] model, there is an affine transformation between the latent code and the internal activation. Thus the manipulation model can be simplified as

$$\mathbf{y}' \triangleq G_1(\mathbf{z}') = G_1(\mathbf{z} + \alpha\mathbf{n}) = \mathbf{A}\mathbf{z} + \mathbf{b} + \alpha\mathbf{A}\mathbf{n} = \mathbf{y} + \alpha\mathbf{A}\mathbf{n}, \qquad (3)$$

where $\mathbf{y}$ is the original projected code and $\mathbf{y}'$ is the projected code after manipulation by $\mathbf{n}$. From Eq. (3) we can see that the manipulation process is instance independent. In other words, given any latent code $\mathbf{z}$ together with a particular latent direction $\mathbf{n}$, the editing can always be achieved by adding the term $\alpha\mathbf{A}\mathbf{n}$ onto the projected code after the first step. From this perspective, the weight parameter $\mathbf{A}$ should contain the essential knowledge of the image variation. Thus we aim to discover important latent directions by decomposing $\mathbf{A}$ in an unsupervised manner. We propose to solve the following optimization problem:

$$\mathbf{N}^* = \underset{\{\mathbf{N} \in R^{d \times k}:\mathbf{n}_i^T\mathbf{n}_i=1 \ \forall i=1,\cdots,k\}}{\arg\max} \sum_{i=1}^{k} \|\mathbf{A}\mathbf{n}_i\|_2^2, \qquad (4)$$

where $\mathbf{N} = [\mathbf{n}_1, \mathbf{n}_2, \cdots, \mathbf{n}_k]$ correspond to the top k semantics sorted by their eigenvalues, and $\mathbf{A}$ is the learned weight in the affine transform between the latent code and the internal activation. This objective aims at finding the directions that can cause large variations after the projection of $\mathbf{A}$. The resulting solution becomes the eigenvectors of the matrix $\mathbf{A}^T\mathbf{A}$. Those resulting directions at different layers control different attributes of the output image, thus pushing the latent code $\mathbf{z}$ on the important directions $\{\mathbf{n}_1, \mathbf{n}_2, \cdots, \mathbf{n}_k\}$ facilitates the interactive image editing. Figure 4 shows some editing result.

Fig. 4. Manipulation results from SeFa [19] on the left and the interface for interactive image editing on the right. On the left, each attribute corresponds to some $\mathbf{n}_i$ in the latent space of the generator. In the interface, user can simply drag each slider bar associating with certain attribute to edit the output image

Many other methods have been developed for the unsupervised discovery of interpretable latent representation. Härkönen et al. [7] perform PCA on the sampled data to find primary directions in the latent space. Voynov and Babenko [20] jointly learn a candidate matrix and a classifier such that the classifier can properly recognize the semantic directions in the matrix. Peebles et al. [14] develops a Hessian penalty as a regularizer for improving disentanglement in training. He et al. [8] designs a linear subspace with an orthogonal basis in each layer of the generator to encourage the decomposition of attributes. Many challenges remain for the unsupervised approach, such as how to evaluate the result from unsupervised learning, annotate each discovered dimension, and improve the disentanglement in the GAN training process.

4   Embedding-Guided Approach

The embedding-guided approach aligns language embedding with generative representations. It allows users to use any free-form text to guide the image generation. The difference between the embedding-guided approach and the previous unsupervised approach is that the embedding-guided approach is conditioned on the given text to manipulate the image to be more flexible, while the unsupervised approach discovers the steerable dimensions in a bottom-up way thus it lacks fine-grained control.

Recent work on StyleCLIP [13] combines the pretrained language-image embedding CLIP [15] and StyleGAN generator [12] for free-form text-driven image editing. CLIP is a pretrained embedding model from 400 million image-text pairs. Given an image I_s, it first projects it back into the latent space as $\mathbf{w}_s$ using existing GAN inversion method. Then StyleCLIP designs the following optimization objective

$$\mathbf{w}^* = \arg\min D_{CLIP}(G(\mathbf{w}), t) + \lambda_{L2}||\mathbf{w} - \mathbf{w}_s||_2 + \lambda_{ID}L_{ID}(\mathbf{w}, \mathbf{w}_s), \quad (5)$$

where $D_{CLIP}(.,.)$ measure the distance between an image and a text using the pre-trained CLIP model, the second and the third terms are some regularizers to keep the similarity and identity with the original input image. Thus this optimization objective results in a latent code $\mathbf{w}^*$ that generates an image close to the given text in the CLIP embedding space as well as similar to the original input image. StyleCLIP further develops some architecture design to speed up the iterative optimization. Figure 5 shows the text driven image editing results.

Some concurrent work called Paint by Word from Bau et al. [2] combines CLIP embedding with region-based image editing. It has a masked optimization objective that allows the user to brush the image to provide the input mask.

Fig. 5. Text driven image editing results from a) StyleCLIP [13] and b) Paint by Word [2].

5 Concluding Remarks

Interpreting deep generative models leads to a deeper understanding of how the learned representations decompose images to generate them. Discovering the human-understandable concepts and steerable dimensions in the deep generative representations also facilitates the promising applications of interactive image generation and editing. We have introduced representative methods from three approaches: the supervised approach, the unsupervised approach, and the embedding-guided approach. The supervised approach can achieve the best image editing quality when the labels or classifiers are available. It remains challenging for the unsupervised and embedding-guided approaches to achieve disentangled manipulation. More future works are expected on the accurate inversion of the real images and the precise local and global image editing.

References

1. Abdal, R., Zhu, P., Mitra, N.J., Wonka, P.: StyleFlow: attribute-conditioned exploration of StyleGAN-generated images using conditional continuous normalizing flows. ACM Trans. Graph. (TOG) **40**(3), 1–21 (2021)
2. Bau, D., et al.: Paint by word. arXiv preprint arXiv:2103.10951 (2021)
3. Bau, D., Zhou, B., Khosla, A., Oliva, A., Torralba, A.: Network dissection: quantifying interpretability of deep visual representations. In: Proceedings of the IEEE/CVF Conference on Computer Vision and Pattern Recognition, pp. 6541–6549 (2017)

4. Bau, D., et al.: Gan dissection: visualizing and understanding generative adversarial networks. In: International Conference on Learning Representations (2018)
5. Goetschalckx, L., Andonian, A., Oliva, A., Isola, P.: GANalyze: toward visual definitions of cognitive image properties. In: Proceedings of International Conference on Computer Vision (ICCV), pp. 5744–5753 (2019)
6. Goodfellow, I.J., et al.: Generative adversarial networks. In: Advances in Neural Information Processing Systems (2014)
7. Härkönen, E., Hertzmann, A., Lehtinen, J., Paris, S.: GANSpace: discovering interpretable GAN controls. In: Advances in Neural Information Processing Systems (2020)
8. He, Z., Kan, M., Shan, S.: EigenGAN: layer-wise eigen-learning for GANs. In: Proceedings of International Conference on Computer Vision (ICCV) (2021)
9. Jahanian, A., Chai, L., Isola, P.: On the "steerability" of generative adversarial networks. In: International Conference on Learning Representations (2019)
10. Karras, T., Aila, T., Laine, S., Lehtinen, J.: Progressive growing of GANs for improved quality, stability, and variation. In: International Conference on Learning Representations (2018)
11. Karras, T., et al.: Alias-free generative adversarial networks. In: Advances in Neural Information Processing Systems (2021)
12. Karras, T., Laine, S., Aila, T.: A style-based generator architecture for generative adversarial networks. In: Proceedings of the IEEE/CVF Conference on Computer Vision and Pattern Recognition, pp. 4401–4410 (2019)
13. Patashnik, O., Wu, Z., Shechtman, E., Cohen-Or, D., Lischinski, D.: StyleCLIP: text-driven manipulation of StyleGAN imagery. In: Proceedings of International Conference on Computer Vision (ICCV) (2021)
14. Peebles, W., Peebles, J., Zhu, J.-Y., Efros, A., Torralba, A.: The hessian penalty: a weak prior for unsupervised disentanglement. In: Vedaldi, A., Bischof, H., Brox, T., Frahm, J.-M. (eds.) ECCV 2020. LNCS, vol. 12351, pp. 581–597. Springer, Cham (2020). https://doi.org/10.1007/978-3-030-58539-6_35
15. Radford, A., et al.: Learning transferable visual models from natural language supervision. arXiv preprint arXiv:2103.00020 (2021)
16. Radford, A., Metz, L., Chintala, S.: Unsupervised representation learning with deep convolutional generative adversarial networks. In: International Conference on Learning Representations (2015)
17. Shen, Y., Gu, J., Tang, X., Zhou, B.: Interpreting the latent space of GANs for semantic face editing. In: Proceedings of the IEEE/CVF Conference on Computer Vision and Pattern Recognition, pp. 9243–9252 (2020)
18. Shen, Y., Yang, C., Tang, X., Zhou, B.: InterFaceGAN: interpreting the disentangled face representation learned by GANs. IEEE Trans. Pattern Anal. Mach. Intell. **44**(4), 2004–2018 (2020)
19. Shen, Y., Zhou, B.: Closed-form factorization of latent semantics in GANs. In: Proceedings of the IEEE/CVF Conference on Computer Vision and Pattern Recognition, pp. 1532–1540 (2021)
20. Voynov, A., Babenko, A.: Unsupervised discovery of interpretable directions in the GAN latent space. In: International Conference on Machine Learning (2020)
21. Xia, W., Zhang, Y., Yang, Y., Xue, J.H., Zhou, B., Yang, M.H.: Gan inversion: a survey. IEEE Trans. Pattern Anal. Mach. Intell. (2022)
22. Yang, C., Shen, Y., Zhou, B.: Semantic hierarchy emerges in deep generative representations for scene synthesis. Int. J. Comput. Vis. **129**(5), 1451–1466 (2021)

23. Zhang, Y., et al.: Image GANs meet differentiable rendering for inverse graphics and interpretable 3D neural rendering. In: International Conference on Learning Representations (2021)
24. Zhou, B., Zhao, H., Puig, X., Fidler, S., Barriuso, A., Torralba, A.: Scene parsing through ADE20K dataset. In: Proceedings of the IEEE/CVF Conference on Computer Vision and Pattern Recognition, pp. 633–641 (2017)

XAI and Strategy Extraction via Reward Redistribution

Marius-Constantin Dinu[1,2,6,7], Markus Hofmarcher[1,2(✉)], Vihang P. Patil[1,2],
Matthias Dorfer[4,7], Patrick M. Blies[4], Johannes Brandstetter[1,2],
Jose A. Arjona-Medina[1,2,6,7], and Sepp Hochreiter[1,2,3,5,7]

[1] Institute for Machine Learning, Johannes Kepler University Linz, Linz, Austria
{dinu,hofmarcher}@ml.jku.at
[2] LIT AI Lab, Johannes Kepler University Linz, Linz, Austria
[3] ELLIS Unit Linz, Johannes Kepler University Linz, Linz, Austria
[4] enliteAI, Vienna, Austria
[5] Institute of Advanced Research in Artificial Intelligence (IARAI), Vienna, Austria
[6] Dynatrace Research, Linz, Austria
[7] AI Austria, Reinforcement Learning Community, Vienna, Austria

Abstract. In reinforcement learning, an agent interacts with an environment from which it receives rewards, that are then used to learn a task. However, it is often unclear what strategies or concepts the agent has learned to solve the task. Thus, interpretability of the agent's behavior is an important aspect in practical applications, next to the agent's performance at the task itself. However, with the increasing complexity of both tasks and agents, interpreting the agent's behavior becomes much more difficult. Therefore, developing new interpretable RL agents is of high importance. To this end, we propose to use Align-RUDDER as an interpretability method for reinforcement learning. Align-RUDDER is a method based on the recently introduced RUDDER framework, which relies on contribution analysis of an LSTM model, to redistribute rewards to key events. From these key events a strategy can be derived, guiding the agent's decisions in order to solve a certain task. More importantly, the key events are in general interpretable by humans, and are often sub-tasks; where solving these sub-tasks is crucial for solving the main task. Align-RUDDER enhances the RUDDER framework with methods from multiple sequence alignment (MSA) to identify key events from demonstration trajectories. MSA needs only a few trajectories in order to perform well, and is much better understood than deep learning models such as LSTMs. Consequently, strategies and concepts can be learned from a few expert demonstrations, where the expert can be a human or an agent trained by reinforcement learning. By substituting RUDDER's LSTM with a profile model that is obtained from MSA of demonstration trajectories, we are able to interpret an agent at three stages: First, by extracting common strategies from demonstration trajectories with MSA. Second, by encoding the most prevalent strategy via the MSA profile model and therefore explaining the expert's behavior. And third,

---

M.-C. Dinu and M. Hofmarcher—Equal contribution.

A. Holzinger et al. (Eds.): xxAI 2020, LNAI 13200, pp. 177–205, 2022.
https://doi.org/10.1007/978-3-031-04083-2_10

by allowing the interpretation of an arbitrary agent's behavior based on its demonstration trajectories.

Keywords: Explainable AI · Contribution analysis · Reinforcement learning · Credit assignment · Reward redistribution

1 Introduction

With recent advances in computing power together with increased availability of large datasets, machine learning has emerged as a key technology for modern software systems. Especially in the fields of computer vision [34,52] and natural language processing [14,71] vast improvements have been made using machine learning.

In contrast to computer vision and natural language processing, which are both based on supervised learning, reinforcement learning is more general as it constructs agents for planning and decision-making. Recent advances in reinforcement learning have resulted in impressive models that are capable of surpassing humans in games [39,58,73]. However, reinforcement learning is still waiting for its breakthrough in real world applications, not least because of two issues. First, the amount of human effort and computational resources required to develop and train reinforcement learning systems is prohibitively expensive for widespread adoption. Second, machine learning and in particular reinforcement learning produces black box models, which do not allow explaining model outcomes and to build trust in these models. The insufficient explainability limits the application of reinforcement learning agents, therefore reinforcement learning is often limited to computer games and simulations.

Advances in the field of explainable AI (XAI) have introduced methods and techniques to alleviate the problem of insufficient explainability for supervised machine learning [3–5,41,42,64]. However, these XAI methods cannot explain the behavior of the more complex reinforcement learning agents. Among other problems, delayed and sparse rewards or hand-crafted reward functions make it hard to explain an agent's final behavior. Therefore, interpreting and explaining agents trained with reinforcement learning is an integral component for viably moving towards real-world reinforcement learning applications.

We explore the current state of explainability methods and their applicability in the field of reinforcement learning and introduce a method, Align-RUDDER [45], which is intrinsically explainable by exposing the global strategy of the trained agent. The paper is structured as follows: In Sect. 2.1 we review explainability methods and how they can be categorized. Sect. 2.2 defines the setting of reinforcement learning. In Sect. 2.3, Sect. 2.4 and Sect. 2.5 we explore the problem of credit assignment and potential solutions from the field of explainable AI. In Sect. 2.6 we review the concept of reward redistribution as a solution for credit assignment. Section 3 introduces the concept of strategy extraction and explores its potential for training reinforcement learning agents (Sect. 3.1) as well as its intrinsic explainability (Sect. 3.2) and finally its usage for explaining arbitrary agent behaviors in Sect. 4. Finally, in Sect. 5 we explore limitations of this approach before concluding in Sect. 6.

2   Background

2.1   Explainability Methods

The importance of explainability methods to provide insights into black box machine learning methods such as deep neural networks has significantly increased in recent years [72]. These methods can be categorized based on multiple factors [15].

First, we can distinguish local and global methods, where global methods explain the general model behavior, while local models focus on explaining specific decisions (e.g. explain the classification of a specific sample) or the influence of individual features on the model output [1,15]. Second, we distinguish between intrinsically explainable models and post-hoc methods [15]. Intrinsically explainable models are designed to provide explanations as well as model predictions. Examples for such models are decision trees [35], rule-based models [75], linear models [22] or attention models [13]. Post-hoc methods are applied to existing models and often require a second model to provide explanations (e.g. approximate an existing model with a linear model that can be interpreted) or provide limited explanations (e.g. determine important input features but no detailed explanations of the inner workings of a model). While intrinsically explainable models offer more detailed explanations and insights, they often sacrifice predictive performance. Post-hoc methods, in contrast, have little to no influence on predictive performance but lack detailed explanations of the model.

Post-hoc explainability methods often provide insights in the form of *attributions*, i.e. a measure of how important certain features are with regard to the model's output. In Fig. 1 we illustrate the model attribution from input towards its prediction. We further categorize attribution methods into *sensitivity analysis* and *contribution analysis*.

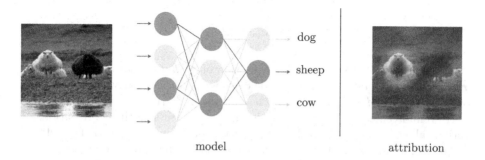

Fig. 1. Illustration of model input attributions towards its prediction [46].

Sensitivity analysis methods, or *"backpropagation through a model"* [8,43,50, 51], provide attributions by calculating the gradient of the model with respect to its input. The magnitude of these gradients is then used to assign a measure of importance to individual features of the input. While sensitivity analysis is typically simple to implement, these methods have several problems such as susceptibility to local minima, instabilities, exploding or vanishing gradients and proper exploration [28,54]. The major drawback, however, is that the relevance of features can be missed since it does not consider their *contribution* to the output but only how small perturbations of features change the output. Therefore, important features can receive low attribution scores as small changes would not result in a significant change of the model's output, but removing them would completely change the output. A prominent example for sensitivity analysis methods are *saliency maps* [59].

Contribution analysis methods provide attributions based on the contribution of individual features to the model output, and therefore do not suffer from the drawbacks of sensitivity analysis methods. This can be achieved in a variety of ways, prominent examples are *integrated gradients* [64] or *layer-wise relevance propagation* (ϵ-LRP) [11].

To illustrate the differences between sensitivity analysis and contribution analysis, we can consider a model $y = f(x)$ that takes an n-dimensional input vector $x = \{x_1, \ldots, x_n\} \in \mathbb{R}^n$ and predicts a k-dimensional output vector $y = \{y_1, \ldots, y_k\} \in \mathbb{R}^k$. We then define an n-dimensional attribution vector $R^k = \{R_1^k, \ldots, R_n^k\} \in \mathbb{R}^n$ for the k-th output unit, which provides the *relevance* of each input value towards its final prediction. The attribution is obtained through the model gradient:

$$R_i(x) = \frac{\partial f(x)}{\partial x_i}, \tag{1}$$

although this is not the only option for attribution through gradients. Alternatively, the attribution can be defined by multiplying the input vector with the model gradient [6]:

$$R_i(x) = x_i \frac{\partial f(x)}{\partial x_i}. \tag{2}$$

Considering Eq. 1 we answer the question of *"What do we need to change in x to get a certain outcome y_k?"*, while considering Eq. 2 we answer the question of *"How much did x_i contribute to the outcome y_k?"* [1].

In reinforcement learning, we are interested in assessing the contributions of actions along a sequence which were relevant for achieving a particular return. Therefore, we are interested in contribution analysis methods rather than sensitivity analysis. We point out that this is closely related to the credit assignment problem, which we will further elaborate in the following sections.

2.2   Reinforcement Learning

In reinforcement learning, an agent is trained to take a sequence of actions by interacting with an environment and by learning from the feedback provided by the environment. The agent selects actions based on its policy, which are executed in the environment. The environment then transitions into its next state based on state-transition probabilities, and the agent receives feedback in the form of the next state and a reward signal. The objective of reinforcement learning is to learn a policy that maximizes the expected cumulative reward, also called return.

More formally, we define our problem setting as a finite Markov decision process (MDP) $\mathcal{P}$ as a 5-tuple $\mathcal{P} = (\mathcal{S}, \mathcal{A}, \mathcal{R}, p, \gamma)$ of finite sets $\mathcal{S}$ with states s (random variable S_t at time t), $\mathcal{A}$ with actions a (random variable A_t), and $\mathcal{R}$ with rewards r (random variable R_{t+1}) [47]. Furthermore, $\mathcal{P}$ has transition-reward distributions $p(S_{t+1} = s', R_{t+1} = r \mid S_t = s, A_t = a)$ conditioned on state-actions, a policy given as action distributions $\pi(A_{t+1} = a' \mid S_{t+1} = s')$ conditioned on states, and a discount factor $\gamma \in [0, 1]$. The return G_t is $G_t = \sum_{k=0}^{\infty} \gamma^k R_{t+k+1}$. We often consider finite horizon MDPs with sequence length T and $\gamma = 1$ giving $G_t = \sum_{k=0}^{T-t} R_{t+k+1}$. The state-value function $V^\pi(s)$ for a policy π is

$$V^\pi(s) = \mathrm{E}_\pi \left[G_t \mid S_t = s \right]$$

and its respective action-value function $Q^\pi(s, a)$ is

$$Q^\pi(s, a) = \mathrm{E}_\pi \left[G_t \mid S_t = s, A_t = a \right].$$

The goal of reinforcement learning is to maximize the expected return at time $t = 0$, that is $v_0^\pi = \mathrm{E}_\pi [G_0]$. The optimal policy π^* is $\pi^* = \mathrm{argmax}_\pi [v_0^\pi]$. We consider the difficult task of learning a policy when the reward given by the environment is sparse or delayed. An integral part to facilitate learning in this challenging setting is credit assignment, i.e. to determine the contribution of states and actions towards the return.

2.3   Credit Assignment in Reinforcement Learning

In reinforcement learning, we face two fundamental problems. First, the trade-off between exploring actions that lead to promising new states and exploiting actions that maximize the return. Second, the credit assignment problem, which involves correctly attributing credit to actions in a sequence that led to a certain return or outcome [2, 66, 67]. Credit assignment becomes more difficult as the delay between selected actions and their associated rewards increases [2, 45]. The study of credit assignment in sequences is a long-standing challenge and has been around since the start of artificial intelligence research [38]. Chess is an example of a sparse and delayed reward problem, where the reward is given at the end of the game. Assigning credit to the large number of decisions taken in a game of chess is quite difficult when the feedback is received only at the end of the game (i.e. win, lose or draw). It is difficult for the learning system to identify which actions were more or less important for the resulting outcome. As a result, the notion of winning or losing alone is often not informative enough for learning systems [38]. This motivates the need to improve credit assignment methods, especially for problems with sparse and delayed rewards. We further elaborate on various credit assignment methods in the next section.

2.4   Methods for Credit Assignment

Credit assignment in reinforcement learning can be classified into two different classes: 1) *Structural credit assignment*, and 2) *Temporal credit assignment* [66]. Structural credit assignment is related to the internals of the learning system that lead to choosing a particular action. Backpropagation [27] is quite popular for such structural credit assignment in Deep Reinforcement Learning. In contrast, temporal credit assignment is related to the events (states and/or actions) which led to a particular outcome in a sequence. In this work, we examine temporal credit assignment methods in detail.

Temporal credit assignment methods are used to obtain policies which maximize future rewards. Temporal difference (TD) learning [67] is a temporal credit assignment method which has close ties to dynamic programming and the Bellman operator [10]. It combines policy evaluation and improvement in a single step, by using the maximum action-value estimate at the next state to improve the action-value estimate at the current state. However, TD learning suffers from high bias and slows down learning when the rewards are sparse and delayed. *Eligibility traces* and TD(λ) [60] were introduced to ameliorate the performance of TD. Instead of looking one step into the future, information from n-steps in the future or past are used to update the current estimate of the action-value function. However, the performance of the algorithm is highly dependent on how much further in the future or in the past it looks into. In TD learning, one tries to find the action-value which maximizes the future return. In contrast, there exist direct policy optimization methods like policy gradient [65] and related methods like actor-critic [40,56].

More recent attempts to tackle credit assignment for delayed and sparse rewards have been made in RUDDER: Return Decomposition for Delayed Rewards (RUDDER) [2] and Hindsight Credit Assignment (HCA) [21]. RUDDER aims to identify actions which increase or decrease the expected future return. These actions are assigned credit directly by RUDDER, which makes learning faster by reducing the delay. We discuss RUDDER in detail in Sect. 2.6.

Unlike RUDDER, HCA assigns credit by estimating the likelihood of past actions having led to the observed outcome and consequently uses hindsight information to assign credit to past decisions. Both methods have in common, that the credit assignment problem is framed as a supervised learning task. In the next section, we look at credit assignment from the lens of explainability methods.

2.5 Explainability Methods for Credit Assignment

We have established that assigning credit to individual states, actions or state-action events along a sequence, which is also known as a trajectory or episode in reinforcement learning terminology, can tremendously simplify the task of learning an optimal policy. Therefore, if a method is able to determine which events were important for a certain outcome, it can be used to study sequences generated by a policy. As explainability methods were designed for this purpose, we can employ them to assign credit to important events and therefore speed up learning. As we have explored in Sect. 2.1, there are several methods we can choose from. The choice between intrinsically explainable models and post-hoc methods depends on whether a method can be combined with a reinforcement learning algorithm and is able to solve the task. In most cases, post-hoc methods are preferable, as they do not restrict the learning algorithm and model class. Since we are mainly interested in temporal credit assignment, we will look at explainability methods with a global scope. Sensitivity analysis methods have many drawbacks (see Sect. 2.1) and are therefore not suited for this purpose. Thus, we want to use contribution analysis methods.

2.6 Credit Assignment via Reward Redistribution

RUDDER [2] demonstrates how contribution analysis methods can be applied to target the credit assignment problem. RUDDER redistributes the return to relevant events and therefore sets future reward expectations to zero. The reward redistribution is achieved through return decomposition, which reduces high variance compared to Monte Carlo methods and high biases compared to TD methods [2]. This is possible because the state-value estimates are simplified to compute averages of immediate rewards.

In a common reinforcement learning setting, one can assign credit to an action a when receiving a reward r by updating a policy $\pi(a|s)$ according to its respective Q-function estimates. However, one fails when rewards are delayed, since the value network has to average over a large number of probabilistic future state-action paths that increase exponentially with the delay of the reward [36,48]. In contrast to using a forward view, a backward view approach based on a backward analysis of a forward model avoids problems with unknown future state-action paths, since the sequence is already completed and known. Backward analysis transforms the forward view approach into a regression task, at which deep learning methods excel. As a forward model, an LSTM can be trained to predict the final return, given a sequence of state-actions. LSTM was already used in reinforcement learning [55] for advantage learning [7] and learning policies [23,24,40]. Using contribution analysis, RUDDER can decompose the return prediction (the output relevance) into contributions of single state-action pairs along the observed sequence, obtaining a redistributed reward (the relevance redistribution). As a result, a new MDP is created with the same optimal policies and, in the optimal case, with no delayed rewards (expected future rewards equal zero) [2]. Indeed, for MDPs the Q-value is equal to the expected immediate reward plus the expected future rewards. Thus, if the expected future rewards are zero, the Q-value estimation simplifies to computing the mean of the immediate rewards.

Therefore, in the context of explainable AI, RUDDER uses contribution analysis to decompose the return prediction (the output relevance) into contributions of single state-action pairs along the observed sequence. RUDDER achieves this by training an LSTM model to predict the final return of a sequence of state-actions as early as possible. By taking the difference of the predicted returns from two consecutive state-actions, the contribution to the final return can be inferred [2].

Sequence-Markov Decision Processes (SDPs). An optimal reward redistribution should transform a delayed reward MDP into a return-equivalent MDP with zero expected future rewards. However, given an MDP, setting future rewards equal to zero is in general not possible. Therefore, RUDDER introduces *sequence-Markov decision processes* (SDPs), for which reward distributions are not required to be Markovian. An SDP is defined as a decision process which is equipped with a Markov policy and has Markov transition probabilities but a reward that is not required to be Markovian. Two SDPs $\tilde{\mathcal{P}}$ and $\mathcal{P}$ are *return-equivalent*, if (i) they differ only in their reward distribution and (ii) they have the same expected return at $t = 0$ for each policy π: $\tilde{v}_0^\pi = v_0^\pi$. RUDDER constructs a reward redistribution that leads to a return-equivalent SDP with a second-order Markov reward distribution and expected future rewards that are equal to zero. For these return-equivalent SDPs, Q-value estimation simplifies to computing the mean.

Return Equivalence. Strictly return-equivalent SDPs $\tilde{\mathcal{P}}$ and $\mathcal{P}$ can be constructed by reward redistributions. Given an SDP $\tilde{\mathcal{P}}$, a *reward redistribution* is a procedure that redistributes for each sequence $s_0, a_0, \ldots, s_T, a_T$ the realization of the sequence-associated return variable $\tilde{G}_0 = \sum_{t=0}^T \tilde{R}_{t+1}$ or its expectation along the sequence. The reward redistribution creates a new SDP $\mathcal{P}$ with the redistributed reward R_{t+1} at time $(t + 1)$ and the return variable $G_0 = \sum_{t=0}^T R_{t+1}$. A reward redistribution is second-order Markov if the redistributed reward R_{t+1} depends only on $(s_{t-1}, a_{t-1}, s_t, a_t)$. If the SDP $\mathcal{P}$ is obtained from the SDP $\tilde{\mathcal{P}}$ by reward redistribution, then $\tilde{\mathcal{P}}$ and $\mathcal{P}$ are strictly *return-equivalent*. Theorem 1 in RUDDER states that the optimal policies remain the same for $\tilde{\mathcal{P}}$ and $\mathcal{P}$ [2].

Reward Redistribution. We consider that a delayed reward MDP $\tilde{\mathcal{P}}$, with a particular policy π, can be transformed into a return-equivalent SDP $\mathcal{P}$ with an optimal reward redistribution and no delayed rewards:

Definition 1 ([2]). *For $1 \leqslant t \leqslant T$ and $0 \leqslant m \leqslant T - t$, the expected sum of delayed rewards at time $(t - 1)$ in the interval $[t + 1, t + m + 1]$ is defined as $\kappa(m, t - 1) = \mathrm{E}_\pi \left[\sum_{\tau=0}^m R_{t+1+\tau} \mid s_{t-1}, a_{t-1} \right]$.*

Theorem 2 ([2]). *We assume a delayed reward MDP $\tilde{\mathcal{P}}$, where the accumulated reward is given at sequence end. A new SDP $\mathcal{P}$ is obtained by a second-order Markov reward redistribution, which ensures that $\mathcal{P}$ is return-equivalent to $\tilde{\mathcal{P}}$. For a specific π, the following two statements are equivalent:*

(I) $\kappa(T - t - 1, t) = 0$, *i.e. the reward redistribution is optimal,*
(II) $\mathrm{E}\left[R_{t+1} \mid s_{t-1}, a_{t-1}, s_t, a_t\right] = \tilde{q}^\pi(s_t, a_t) - \tilde{q}^\pi(s_{t-1}, a_{t-1})$.

An optimal reward redistribution fulfills for $1 \leqslant t \leqslant T$ and $0 \leqslant m \leqslant T - t$:
$\kappa(m, t - 1) = 0$.

Theorem 2 shows that an optimal reward redistribution can be obtained by a second-order Markov reward redistribution for a given policy. It is an existence proof which explicitly gives the expected redistributed reward. In addition, higher-order Markov reward redistributions can also be optimal. In case of higher-order Markov reward redistribution, Equation (II) in Theorem 2 can have random variables R_{t+1} that depend on arbitrary states that are visited in the trajectory. Then Equation (II) averages out all states except s_t and s_{t-1} and averages out all randomness. In particular, this is also interesting for Align-RUDDER, since it can achieve an optimal reward redistribution. Therefore, although Align-RUDDER is in general not second-order Markov, Theorem 2 still holds in case of optimality.

For RUDDER, reward redistribution as in Theorem 2 can be achieved through return decomposition by predicting $\tilde{r}_{T+1} \in \tilde{R}_{T+1}$ of the original MDP $\tilde{P}$ by a function g from the state-action sequence. RUDDER determines for each sequence element its contribution to the prediction of $\tilde{r}_{T+1}$ at the end of the sequence. Therefore, it performs backward analysis through contribution analysis. Contribution analysis computes the contribution of the current input to the final prediction, i.e. the information gain by the current input on the final prediction. In principle, RUDDER could use any contribution analysis method. However, RUDDER prefers three methods: (A) differences of return predictions, (B) integrated gradients (IG) [64], and (C) layer-wise relevance propagation (LRP) [5]. For contribution method (A), RUDDER ensures that g predicts the final reward $\tilde{r}_{T+1}$ at every time step. Hence, the change in prediction is a measure of the contribution of an input to the final prediction and assesses the information gain by this input. The redistributed reward is given by the difference of consecutive predictions. In contrast to method (A), methods (B) and (C) use information from later on in the sequence for determining the contribution of the current input. Thus, a non-Markovian reward is introduced, as it depends on later sequence elements. However, the non-Markovian reward must be viewed as probabilistic reward, which is prone to have high variance. Therefore, RUDDER prefers method (A).

A principle insight on which RUDDER is based, is that the Q-function of optimal policies for complex tasks resembles a step function as they are hierarchical and composed of sub-tasks (blue curve, row 1 of Fig. 2, right panel). Completing such a sub-task is then reflected by a step in the Q-function. Therefore, a step in the Q-function is a change in return expectation, that is, the expected amount of the return or the probability to obtain the return changes. With return decomposition one identifies the steps of the Q-function (green arrows in Fig. 2, right panel), and an LSTM can therefore predict the expected return (red arrow, row 1 of Fig. 2, right panel), given the state-action sub-sequence to

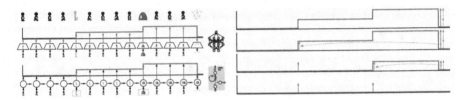

Fig. 2. Basic insight into reward redistribution [45]. Left panel, Row 1: An agent has to take a key to unlock a door. Both events increase the probability of receiving the treasure, which the agent always gets as a delayed reward, when the door is unlocked at sequence end. **Row 2:** The Q-function approximation typically predicts the expected return at every state-action pair (red arrows). **Row 3:** However, the Q-function approximation requires only to predict the steps (red arrows). **Right panel, Row 1:** The Q-function is the future-expected return (blue curve). Green arrows indicate Q-function steps and the big red arrow the delayed reward at sequence end. **Row 2 and 3:** The redistributed rewards correspond to steps in the Q-function (small red arrows). **Row 4:** After redistributing the reward, only the redistributed immediate reward remains (red arrows). Reward is no longer delayed. (Color figure online)

redistribute the reward. The prediction is decomposed into single steps of the Q-function (green arrows in Fig. 2). The redistributed rewards (small red arrows in second and third row of right panel of Fig. 2) remove the steps. Thus, the expected future reward is equal to zero (blue curve at zero in last row in right panel of Fig. 2). Future rewards of zero means that learning the Q-values simplifies to estimating the expected immediate rewards (small red arrows in right panel of Fig. 2), since delayed rewards are no longer present. Also, Hindsight Credit Assignment [21] identifies such Q-function steps that stem from actions alone. Figure 2 further illustrates how a Q-function predicts the expected return from every state-action pair, and how it is prone to prediction errors that hamper learning (second row, left panel). Since the Q-function is mostly constant, it is not necessary to predict the expected return for every state-action pair. It is sufficient to *identify relevant state-actions* across the whole episode and use them for predicting the expected return. This is achieved by computing the difference of two subsequent predictions of the LSTM model. If a state-action pair increases the prediction of the return, it is immediately rewarded. Using state-action sub-sequences $(s, a)_{0:t} = (s_0, a_0, \ldots, s_t, a_t)$, the redistributed reward is $R_{t+1} = g((s, a)_{0:t}) - g((s, a)_{0:t-1})$, where g is the return decomposition function, which is represented by an LSTM model and predicts the return of the episode. The LSTM model first learns to approximate the largest steps of the Q-function, since they reduce the prediction error the most. Therefore, the LSTM model extracts first the relevant state-actions pairs (events). Furthermore, the LSTM network [29–32] can store the relevant state-actions in its memory cells and subsequently, only updates its states to change its return prediction, when a new relevant state-action pair is observed. Thus, the LSTM return prediction is constant at most time points and does not have to be learned. The basic insight that Q-functions are step functions is the motivation for identifying these steps via return decomposition to speed up learning through reward redistribution, and furthermore enhance explainability through its state-action contributions.

In conclusion, redistributed reward serves as reward for a subsequent learning method [2]: (A) The Q-values can be directly estimated [2], which is also shown in Sect. 3 for the artificial tasks and Behavioral Cloning (BC) [70] pre-training for the Minecraft environment [19]. (B) The redistributed rewards can serve for learning with policy gradients like Proximal Policy Optimization (PPO) [57], which is also used in the Minecraft experiments for full training. (C) The redistributed rewards can serve for temporal difference learning, like Q-learning [74].

3   Strategy Extraction via Reward Redistribution

A strategy is a sequence of events which leads to a desirable outcome. Assuming a sequence of events is provided, the extraction of a strategy is the process of extracting events which are important for the desired outcome. This outcome could be a common state or return achieved at the end of the sequences. For example, if the desired outcome is to construct a wooden pickaxe in Minecraft, a strategy extracted from human demonstrations might contain event sequences for collecting a log, making planks, crafting a crafting table and finally a wooden pickaxe.

Strategy extraction is useful to study policies and also demonstration sequences. High return episodes can be studied to extract a strategy achieving such high returns. For example, Minecraft episodes where a stone pickaxe is obtained will include a strategy to make a wooden pickaxe, followed by collecting stones and finally the stone pickaxe. Similarly, strategies can be extracted from low return episodes, which can be helpful in learning which events to avoid. Extracted strategies explain the behavior of underlying policies or demonstrations. Furthermore, by comparing new trajectories to a strategy obtained from high return episodes, the reward signal can be redistributed to those events that are necessary for following the strategy and therefore are important.

However, current exploration strategies struggle with discovering episodes with high rewards in complex environments with delayed rewards. Therefore, episodes with high rewards are assumed and are given as demonstrations, such that they do not have to be discovered by exploration. Unfortunately, the number of demonstrations is typically small, as obtaining them is often costly and time-consuming. Therefore, deep learning methods that require a large amount of data, such as RUDDER's LSTM model, will not work well for this task while Align-RUDDER can learn a good strategy from as few as two demonstrations.

Reward redistribution identifies events which lead to an increase (or decrease) in expected return. The sequence of *important* events is the strategy. Thus, reward redistribution can be used to extract strategies. We illustrate this on the example of profile models in Sect. 3.1. Furthermore, a strategy can be used to redistribute reward by comparing a new sequence to an already given strategy. This results in faster learning, and is explained in detail in Sect. 3.2. Finally, we study expert episodes for the complex task of mining a diamond in Minecraft in Sect. 4.2.

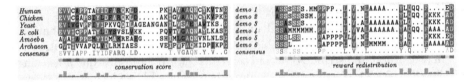

Fig. 3. The function of a protein is largely determined by its structure [45]. The relevant regions of this structure are even conserved across organisms, as shown in the left panel. Similarly, solving a task can often be decomposed into sub-tasks which are conserved across multiple demonstrations. This is shown in the right panel, where events are mapped to the letter code for amino acids. Sequence alignment makes those conserved regions visible and enables redistribution of reward to important events.

3.1   Strategy Extraction with Profile Models

Align-RUDDER introduced techniques from sequence alignment to replace the LSTM model from RUDDER by a profile model for reward redistribution. The profile model is the result of a multiple sequence alignment of the demonstrations and allows aligning new sequences to it. Both the sub-sequences $(s, a)_{0:t-1}$ and $(s, a)_{0:t}$ are mapped to sequences of events and are then aligned to the profile model. Thus, both sequences receive an alignment score S, which is proportional to the return decomposition function g. Similar to the LSTM model, Align-RUDDER identifies the largest steps in the Q-function via relevant events determined by the profile model. The redistributed reward is again $R_{t+1} = g((s, a)_{0:t}) - g((s, a)_{0:t-1})$ (see Eq. (3)). Therefore, redistributing the reward by sequence alignment fits into the RUDDER framework with all its theoretical guarantees. RUDDER is valid and works if its LSTM is replaced by other recurrent networks, attention mechanisms, or, as in case of Align-RUDDER, sequence and profile models [2].

Reward Redistribution by Sequence Alignment. In bioinformatics, sequence alignment identifies similarities between biological sequences to determine their evolutionary relationship [44,62]. The result of the alignment of multiple sequences is a profile model. The profile model is a consensus sequence, a frequency matrix, or a Position-Specific Scoring Matrix (PSSM) [63]. New sequences can be aligned to a profile model and receive an alignment score that indicates how well the new sequences agree to the profile model.

Align-RUDDER uses such alignment techniques to align two or more high return demonstrations. For the alignment, Align-RUDDER assumes that the demonstrations follow the same underlying strategy, therefore they are similar to each other analogous to being evolutionary related. Figure 3 shows an alignment of biological sequences and an alignment of demonstrations where events are mapped to letters. If the agent generates a state-action sequence $(s, a)_{0:t-1}$, then this sequence is aligned to the profile model g giving a score $g((s, a)_{0:t-1})$. The next action of the agent extends the state-action sequence by one state-action pair (s_t, a_t). The extended sequence $(s, a)_{0:t}$ is also aligned to the profile model g, giving another score $g((s, a)_{0:t})$. The redistributed reward R_{t+1} is the difference of these scores: $R_{t+1} = g((s,a)_{0:t}) - g((s,a)_{0:t-1})$ (see Eq. (3)). This difference indicates how much of the return is gained or lost by adding another sequence element. Align-RUDDER scores how close an agent follows an underlying strategy, which has been extracted by the profile model.

The new reward redistribution approach consists of five steps, see Fig. 4: (I) Define events to turn episodes of state-action sequences into sequences of events. (II) Determine an alignment scoring scheme, so that relevant events are aligned to each other. (III) Perform a multiple sequence alignment (MSA) of the demonstrations. (IV) Compute the profile model like a PSSM. (V) Redistribute the reward: Each sub-sequence τ_t of a new episode τ is aligned to the profile. The redistributed reward R_{t+1} is proportional to the difference of scores S based on the PSSM given in step (IV), i.e. $R_{t+1} \propto S(\tau_t) - S(\tau_{t-1})$.

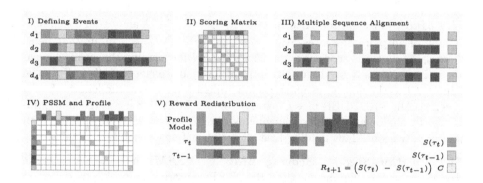

Fig. 4. The five steps of Align-RUDDER's reward redistribution [45]. **(I)** Define events and turn demonstrations into sequences of events. Each block represent an event to which the original state is mapped. **(II)** Construct a scoring matrix using event probabilities from demonstrations for diagonal elements and setting off-diagonal to a constant value. **(III)** Perform an MSA of the demonstrations. **(IV)** Compute a PSSM. Events with the highest column scores are indicated at the top row. **(V)** Redistribute reward as the difference of scores of sub-sequences aligned to the profile.

In the following, the five steps of Align-RUDDER's reward redistribution are explained in detail.

(I) Defining Events. Align-RUDDER considers differences of consecutive states to detect a change caused by an important event like achieving a subtask[1]. An *event* is defined as a cluster of state differences, where similarity-based clustering like affinity propagation (AP) [18] is used. If states are only enumerated, it is suggested to use the "successor representation" [12] or "successor features" [9]. In Align-RUDDER, the demonstrations are combined with state-action sequences generated by a random policy to construct the successor representation.

---

[1] Any sequence of events can be used for clustering and reward redistribution, and consequently for sub-task extraction.

A sequence of events is obtained from a state-action sequence by mapping states s to its cluster identifier e (the event) and ignoring the actions. Alignment techniques from bioinformatics assume sequences composed of a few events, e.g. 20 events. If there are too many events, good fitting alignments cannot be distinguished from random alignments. This effect is known in bioinformatics as "Inconsistency of Maximum Parsimony" [16].

(II) Determining the Alignment Scoring System. A scoring matrix $\mathbb{S}$ with entries $s_{i,j}$ determines the score for aligning event i with j. A priori, we only know that a relevant event should be aligned to itself but not to other events. Therefore, we set $s_{i,j} = 1/p_i$ for $i = j$ and $s_{i,j} = \alpha$ for $i \neq j$. Here, p_i is the relative frequency of event i in the demonstrations. α is a hyperparameter, which is typically a small negative number. This scoring scheme encourages alignment of rare events, for which p_i is small.

(III) Multiple Sequence Alignment (MSA). An MSA algorithm maximizes the sum of all pairwise scores $S_{\text{MSA}} = \sum_{i,j,i<j} \sum_{t=0}^{L} s_{i,j,t_i,t_j,t}$ in an alignment, where $s_{i,j,t_i,t_j,t}$ is the score at alignment column t for aligning the event at position t_i in sequence i to the event at position t_j in sequence j. $L \geq T$ is the alignment length, since gaps make the alignment longer than the length of each sequence. Align-RUDDER uses ClustalW [69] for MSA. MSA constructs a guiding tree by agglomerative hierarchical clustering of pairwise alignments between all demonstrations. This guiding tree allows identifying multiple strategies.

(IV) Position-Specific Scoring Matrix (PSSM) and MSA Profile Model. From the alignment, Align-RUDDER constructs a profile model as a) column-wise event probabilities and b) a PSSM [63]. The PSSM is a column-wise scoring matrix to align new sequences to the profile model.

(V) Reward Redistribution. The reward redistribution is based on the profile model. A sequence $\tau = e_{0:T}$ (e_t is event at position t) is aligned to the profile, which gives the score $S(\tau) = \sum_{l=0}^{L} s_{l,t_l}$. Here, s_{l,t_l} is the alignment score for the event e_{t_l} at position l in the alignment. Alignment gaps are columns to which no event was aligned, which have $t_l = T+1$ with gap penalty $s_{l,T+1}$. If $\tau_t = e_{0:t}$ is the prefix sequence of τ of length $t + 1$, then the reward redistribution R_{t+1} for $0 \leqslant t \leqslant T$ is

$$R_{t+1} = (S(\tau_t) - S(\tau_{t-1})) C$$
$$= g((s,a)_{0:t}) - g((s,a)_{0:t-1}), \tag{3}$$
$$R_{T+2} = \tilde{G}_0 - \sum_{t=0}^{T} R_{t+1},$$

where $C = \mathrm{E}_{\text{demo}}\left[\tilde{G}_0\right] / \mathrm{E}_{\text{demo}}\left[\sum_{t=0}^{T} S(\tau_t) - S(\tau_{t-1})\right]$ with $S(\tau_{-1}) = 0$. The original return of the sequence τ is $\tilde{G}_0 = \sum_{t=0}^{T} \tilde{R}_{t+1}$, and the expectation of the return over demonstrations is E_{demo}. The constant C scales R_{t+1} to the range of $\tilde{G}_0$. R_{T+2} is the correction of the redistributed reward [2], with zero expectation for demonstrations: $\text{E}_{\text{demo}}[R_{T+2}] = 0$. Since $\tau_t = e_{0:t}$ and $e_t = f(s_t, a_t)$, then $g((s,a)_{0:t}) = S(\tau_t)C$. Strict return-equivalence [2] is ensured by $G_0 = \sum_{t=0}^{T+1} R_{t+1} = \tilde{G}_0$. The redistributed reward depends only on the past: $R_{t+1} = h((s,a)_{0:t})$.

Higher-Order Markov Reward Redistribution. Align-RUDDER may lead to higher-order Markov redistribution. However, Corollary 1 in the Appendix of [45] states that the optimality criterion from Theorem 2 in Arjona-Medina et al. [2] also holds for higher-order Markov reward redistribution, if the expected redistributed higher-order Markov reward is the difference of Q-values. In that case, the redistribution is optimal, and there is no delayed reward. Furthermore, the optimal policies are the same as for the original problem. This corollary is the motivation for redistributing the reward to the steps in the Q-function. Furthermore, Corollary 2 in the Appendix of [45] states that under a condition, an optimal higher-order reward redistribution can be expressed as the difference of Q-values.

3.2 Explainable Agent Behavior via Strategy Extraction

The reward redistribution identifies sub-tasks as alignment positions with high redistributed rewards. These sub-tasks are indicated by high scores s in the PSSM. Reward redistribution also determines the terminal states of sub-tasks, since it assigns rewards for solving the sub-tasks. As such, the strategy for solving a given task is extracted from those demonstrations used for alignment and represented as a sequence of sub-tasks. By assigning rewards to these sub-tasks with Align-RUDDER, a policy can be learned that is also able to achieve these sub-tasks and therefore high returns.

While RUDDER with an LSTM model for reward redistribution is also able to assign reward to important events, in practice it is not easy to identify sub-tasks. Changes in predicted reward from one event to the next are often small, as it is difficult for an LSTM model to learn sharp increases or decreases. Furthermore, it would be necessary to inspect a relatively large number of episodes to identify common sub-tasks. In contrast, the sub-tasks extracted via sequence alignment are often easy to interpret and can be obtained from only a few episodes. The strategy of agents trained via Align-RUDDER can easily be explained by inspecting the alignment and visualizing the sequence of aligned events. As the strategy represents the global long-term behavior of an agent, its behavior can be interpreted through the strategy.

4   Experiments

Using several examples we show how reward redistribution with Align-RUDDER enables learning a policy with only a few demonstrations, even in highly complex environments. Furthermore, the strategy these policies follow is visualized, highlighting the ability of Align-RUDDER's alignment-based approach to interpret agent behavior.

4.1 Gridworld

First, we analyze Align-RUDDER on two artificial tasks. The tasks are variations of the gridworld *rooms example* [68], where cells (locations) are the MDP states. The *FourRooms* environment is a 12×12 gridworld with four rooms. The target is in room four, and the start is in room one (from bottom left, to bottom right) with 20 portal entry locations. *EightRooms* is a larger variant with a 12×24 gridworld divided into eight rooms. Here, the target is in room eight, and the starting location in room one, again with 20 portal entry locations. We show the two artificial tasks with sample trajectories in Fig. 5.

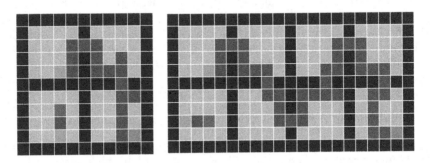

Fig. 5. Examples of trajectories in the two artificial task environments with four (left) and eight (right) rooms. The initial position is indicated in red, the portal between the first and second room in yellow and the goal in green [45]. Blue squares indicate the path of the trajectory. (Color figure online)

In this setting, the states do not have to be time-aware for ensuring stationary optimal policies but the unobserved used-up time introduces a random effect. The grid is divided into rooms. The agent's goal is to reach a target from an initial state with the lowest number of steps. It has to cross different rooms, which are connected by doors, except for the first room, which is only connected to the second room by a *portal*. If the agent is at the portal entry cell of the first room, then it is teleported to a fixed portal arrival cell in the second room. The location of the portal entry cell is random for each episode, while the portal arrival cell is fixed across episodes. The portal entry cell location is given in the state for the first room. The portal is introduced to ensure that initialization with behavioral cloning (BC) alone is not sufficient for solving the task. It enforces that going to the portal entry cells is learned, even when they are at positions not observed in demonstrations. At every location, the agent can move *up, down, left, right*. The state transitions are stochastic. An episode ends after $T = 200$ time steps. If the agent arrives at the target, then at the next step it goes into an absorbing state, where it stays until $T = 200$ without receiving further rewards. Reward is only given at the end of the episode. Demonstrations are generated by an optimal policy with an exploration rate of 0.2.

The five steps of Align-RUDDER's reward redistribution for these experiments are:

(i) Defining Events. Events are clusters of states obtained by Affinity Propagation using the successor representation based on demonstrations as similarity. Figure 6 shows examples of clusters for the two versions of the environment.

(ii) Determining the Alignment Scoring System. The scoring matrix is obtained according to (II), using $\epsilon = 0$ and setting all off-diagonal values of the scoring matrix to -1.

(iii) Multiple sequence alignment (MSA). ClustalW is used for the MSA of the demonstrations with zero gap penalties and no biological options.

(iv) Position-Specific Scoring Matrix (PSSM) and MSA profile model. The MSA supplies a profile model and a PSSM, as in (IV).

(v) Reward Redistribution. Sequences generated by the agent are mapped to sequences of events according to (I). Reward is redistributed via differences of profile alignment scores of consecutive sub-sequences according to Eq. (3) using the PSSM.

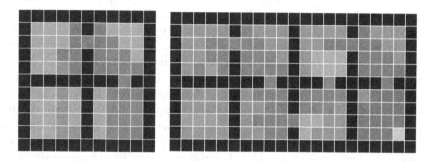

Fig. 6. Examples of different clusters in the FourRooms (left) and EightRooms (right) environment with 1% stochasticity on the transitions after performing clustering with Affinity Propagation using the successor representation with 25 demonstrations. Different colors represent different clusters [45].

The reward redistribution determines sub-tasks like doors or portal arrival. Some examples are shown in Fig. 7. In these cases, three sub-tasks emerged. One for entering the portal and going to the first room, one for travelling from the entrance of one room to the exit of the next room, and finally going to the goal in the last room. The sub-tasks partition the Q-table into sub-tables that represent a sub-agent. The emerging set of sub-agents describe the global behavior of the Align-RUDDER method and can be directly used to explain the decision-making for specific tasks.

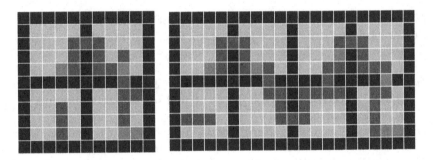

Fig. 7. Reward redistribution for the above trajectories in the FourRooms (left) and EightRooms (right) environments [45]. Here, sub-tasks emerged via reward redistribution for entering the portal, travelling from the entrance of one room to the exit of the next and finally for reaching the goal.

Results. In addition to enabling an interpretation of the strategy for solving a task, the redistributed reward signal speeds up the learning process of existing methods and requires fewer examples when compared to related approaches. All compared methods learn a Q-table and use an ϵ-greedy policy with $\epsilon = 0.2$. The Q-table is initialized by behavioral cloning (BC). The state-action pairs which are not initialized, since they are not visited in the demonstrations, get an initialization by drawing a sample from a normal distribution with mean 1 and standard deviation 0.5 (avoiding equal Q-values). Align-RUDDER learns the Q-table via RUDDER's Q-value estimation (learning method (A) from above). For BC+Q, RUDDER (LSTM), SQIL [49], and DQfD [26] a Q-table is learned by Q-learning. Hyperparameters are selected via grid search with a similar computational budget for each method. For different numbers of demonstrations, performance is measured by the number of episodes to achieve 80% of the average return of the demonstrations. A Wilcoxon rank-sum test determines the significance of performance differences between Align-RUDDER and the other methods.

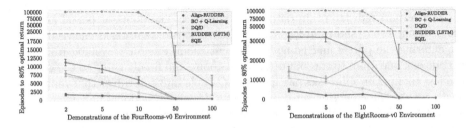

Fig. 8. Comparison of Align-RUDDER and other methods in the FourRooms (left) and EightRooms (right) environments with respect to the number of episodes required for learning on different numbers of demonstrations. Results are the average over 100 trials. Align-RUDDER significantly outperforms all other methods [45].

Figure 8 shows the number of episodes required for achieving 80% of the average reward of the demonstrations for different numbers of demonstrations. In both environments, Align-RUDDER significantly outperforms all other methods, for $\leqslant 10$ demonstrations (with p-values of $< 10^{-10}$ and $< 10^{-19}$ for Task (I) and (II), respectively).

4.2   Minecraft

To demonstrate the effectiveness of Align-RUDDER even in highly complex environments, it was applied to the complex high-dimensional problem of obtaining a diamond in Minecraft with the *MineRL* environment [19]. This task requires an agent to collect a diamond by exploring the environment, gathering resources and building necessary tools. To obtain a diamond the agent needs to collect resources (log, cobblestone, etc.) and craft tools (table, pickaxe, etc.). Every episode of the environment is procedurally generated, and the agent is placed at a random location. This is a challenging environment for reinforcement learning as episodes are typically very long, the reward signal is sparse and exploration difficult. By using demonstrations from human players, Align-RUDDER can circumvent the exploration problem and with reward redistribution can ameliorate the sparse reward problem. Furthermore, by identifying sub-tasks, individual agents can be trained to solve simpler tasks, and help divide the complex long time-horizon task in more approachable sub-problems. In complement to that, we can also inspect and interpret the behavior of expert policies using Align-RUDDER's alignment method. In our example, the expert policies are presented in the form of human demonstrations that successfully obtained a diamond. Align-RUDDER is able to extract a strategy from as few as ten trajectories. In the following, we outline the five steps of Align-RUDDER in the Minecraft environment. Furthermore, we inspect the alignment-based reward redistribution and show how it enables interpretation of both the expert policies and the trained agent.

(i) Defining Events. A state consists of a visual input and an inventory. Both inputs are normalized and then the difference of consecutive states is clustered, obtaining 19 clusters corresponding to events. Upon inspection these clusters correspond to inventory changes, i.e. gaining a particular item. Finally, the demonstration trajectories are mapped to sequences of events. This is shown in Fig. 9.

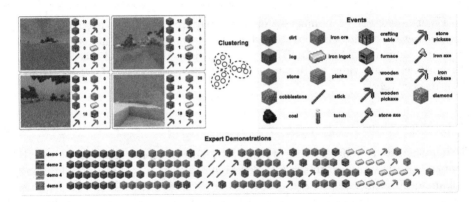

Fig. 9. Step **(I)**: Define events and map demonstrations into sequences of events.

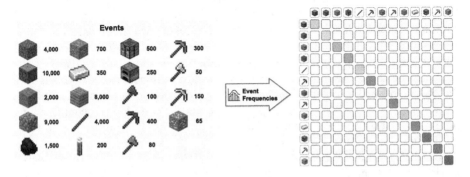

Fig. 10. Step **(II)**: Construct a scoring matrix using event probabilities from demonstrations for diagonal elements and setting off-diagonal to a constant value. Darker colors signify higher score values. For illustration, only a subset of events is shown. (Color figure online)

(ii) Determining the Alignment Scoring System. The scoring matrix is computed according to (II). Since there is no prior knowledge on how the individual events are related to each other, the scoring matrix has the inverse frequency of an event occurring in the expert trajectories on the diagonal and a small constant value on the off-diagonal entries. As can be seen in Fig. 10, this results in lower scores for clusters corresponding to earlier events as they occur more often and high values for rare events such as building a pickaxe or mining the diamond.

(iii) Multiple Sequence Alignment (MSA). The 10 expert episodes that obtained a diamond in the shortest amount of time are aligned using ClustalW with zero gap penalties and no biological options (i.e. arguments to ClustalW related to biological sequences). The MSA algorithm maximizes the pairwise sum of scores of all alignments using the scoring matrix from (II). Figure 11 shows an example of a such an alignment.

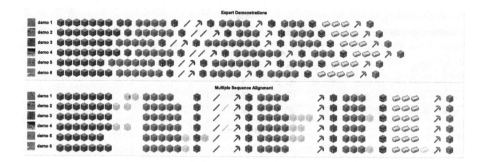

Fig. 11. Step **(III)**: Perform multiple sequence alignment (MSA) of the demonstrations.

(iv) Position-Specific Scoring Matrix (PSSM) and MSA Profile Model. The multiple alignment gives a profile model and a PSSM. In Fig. 12 an example of a PSSM is shown, resulting from an alignment of the previous example sequences. The PSSM contains for each position in the alignment the frequency of each event occurring in the trajectories used for the alignment. At this point, the strategy followed by the majority of experts is already visible.

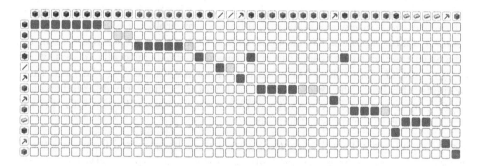

Fig. 12. Step **(IV)**: Compute a position-specific scoring matrix (PSSM). The score at a position from the MSA (column) and for an event (row) depends on the frequency of that event at that position in the MSA. For example, the event in the last position is present in all the sequences, and thus gets a high score at the last position. But it is absent in the remaining position, and thus gets a score of zero elsewhere.

(v) Reward Redistribution. The reward is redistributed via differences of profile alignment scores of consecutive sub-sequences according to Eq. (3) using the PSSM. Figure 13 illustrates this on the example of an incomplete trajectory. In addition to aligning trajectories generated by an agent, we can use demonstrations from human players that were not able to obtain the diamond and therefore highlight problems those players have encountered.

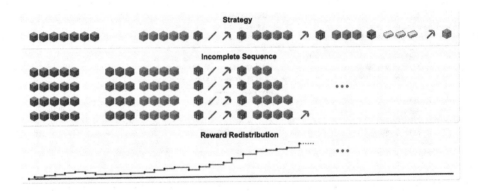

Fig. 13. Step (**V**): A new sequence is aligned step by step to the profile model using the PSSM, resulting in an alignment score for each sub-sequence. The redistributed reward is then proportional to the difference of scores of subsequent alignments.

Interpreting Agent Behavior. The strategy for obtaining a diamond, an example of which is shown in Fig. 13, is a direct result of Align-RUDDER. If it is possible to map event clusters to a meaningful representation, as is the case here by mapping the clusters to changes in inventory states, the strategy describes the behavior of the expert policies in a very intuitive and interpretable fashion. Furthermore, new trajectories generated by the learned agent can be aligned to the strategy, highlighting differences or problems where the trained agent is unable to follow the expert strategy. Inspecting the strategy it can be seen that random events, such as collecting dirt which naturally occurs when digging, are not present as they are not important for solving the task. Surprisingly, also items that seem helpful such as torches for providing light when digging are not used by the majority of experts even though they have to operate in near complete darkness without them.

Results. Sub-agents can be trained for the sub-tasks extracted from the expert episodes. The sub-agents are first pre-trained on the expert episodes for the sub-tasks using BC, and further trained in the environment using Proximal Policy Optimization (PPO) [57]. Using only 10 expert episodes, Align-RUDDER is able to learn to mine a diamond. A diamond is obtained in 0.1% of the cases, and to the best of our knowledge, no pure learning method[2] has yet mined a diamond [53]. With a 0.5 success probability for each of the 31 extracted sub-tasks[3], the resulting success rate for mining the diamond would be 4.66×10^{-10}. Table 1 shows a comparison of methods on the Minecraft MineRL dataset by the maximum item score [37]. Results are taken from [37], in particular from Fig. 2, and completed by [33,53,61]. Align-RUDDER was not evaluated during

---

[2] This includes not only learning to extract the sub-tasks, but also learning to solve the sub-tasks themselves.

[3] A 0.5 success probability already defines a very skilled agent in the MineRL environment.

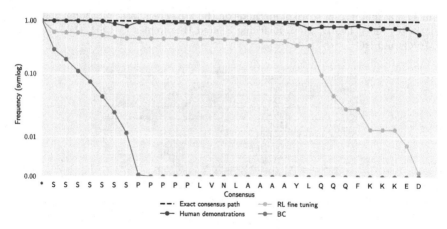

Fig. 14. Comparing the consensus frequencies between behavioral cloning (BC, green), where fine-tuning starts, the fine-tuned model (orange), and human demonstrations (blue) [45]. The plot is in symmetric log scale (symlog in matplotlib). The mapping from the letters on the x-axis to items is as follows: **S:** log, **P:** plank, **L:** crafting table, **V:** stick, **N:** wooden pickaxe, **A:** cobblestone, **Y:** stone pickaxe, **Q:** iron ore, **F:** furnace, **K:** iron ingot, **E:** iron pickaxe, **D:** diamond ore. (Color figure online)

the challenge, and may therefore have advantages. However, it did not receive the intermediate rewards provided by the environment that hint at sub-tasks, but self-discovered such sub-tasks, which demonstrates its efficient learning. Furthermore, Align-RUDDER is capable of extracting a common strategy from only a few demonstrations and train globally explainable models based on this strategy (Fig. 14).

5   Limitations

While Align-RUDDER can extract strategies and speed up learning even in complex environments, the resulting performance depends on the quality of the alignment model. A low quality alignment model can be a result of multiple factors, one of which is having many distinct events ($\gg$20). Clustering can be used to reduce the number of events, which could also lead to a low quality alignment model if too many relevant events are clustered together. While the optimal policy does not change due to a poor alignment of expert episodes, the benefit of employing reward redistribution based on such an alignment diminishes.

The alignment could fail if all expert episodes have different underlying strategies, i.e. no events are common in the expert episodes. We assume that the expert episodes follow the same underlying strategy, therefore they are similar to each other and can be aligned. However, if an underlying strategy does not exist, then the alignment may fail to identify relevant events that should receive high redistributed rewards. In this case, reward is given at sequence end, when the redistributed reward is corrected, which leads to an episodic reward without

Table 1. Maximum item score of methods on the Minecraft task. Methods: Soft-Actor Critic (SAC, [20]), DQfD, Meta Learning Shared Hierarchies (MLSH, [17]), Rainbow [25], PPO, and BC.

Method	Team Name									
Align-RUDDER	Ours									
DQfD	CDS									
BC	MC_RL									
CLEAR	I4DS									
Options&PPO	CraftRL									
BC	UEFDRL									
SAC	TD240									
MLSH	LAIR									
Rainbow	Elytra									
PPO	karolisram									

reducing the delay of the rewards and speeding up learning. This is possible, as there can be many distinct paths to the same end state. This problem can be resolved if there are at least two demonstrations of each of these different strategies. This helps with identifying events for all different strategies, such that the alignment will not fail.

Align-RUDDER has the potential to reduce the cost for training and deploying agents in real world applications, and therefore enable systems that have not been possible until now. However, the method relies on expert episodes and thereby expert decisions, which are usually strongly biased. Therefore, the responsible use of Align-RUDDER depends on a careful selection of the training data and awareness of the potential biases within those.

6   Conclusion

We have analyzed Align-RUDDER, which solves highly complex tasks with delayed and sparse rewards. The global behavior of agents trained by Align-RUDDER can easily be explained by inspecting the alignment of events. Furthermore, the alignment step of Align-RUDDER can be employed to explain arbitrary agents' behavior, so long as episodes generated with this agent are available or can be generated.

Furthermore, we have shown that Align-RUDDER outperforms state-of-the-art methods designed for learning from demonstrations in the regime of few demonstrations. On the Minecraft ObtainDiamond task, Align-RUDDER is, to the best of our knowledge, the first pure learning method to mine a diamond.

Acknowledgements. The ELLIS Unit Linz, the LIT AI Lab, the Institute for Machine Learning, are supported by the Federal State Upper Austria. IARAI is supported by Here Technologies. We thank the projects AI-MOTION (LIT-2018-6-YOU-212), AI-SNN (LIT-2018-6-YOU-214), DeepFlood (LIT-2019-8-YOU-213), Medical Cognitive Computing Center (MC3), INCONTROL-RL (FFG-881064), PRIMAL (FFG-873979), S3AI (FFG-872172), DL for GranularFlow (FFG-871302), AIRI

FG 9-N (FWF-36284, FWF-36235), ELISE (H2020-ICT-2019-3 ID: 951847), AIDD (MSCA-ITN-2020 ID: 956832). We thank Janssen Pharmaceutica (MaDeSMart, HBC.2018.2287), Audi.JKU Deep Learning Center, TGW LOGISTICS GROUP GMBH, Silicon Austria Labs (SAL), FILL Gesellschaft mbH, Anyline GmbH, Google, ZF Friedrichshafen AG, Robert Bosch GmbH, UCB Biopharma SRL, Merck Healthcare KGaA, Verbund AG, Software Competence Center Hagenberg GmbH, TÜV Austria, Frauscher Sensonic and the NVIDIA Corporation.

References

1. Ancona, M., Ceolini, E., Öztireli, C., Gross, M.: Gradient-based attribution methods. In: Samek, W., Montavon, G., Vedaldi, A., Hansen, L.K., Müller, K.-R. (eds.) Explainable AI: Interpreting, Explaining and Visualizing Deep Learning. LNCS (LNAI), vol. 11700, pp. 169–191. Springer, Cham (2019). https://doi.org/10.1007/978-3-030-28954-6_9. ISBN 978-3-030-28954-6

2. Arjona-Medina, J.A., Gillhofer, M., Widrich, M., Unterthiner, T., Brandstetter, J., Hochreiter, S.: RUDDER: return decomposition for delayed rewards. In: Advances in Neural Information Processing Systems, vol. 32, pp. 13566–13577 (2019)

3. Arras, L., Montavon, G., Müller, K.-R., Samek, W.: Explaining recurrent neural network predictions in sentiment analysis. arXiv, abs/1706.07206 (2017)

4. Arras, L., et al.: Explaining and interpreting LSTMs. In: Samek, W., Montavon, G., Vedaldi, A., Hansen, L.K., Müller, K.-R. (eds.) Explainable AI: Interpreting, Explaining and Visualizing Deep Learning. LNCS (LNAI), vol. 11700, pp. 211–238. Springer, Cham (2019). https://doi.org/10.1007/978-3-030-28954-6_11. ISBN978-3-030-28954-6

5. Bach, S., Binder, A., Montavon, G., Klauschen, F., Müller, K.-R., Samek, W.: On pixel-wise explanations for non-linear classifier decisions by layer-wise relevance propagation. PLoS One **10**(7), e0130140 (2015). https://doi.org/10.1371/journal.pone.0130140

6. Baehrens, D., Schroeter, T., Harmeling, S., Kawanabe, M., Hansen, K., Müller, K.-R.: How to explain individual classification decisions. J. Mach. Learn. Res. **11**, 1803–1831 (2010). ISSN 1532-4435

7. Bakker, B.: Reinforcement learning with long short-term memory. In: Dietterich, T.G., Becker, S., Ghahramani, Z. (eds.) Advances in Neural Information Processing Systems, vol. 14, pp. 1475–1482. MIT Press (2002)

8. Bakker, B.: Reinforcement learning by backpropagation through an LSTM model/critic. In: IEEE International Symposium on Approximate Dynamic Programming and Reinforcement Learning, pp. 127–134 (2007). https://doi.org/10.1109/ADPRL.2007.368179

9. Barreto, A., et al.: Successor features for transfer in reinforcement learning. In: Guyon, I., et al. (eds.) Advances in Neural Information Processing Systems, vol. 30. Curran Associates Inc. (2017)

10. Bellman, R.E.: Adaptive Control Processes. Princeton University Press, New Jersey (1961)

11. Binder, A., Bach, S., Montavon, G., Müller, K.-R., Samek, W.: Layer-wise relevance propagation for deep neural network architectures. In: Information Science and Applications (ICISA) 2016. LNEE, vol. 376, pp. 913–922. Springer, Singapore (2016). https://doi.org/10.1007/978-981-10-0557-2_87. ISBN 978-981-10-0557-2

12. Dayan, P.: Improving generalization for temporal difference learning: the successor representation. Neural Comput. **5**(4), 613–624 (1993)

13. Correia, A.D.S., Colombini, E.L.: Attention, please! a survey of neural attention models in deep learning. arXiv, abs/2103.16775 (2021)
14. Devlin, J., Chang, M., Lee, K., Toutanova, K.: BERT: pre-training of deep bidirectional transformers for language understanding. arXiv, abs/1810.04805 (2019)
15. Du, M., Liu, N., Hu, X.: Techniques for interpretable machine learning. Commun. ACM **63**(1), 68–77 (2019). https://doi.org/10.1145/3359786. ISSN 0001-0782
16. Felsenstein, J.: Cases in which parsimony or compatibility methods will be positively misleading. Syst. Zool. **27**(4), 401–410 (1978). https://doi.org/10.2307/2412923
17. Frans, K., Ho, J., Chen, X., Abbeel, P., Schulman, J.: Meta learning shared hierarchies. In: International Conference on Learning Representations (2018). arXiv abs/1710.09767
18. Frey, B.J., Dueck, D.: Clustering by passing messages between data points. Science **315**(5814), 972–976 (2007). https://doi.org/10.1126/science.1136800
19. Guss, W.H., et al.: MineRL: a large-scale dataset of minecraft demonstrations. In: Proceedings of the 28th International Joint Conference on Artificial Intelligence (IJCAI 2019) (2019)
20. Haarnoja, T., Zhou, A., Abbeel, P., Levine, S.: Soft actor-critic: off-policy maximum entropy deep reinforcement learning with a stochastic actor. In: Dy, J., Krause, A. (eds.) Proceedings of Machine Learning Research, vol. 80, pp. 1861–1870. PMLR (2018). arXiv abs/1801.01290
21. Harutyunyan, A., et al.: Hindsight credit assignment. In: Advances in Neural Information Processing Systems, vol. 32, pp. 12467–12476 (2019)
22. Hastie, T., Tibshirani, R.: Generalized additive models. Stat. Sci. **1**(3), 297–310 (1986). https://doi.org/10.1214/ss/1177013604
23. Hausknecht, M.J., Stone, P.: Deep recurrent Q-learning for partially observable MDPs. arXiv, abs/1507.06527 (2015)
24. Heess, N., Wayne, G., Tassa, Y., Lillicrap, T.P., Riedmiller, M.A., Silver, D.: Learning and transfer of modulated locomotor controllers. arXiv, abs/1610.05182 (2016)
25. Hessel, M., et al.: Rainbow: combining improvements in deep reinforcement learning. arXiv, abs/1710.02298 (2017)
26. Hester, T., et al.: Deep Q-learning from demonstrations. In: The Thirty-Second AAAI Conference on Artificial Intelligence (AAAI-18). Association for the Advancement of Artificial Intelligence (2018)
27. Hinton, G.E., Sejnowski, T.E.: Learning and relearning in Boltzmann machines. In: Parallel Distributed Processing, vol. 1, pp. 282–317. MIT Press, Cambridge (1986)
28. Hochreiter, S.: Implementierung und Anwendung eines 'neuronalen' Echtzeit-Lernalgorithmus für reaktive Umgebungen. Practical work, Supervisor: J. Schmidhuber, Institut für Informatik, Technische Universität München (1990)
29. Hochreiter, S.: Untersuchungen zu dynamischen neuronalen Netzen. Master's thesis, Technische Universität München (1991)
30. Hochreiter, S., Schmidhuber, J.: Long short-term memory. Technical report FKI-207-95, Fakultät für Informatik, Technische Universität München (1995)
31. Hochreiter, S., Schmidhuber, J.: Long short-term memory. Neural Comput. **9**(8), 1735–1780 (1997)
32. Hochreiter, S., Schmidhuber, J.: LSTM can solve hard long time lag problems. In: Mozer, M.C., Jordan, M.I., Petsche, T. (eds.) Advances in Neural Information Processing Systems, vol. 9, pp. 473–479. MIT Press, Cambridge (1997)

33. Kanervisto, A., Karttunen, J., Hautamäki, V.: Playing Minecraft with behavioural cloning. In: Escalante, H.J., Hadsell, R. (eds.) Proceedings of Machine Learning Research (PMLR), vol. 123, pp. 56–66. PMLR (2020)

34. Lin, T.-Y., et al.: Microsoft COCO: common objects in context. In: Fleet, D., Pajdla, T., Schiele, B., Tuytelaars, T. (eds.) ECCV 2014. LNCS, vol. 8693, pp. 740–755. Springer, Cham (2014). https://doi.org/10.1007/978-3-319-10602-1_48. ISBN 978-3-319-10602-1

35. Lundberg, S.M., et al.: From local explanations to global understanding with explainable AI for trees. Nat. Mach. Intell. **2**(1), 56–67 (2020). https://doi.org/10.1038/s42256-019-0138-9. ISSN 2522-5839

36. Luoma, J., Ruutu, S., King, A.W., Tikkanen, H.: Time delays, competitive interdependence, and firm performance. Strateg. Manag. J. **38**(3), 506–525 (2017). https://doi.org/10.1002/smj.2512

37. Milani, S., et al.: Retrospective analysis of the 2019 MineRL competition on sample efficient reinforcement learning. arXiv, abs/2003.05012 (2020)

38. Minsky, M.: Steps towards artificial intelligence. Proc. IRE **49**(1), 8–30 (1961). https://doi.org/10.1109/JRPROC.1961.287775

39. Mnih, V., et al.: Human-level control through deep reinforcement learning. Nature **518**(7540), 529–533 (2015). https://doi.org/10.1038/nature14236

40. Mnih, V., et al.: Asynchronous methods for deep reinforcement learning. In: Proceedings of the 33rd International Conference on Machine Learning (ICML), Volume 48 of Proceedings of Machine Learning Research, pp. 1928–1937. PMLR.org (2016)

41. Montavon, G., Lapuschkin, S., Binder, A., Samek, W., Müller, K.-R.: Explaining nonlinear classification decisions with deep Taylor decomposition. Pattern Recogn. **65**, 211–222 (2017). https://doi.org/10.1016/j.patcog.2016.11.008

42. Montavon, G., Samek, W., Müller, K.-R.: Methods for interpreting and understanding deep neural networks. Digit. Signal Process. **73**, 1–15 (2017). https://doi.org/10.1016/j.dsp.2017.10.011

43. Munro, P.W.: A dual back-propagation scheme for scalar reinforcement learning. In: Proceedings of the Ninth Annual Conference of the Cognitive Science Society, Seattle, WA, pp. 165–176 (1987)

44. Needleman, S.B., Wunsch, C.D.: A general method applicable to the search for similarities in the amino acid sequence of two proteins. J. Mol. Biol. **48**(3), 443–453 (1970)

45. Patil, V.P., et al.: Align-rudder: learning from few demonstrations by reward redistribution. arXiv, abs/2009.14108 (2020). CoRR

46. Petsiuk, V., Das, A., Saenko, K.: RISE: randomized input sampling for explanation of black-box models. arXiv, abs/1806.07421 (2018)

47. Puterman, M.L.: Markov Decision Processes, 2nd edn. Wiley (2005). ISBN 978-0-471-72782-8

48. Rahmandad, H., Repenning, N., Sterman, J.: Effects of feedback delay on learning. Syst. Dyn. Rev. **25**(4), 309–338 (2009). https://doi.org/10.1002/sdr.427

49. Reddy, S., Dragan, A.D., Levine, S.: SQIL: imitation learning via regularized behavioral cloning. In: Eighth International Conference on Learning Representations (ICLR) (2020). arXiv abs/1905.11108

50. Robinson, A.J.: Dynamic error propagation networks. PhD thesis, Trinity Hall and Cambridge University Engineering Department (1989)

51. Robinson, T., Fallside, F.: Dynamic reinforcement driven error propagation networks with application to game playing. In: Proceedings of the 11th Conference of the Cognitive Science Society, Ann Arbor, pp. 836–843 (1989)

52. Russakovsky, O., et al.: ImageNet large scale visual recognition challenge. Int. J. Comput. Vis. **115**(3), 211–252 (2015). https://doi.org/10.1007/s11263-015-0816-y
53. Scheller, C., Schraner, Y., Vogel, M.: Sample efficient reinforcement learning through learning from demonstrations in Minecraft. In: Escalante, H.J., Hadsell, R. (eds.) Proceedings of Machine Learning Research (PMLR), vol. 123, pp. 67–76. PMLR (2020)
54. Schmidhuber, J.: Making the world differentiable: On using fully recurrent self-supervised neural networks for dynamic reinforcement learning and planning in non-stationary environments. Technical report FKI-126-90 (revised), Institut für Informatik, Technische Universität München (1990). Experiments by Sepp Hochreiter
55. Schmidhuber, J.: Deep learning in neural networks: an overview. Neural Netw. **61**, 85–117 (2015). https://doi.org/10.1016/j.neunet.2014.09.003
56. Schulman, J., Levine, S., Moritz, P., Jordan, M.I., Abbeel, P.: Trust region policy optimization. In: 32st International Conference on Machine Learning (ICML), Volume 37 of Proceedings of Machine Learning Research, pp. 1889–1897. PMLR (2015)
57. Schulman, J., Wolski, F., Dhariwal, P., Radford, A., Klimov, O.: Proximal policy optimization algorithms. arXiv, abs/1707.06347 (2018)
58. Silver, D., et al.: Mastering the game of Go with deep neural networks and tree search. Nature **529**(7587), 484–489 (2016). https://doi.org/10.1038/nature16961
59. Simonyan, K., Vedaldi, A., Zisserman, A.: Deep inside convolutional networks: visualising image classification models and saliency maps. arXiv, abs/1312.6034 (2014)
60. Singh, S.P., Sutton, R.S.: Reinforcement learning with replacing eligibility traces. Mach. Learn. **22**, 123–158 (1996)
61. Skrynnik, A., Staroverov, A., Aitygulov, E., Aksenov, K., Davydov, V., Panov, A.I.: Hierarchical deep Q-network with forgetting from imperfect demonstrations in Minecraft. arXiv, abs/1912.08664 (2019)
62. Smith, T.F., Waterman, M.S.: Identification of common molecular subsequences. J. Mol. Biol. **147**(1), 195–197 (1981)
63. Stormo, G.D., Schneider, T.D., Gold, L., Ehrenfeucht, A.: Use of the 'Perceptron' algorithm to distinguish translational initiation sites in E. coli. Nucleic Acids Res. **10**(9), 2997–3011 (1982)
64. Sundararajan, M., Taly, A., Yan, Q.: Axiomatic attribution for deep networks. In: Proceedings of the 34th International Conference on Machine Learning, ICML 2017, vol. 70, pp. 3319–3328 (2017)
65. Sutton, R., McAllester, D., Singh, S., Mansour, Y.: Policy gradient methods for reinforcement learning with function approximation. In: Solla, S., Leen, T., Müller, K. (eds.) Advances in Neural Information Processing Systems, vol. 12. MIT Press (2000)
66. Sutton, R.S.: Temporal credit assignment in reinforcement learning. PhD thesis, University of Massachusetts Amherst (1984)
67. Sutton, R.S., Barto, A.G.: Reinforcement Learning: An Introduction, 2nd edn. MIT Press, Cambridge (2018)
68. Sutton, R.S., Precup, D., Singh, S.P.: Between MDPs and Semi-MDPs: a framework for temporal abstraction in reinforcement learning. Artif. Intell. **112**(1–2), 181–211 (1999)

69. Thompson, J.D., Higgins, D.G., Gibson, T.J.: CLUSTAL W: improving the sensitivity of progressive multiple sequence alignment through sequence weighting, position-specific gap penalties and weight matrix choice. Nucleic Acids Res. **22**(22), 4673–4680 (1994)
70. Torabi, F., Warnell, G., Stone, P.: Behavioral cloning from observation (2018)
71. Vaswani, A., et al.: Attention is all you need. In: Advances in Neural Information Processing Systems, vol. 30, pp. 5998–6008. Curran Associates Inc. (2017)
72. Vilone, G., Longo, L.: Explainable artificial intelligence: a systematic review. arXiv, abs/2006.00093 (2020)
73. Vinyals, O., et al.: Grandmaster level in StarCraft II using multi-agent reinforcement learning. Nature **575**(7782), 350–354 (2019)
74. Watkins, C.J.C.H.: Learning from delayed rewards. Ph.D. thesis, King's College (1989)
75. Wei, D., Dash, S., Gao, T., Gunluk, O.: Generalized linear rule models. In: Chaudhuri, K., Salakhutdinov, R. (eds.) Proceedings of the 36th International Conference on Machine Learning, Volume 97 of Proceedings of Machine Learning Research, pp. 6687–6696. PMLR, 09–15 June 2019

Interpretable, Verifiable, and Robust Reinforcement Learning via Program Synthesis

Osbert Bastani[1(✉)], Jeevana Priya Inala[2], and Armando Solar-Lezama[3]

[1] University of Pennsylvania, Philadelphia, PA 19104, USA
obastani@seas.upenn.edu
[2] Microsoft Research, Redmond, WA 98052, USA
jinala@microsoft.com
[3] Massachusetts Institute of Technology, Cambridge, MA 02139, USA
asolar@csail.mit.edu

Abstract. Reinforcement learning is a promising strategy for automatically training policies for challenging control tasks. However, state-of-the-art deep reinforcement learning algorithms focus on training deep neural network (DNN) policies, which are black box models that are hard to interpret and reason about. In this chapter, we describe recent progress towards learning policies in the form of programs. Compared to DNNs, such *programmatic policies* are significantly more interpretable, easier to formally verify, and more robust. We give an overview of algorithms designed to learn programmatic policies, and describe several case studies demonstrating their various advantages.

Keywords: Interpretable reinforcement learning · Program synthesis

1 Introduction

Reinforcement learning is a promising strategy for learning control policies for challenging sequential decision-making tasks. Recent work has demonstrated its promise in applications including game playing [34,43], robotics control [14,31], software systems [13,30], and healthcare [6,37]. A typical strategy is to build a high-fidelity simulator of the world, and then use reinforcement learning to train a control policy to act in this environment. This policy makes decisions (e.g., which direction to walk) based on the current state of the environment (e.g., the current image of the environment captured by a camera) to optimize the cumulative reward (e.g., how quickly the agent reaches its goal).

There has been significant recent progress on developing powerful *deep* reinforcement learning algorithms [33,41], which train a policy in the form of a deep neural network (DNN) by using gradient descent on the DNN parameters to optimize the cumulative reward. Importantly, these algorithms treat the underlying environment as a black box, making them very generally applicable.

© The Author(s) 2022
A. Holzinger et al. (Eds.): xxAI 2020, LNAI 13200, pp. 207–228, 2022.
https://doi.org/10.1007/978-3-031-04083-2_11

A key challenge in many real-world applications is the need to ensure that the learned policy continues to act correctly once it is deployed in the real world. However, DNN policies are typically very difficult to understand and analyze, making it hard to make guarantees about their performance. The reinforcement learning setting is particularly challenging since we need to reason not just about isolated predictions but about sequences of highly connected decisions.

As a consequence, there has been a great deal of recent interest in learning policies in the form of programs, called *programmatic policies*. Such policies include existing interpretable models such as decision trees [9], which are simple programs composed of if-then-else statements, as well as more complex ones such as state machines [26] and list processing programs [27,50]. In general, programs have been leveraged in machine learning to achieve a wide range of goals, such as representing high-level structure in images [16,17,25,46,47,53] and classifying sequence data such as trajectories or text [12,42].

Programmatic policies have a number of advantages over DNN policies that make it easier to ensure they act correctly. For instance, programs tend to be significantly more interpretable than DNNs; as a consequence, human experts can often understand and debug behaviors of a programmatic policy [26,27,50]. In addition, in contrast to DNNs, programs have discrete structure, which make them much more amenable to formal verification [3,9,39], which can be used to prove correctness properties of programmatic policies. Finally, there is evidence that programmatic policies are more robust than their DNN counterparts—e.g., they generalize better to changes in the task or robot configuration [26].

A key challenge with learning programmatic policies is that state-of-the-art reinforcement learning algorithms cannot be applied. In particular, these algorithms are based on the principle of gradient descent on the policy parameters, yet programmatic policies are typically non-differentiable (or at least, their optimization landscape contains many local minima). As a consequence, a common strategy to learning these policies is to first learn the DNN policy using deep reinforcement learning, and then using imitation learning to compress the DNN into a program. Essentially, this strategy reduces the reinforcement learning problem for programmatic policies into a supervised learning problem, for which efficient algorithms often exist—e.g., based on program synthesis [21]. A refinement of this strategy is to adaptively update the DNN policy to mirror the programmatic policy, which reduces the gap between the DNN and the program [26,49].

In this chapter, we provide an overview of recent progress in this direction. We begin by formalizing the reinforcement learning problem (Sect. 2); then, we describe interesting kinds of programmatic policies that have been studied (Sect. 3), algorithms for learning programmatic policies (Sect. 4), and case studies demonstrating the value of programmatic policies (Sect. 5).

2 Background on Reinforcement Learning

We consider a reinforcement learning problem formulated as a *Markov decision process (MDP)* $M = (S, A, P, R)$ [36], where S is the set of states, A is the set of actions, $P(s' \mid a, s) \in [0, 1]$ is the probability of transitioning from state $s \in S$

to state $s' \in S$ upon taking action $a \in A$, and $R(s,a) \in \mathbb{R}$ is the reward accrued by taking action a in state s.

Given an MDP M, our goal is to train an agent that acts in M in a way that accrues high cumulative reward. We represent the agent as a policy $\pi : S \to A$ mapping states to actions. Then, starting from a state $s \in S$, the agent selects action $a = \pi(s)$ according to the policy, observes a reward $R(s,a)$, transitions to the next state $s' \sim P(\cdot \mid s,a)$, and then iteratively continues this process starting from s'. For simplicity, we assume that a deterministic initial state $s_1 \in S$ along with a fixed, finite number of steps $H \in \mathbb{N}$. Then, we formalize the trajectory taken by the agent as a *rollout* $\zeta \in (S \times A \times \mathbb{R})^H$, which is a sequence of state-action-reward tuples $\zeta = ((s_1, a_1, r_1), ..., (s_H, a_H, r_H))$. We can sample a rollout by taking $r_t = R(s_t, a_t)$ and $s_{t+1} \sim P(\cdot \mid s_t, a_t)$ for each $t \in [H] = \{1, ..., H\}$; we let $D^{(\pi)}(\zeta)$ denote the distribution over rollouts induced by using policy π.

Now, our goal is to choose a policy $\pi \in \Pi$ in a given class of policies Π that maximizes the expected reward accrued. In particular, letting $J(\zeta) = \sum_{t=1}^{H} r_t$ be the cumulative reward of rollout ζ, our goal is to compute

$$\hat{\pi} = \arg\max_{\pi \in \Pi} J(\pi) \qquad \text{where} \qquad J(\pi) = \mathbb{E}_{\zeta \sim D^{(\pi)}}[J(\zeta)],$$

i.e., the policy $\pi \in \Pi$ that maximizes the expected cumulative reward over the induced distribution of rollouts $D^{(\pi)}(\zeta)$.

As an example, we can model a robot navigating a room to reach a goal as follows. The state $(x, y) \in S = \mathbb{R}^2$ represents the robot's position, and the action $(v, \phi) \in A = \mathbb{R}^2$ represents the robot's velocity v and direction ϕ. The transition probabilities are $P(s' \mid s, a) = \mathcal{N}(f(s, a), \Sigma)$, where

$$f((x, y), (v, \phi)) = (x + v \cdot \cos \phi \cdot \tau, y + v \cdot \sin \phi \cdot \tau),$$

where $\tau \in \mathbb{R}_{>0}$ is the time increment, and where $\Sigma \in \mathbb{R}^{2 \times 2}$ is the variance in the state transitions due to stochastic perturbations. Finally, the rewards are the distance to the goal—i.e., $R(s, a) = -\|s - g\|_2 + \lambda \cdot \|a\|_2$, where $g \in \mathbb{R}^2$ is the goal and $\lambda \in \mathbb{R}_{>0}$ is a hyperparameter. Intuitively, the optimal policy $\hat{\pi}$ for this MDP takes actions in a way that maximizes the time the robot spends close to the goal g, while avoiding very large (and therefore costly) actions.

3   Programmatic Policies

The main difference in programmatic reinforcement learning compared to traditional reinforcement learning is the choice of policy class Π. In particular, we are interested in cases where Π is a space of programs of some form. In this section, we describe specific choices that have been studied.

3.1   Traditional Interpretable Models

A natural starting point is learning policies in the form of traditional interpretable models, including decision trees [10] and rule lists [52]. In particular, these models can be thought of as simple programs composed of simple primitives such as if-then-else rules and arithmetic operations. For example, in Fig. 1, we

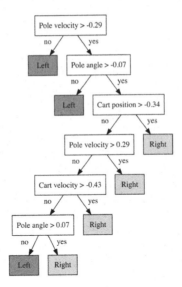

Fig. 1. A decision tree policy trained to control the cart-pole model; it achieves near-perfect performance. Adapted from [9].

show an example of a decision tree policy trained to control the cart-pole robot, which consists of a pole balanced on a cart and the goal is to move the cart back and forth to keep the pole upright [11]. Here, the state consists of the velocity and angle of each the cart and the pole (i.e., $S \subseteq \mathbb{R}^4$), and the actions are to move the cart left or right (i.e., $A = \{\text{left}, \text{right}\}$). As we discuss in Sect. 5, these kinds of policies provide desirable properties such as interpretability, robustness, and verifiability. A key shortcoming is that they have difficulty handling more complex inputs, e.g., sets of other agents, sequences of observations, etc. Thus, we describe programs with more sophisticated components below.

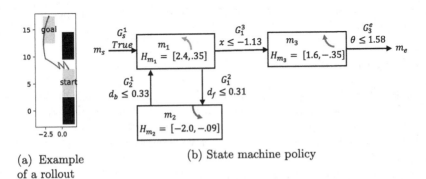

(a) Example of a rollout

(b) State machine policy

Fig. 2. (a) A depiction of the task, which is to drive the blue car (the agent) out from between the two stationary black cars. (b) A state machine policy trained to solve this task. Adapted from [26]. (Color figure online)

3.2   State Machine Policies

A key shortcoming of traditional interpretable models is that they do not possess internal state—i.e., the policy cannot propagate information about the current time step to the next time step. In principle, for an MDP, keeping internal state is not necessary since the state variable contains all information necessary to act optimally. Nevertheless, in many cases, it can be helpful for the policy to keep internal state—for instance, for motions such as walking or swimming that repeat iteratively, it can be helpful to internally keep track of progress within the current iteration. In addition, if the state is partially observed (i.e., the policy only has access to $o = h(s)$ instead of the full state s), then internal state may be necessary to act optimally [28]. In the context of deep reinforcement learning, recurrent neural networks (RNNs) can be used to include internal state [23].

For programmatic policies, a natural analog is to use polices based on finite-state machines. In particular, *state machine policies* are designed to be interpretable while including internal state [26]. Its internal state records one of a finite set of possible *modes*, each of which is annotated with (i) a simple policy for choosing the action when in this mode (e.g., a linear function of the state), and (ii) rules for when to transition to the next mode (e.g., if some linear inequality becomes satisfied, then transition to a given next mode). These policies are closely related to *hybrid automata* [2,24], which are models of a subclass of dynamical systems called *hybrid systems* that include both continuous transitions (modeled by differential equations) and discrete, discontinuous ones (modeled by a finite-state machine). In particular, the closed-loop system consisting of a state-machine policy controlling a hybrid system is also a hybrid system.

As an example, consider Fig. 2; the blue car (the agent) is parked between two stationary black cars, and its goal is to drive out of its parking spot into the goal position while avoiding collisions. The state is $(x, y, \theta, d) \in \mathbb{R}^4$, where (x, y) is the center of the car, θ is its orientation, and d is the distance between the two black cars. The actions are $(v, \psi) \in \mathbb{R}^2$, where v is the velocity and ψ is the steering angle. The transitions are the standard bicycle dynamics [35].

In Fig. 2b, we show the state machine policy synthesized by our algorithm for this task. We use d_f and d_b to denote the distances between the agent and the front and back black cars, respectively. This policy has three different modes (besides a start mode m_s and an end mode m_e). Roughly speaking, it says (i) immediately shift from mode m_s to m_1, and drive the car forward and to the left, (ii) continue until close to the car in front; then, transition to mode m_2, and drive the car backwards and to the right, (iii) continue until close to the car behind; then, transition back to mode m_1, (iv) iterate between m_1 and m_2 until the car can safely exit the parking spot; then, transition to mode m_3, and drive forward and to the right to make the car parallel to the lane.

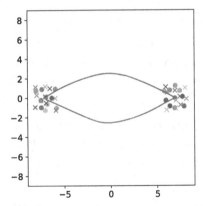

Fig. 3. Two groups of agents (red vs. blue) at their initial positions (circles) trying to reach their goal positions (crosses). The solid line shows the trajectory taken by a single agent in each group. (Color figure online)

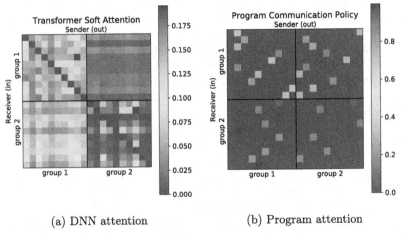

(a) DNN attention   (b) Program attention

$$R_1 : \mathsf{argmax}(\mathsf{map}(-d^{i,j}, \mathsf{filter}(\theta^{i,j} \geq -1.85, \ell))),$$
$$R_2 : \mathsf{random}(\mathsf{filter}(d^{i,j} \geq 3.41, \ell)).$$

(c) Programmatic attention rules

Fig. 4. (a) Soft attention computed by a DNN for the agent along the y-axis deciding whether to focus on the agent along the x-axis. (b) Sparse attention computed by a program. (c) Program used by each agent to select other agents to focus on. Adapted from [27].

3.3   List Processing Programs

Another kind of programmatic policy is list processing programs, which are compositions of components designed to manipulate lists—e.g., the map, filter, and fold operators [18]; the set of possible components can be chosen based on the application. In contrast to state machine policies, list processing programs are designed to handle situations where the state includes lists of elements. For example, in multi-agent systems, the full state consists of a list of states for each individual agent [27]. In this case, the program must compute a single action based on the given list of states. Alternatively, for environments with variable numbers of objects, the set of object positions must be encoded as a list. Finally, they can also be used to choose actions based on the history of the previous k states [50], which achieves a similar goal as state machine policies.

As an example, consider the task in Fig. 3, where agents in group 1 (blue) are navigating from the left to their goal on the right, while agents in group 2 (red) are navigating from the right to their goal on the left. The system state $s \in \mathbb{R}^{2k}$ is a list containing the position (x_i, y_i) of each agent of the k agents. An action $a \in \mathbb{R}^{2k}$ consists of the velocities (v_i, w_i) to be applied by each agent. We consider a strategy where we use a single policy $\pi : S \times [k] \rightarrow \mathbb{R}^2$, which takes as input the system state along with the index of the current agent $i \in [k]$, and produces the action $\pi(s, i)$ to be taken by agent i. This policy is applied to each agent to construct the full list of actions.

To solve this task, each agent must determine which agents to focus on; in the example in Fig. 3, it is useful to attend to the closest neighbor in the same group (to avoid colliding with them), as well as with an arbitrary agent from the opposite group (to coordinate so their trajectories do not collide).

For now, we describe programmatic policies for each agent designed to select a small number of other agents to focus on. This list of agents can in principle be processed by a second programmatic policy to determine the action to choose; however, in Sect. 3.4, we describe a strategy that combines them with a neural network policy to select actions. Figure 4c shows an example of a programmatic policy that each agent can use to choose other agents to focus on for the task in Fig. 3. This program consists of two rules, each of which selects a single agent to focus on; the program returns the set consisting of both selected agents. In each of these rules, agent i is selecting over other agents j in the list ℓ; $d^{i,j}$ is the distance between them and $\theta^{i,j}$ is the angle between them. Intuitively, rule R_1 chooses the nearest other agent j such that $\theta^{i,j} \in [-1.85, \pi]$, which is likely an agent in the same group as agent i that is directly in front of agent i; thus, agent i needs to focus on it to avoid colliding into it. In contrast, R_2 chooses a random agent from the agents that are far away, which is likely an agent in the other group; thus, agent i can use this information to avoid the other group.

3.4   Neurosymbolic Policies

In some settings, we want part of the policy to be programmatic, but other parts of the policy to be DNNs. We refer to policies that combine programs and DNNs as *neurosymbolic policies*. Intuitively, the program handles part of the computation that we would like to be interpretable, whereas the DNN handles the remainder of the computation (potentially the part that cannot be easily approximated by an interpretable model).

One instance of this strategy is to leverage programs as the attention mechanism for a transformer model [27]. At a high level, a transformer [48] is a DNN that operates on a list of inputs. These models operate by first choosing a small subset of other elements of the list to focus on (the attention layer), then uses a fully-connected layer to decide what information from the other agents is useful (the value layer), and finally uses a second fully-connected layer to compute the result (output layer). For example, transformers can be applied to multi-agent systems since it has to reason over the list of other agents.

A neurosymbolic transformer is similar to a transformer but uses programmatic policies for the attention layer; the value layer and the output layer are still neural networks. This architecture makes the attention layer interpretable—e.g., it is easy to understand and visualize why an agent attends to another agent, while still retaining much of the complexity of the original transformer.

For example, the program shown in Fig. 4c can be used to select other agents to attend to in a neurosymbolic transformer; unlike a DNN attention layer, this program is interpretable. An added advantage is that the program produces sparse attention weights; in contrast, a DNN attention layer produces soft attention weights, so every agent needs to attend to every other agent, even if the attention weight is small. Figure 4a shows the soft attention computed by a DNN, and Fig. 4b shows the sparse attention computed by a program.

4   Synthesizing Programmatic Policies

Next, we describe our algorithms for training programmatic policies. We begin by describing the general strategy of first training a deep neural network (DNN) policy using deep reinforcement learning, and then using imitation learning in conjunction with the DNN policy to reduce the reinforcement learning problem for programmatic policies to a supervised learning problem (Sect. 4.1 and 4.2). Then, we describe a refinement of this strategy where the DNN is adaptively updated to better mirror the current programmatic policy (Sect. 4.3). Finally, all of these strategies rely on a subroutine for solving the supervised learning problem; we briefly discuss approaches to doing so (Sect. 4.4).

4.1   Imitation Learning

We focus on the setting of continuous state and action spaces (i.e., $S \subseteq \mathbb{R}^n$ and $A \subseteq \mathbb{R}^m$), but our techniques are applicable more broadly. A number of algorithms have been proposed for computing optimal policies for a given MDP M and policy class Π [44]. For continuous state and action spaces, state-of-the-art deep reinforcement learning algorithms [33,41] consider a parameteric policy class $\Pi = \{\pi_\theta \mid \theta \in \Theta\}$, where the parameters $\Theta \subseteq \mathbb{R}^d$ are real-valued—e.g., π_θ is a DNN and θ are its parameters. Then, they compute π^* by optimizing over θ. One strategy is to use gradient descent on the objective—i.e.,

$$\theta' \leftarrow \theta + \eta \cdot \nabla_\theta J(\pi_\theta).$$

Algorithm 1. Training programmatic policies using imitation learning.

---

procedure IMITATIONLEARN(M, Q^*, m, n)
    Train oracle policy $\pi^* \leftarrow \text{TrainDNN}(M)$
    Initialize training dataset $Z \leftarrow \varnothing$
    Initialize programmatic policy $\hat{\pi}_0 \leftarrow \pi^*$
    for $i \in \{1, ..., n\}$ **do**
        Sample m trajectories to construct $Z_i \leftarrow \{(s, \pi^*(s)) \sim D^{(\hat{\pi}_{i-1})}\}$
        Aggregate dataset $Z \leftarrow Z \cup Z_i$
        Train programmatic policy $\hat{\pi}_i \leftarrow \text{TrainProgram}(Z)$
    end for
    return Best policy $\hat{\pi} \in \{\hat{\pi}_1, ..., \hat{\pi}_N\}$ on cross validation
end procedure

In particular, the policy gradient theorem [45] encodes how to compute an unbiased estimator of this objective in terms of $\nabla_\theta \pi_\theta$. In general, most state-of-the-art approaches rely on gradient descent on the policy parameters θ. However, such approaches cannot be applied to training programmatic policies, since the search space of programs is typically discrete.

Instead, a general strategy is to use imitation learning to reduce the reinforcement learning problem to a supervised learning problem. At a high level, the idea is to first use deep reinforcement learning to learn an high-performing DNN policy π^*, and then train the programmatic policy $\hat{\pi}$ to imitate π^*.

A naïve strategy is to use an imitation learning algorithm called behavioral cloning [4], which uses π^* to explore the MDP, collects state-action pairs $Z = \{(s, a)\}$ pairs occurring in rollouts $\zeta \sim D^{(\pi^*)}$, and then trains $\hat{\pi}$ using supervised learning on the dataset Z—i.e.,

$$\hat{\pi} = \arg\min_{\pi \in \Pi} \sum_{(s,a) \in Z} \mathbb{1}(\pi(s) = a). \tag{1}$$

Intuitively, the key shortcoming with this approach is that if $\hat{\pi}$ makes a mistake compared to the DNN policy π^*, then it might reach a state s that is very different from the states in the dataset Z. Thus, $\hat{\pi}$ may not know the correct action to take in state s, leading to poor performance. As a simple example, consider a self-driving car, and suppose π^* drives perfectly in the center of lane, whereas $\hat{\pi}$ deviates slightly from the center early in the rollout. Then, it reaches a state never seen in the training data Z, which means $\hat{\pi}$ does not know how to act in this state, so it may deviate further.

State-of-the-art imitation learning algorithms are designed to avoid these issues. One simple but effective strategy is the Dataset Aggregation (DAGGER) algorithm [38], which iteratively retrains the programmatic policy based on the distribution of states it visits. The first iteration is the same as behavioral cloning; in particular, it generates an initial dataset Z_0 using π^* and trains an initial programmatic policy $\hat{\pi}_0$. In each subsequent iteration i, it generates a dataset Z_i using the previous programmatic policy $\hat{\pi}_{i-1}$, and then trains $\hat{\pi}_i$ on Z_i.

This strategy is summarized in Algorithm 1; it has been successfully leveraged to train programmatic policies to solve reinforcement learning problems [50].

4.2 Q-Guided Imitation Learning

One shortcoming of Algorithm 1 is that it does not account for the fact that certain actions are more important than others [9]. Instead, the loss function in Eq. 1 treats all state-action pairs in the dataset Z as being equally important. However, in practice, one state-action pair (s, a) may be significantly more consequential than another one (s', a')—i.e., making a mistake $\hat{\pi}(s) \neq a$ might degrade performance by significantly more than a mistake $\hat{\pi}(s') \neq a$.

For example, consider the toy game of Pong in Fig. 8; the goal is to move the paddle to prevent the ball from exiting the screen. Figure 5a shows a state where the action taken is very important; the paddle must be moved to the right, or else the ball cannot be stopped from exiting. In contrast, Fig. 5b shows a state where the action taken is unimportant. Ideally, our algorithm would upweight the former state-action pair and downweight the latter.

One way to address this issue is by leveraging the Q-function, which measures the quality of a state-action pair—in particular, $Q^{(\pi)}(s, a) \in \mathbb{R}$ is the cumulative reward accrued by taking action a in state s, and then continuing with policy π. Traditional imitation learning algorithms do not have access to $Q^{(\pi^*)}$, since π^* is typically a human expert, and it would be difficult to elicit these values. However, the Q function $Q^{(\pi^*)}$ for the DNN policy π^* is computed as a byproduct of many deep reinforcement learning algorithms, so it is typically available in our setting. Given $Q^{(\pi^*)}$, a natural alternative to Eq. 1 is

$$\hat{\pi} = \arg\min_{\pi \in \Pi} \sum_{(s,a) \in Z} (Q^{(\pi^*)}(s, a) - Q^{(\pi^*)}(s, \hat{\pi}(s))). \tag{2}$$

Intuitively, the term $Q^{(\pi^*)}(s, a) - Q^{(\pi^*)}(s, \hat{\pi}(s))$ measures the degradation in performance by taking the incorrect action $\hat{\pi}(s)$ instead of a. Indeed, it can be proven that this objective exactly encodes the gap in performance between $\hat{\pi}$ and π^*—i.e., in the limit of infinite data, it is equivalent to computing

$$\hat{\pi} = \arg\min_{\pi \in \Pi} \{J(\pi^*) - J(\hat{\pi})\}.$$

Finally, a shortcoming of Eq. 2 is that it is not a standard supervised learning problem. To address this issue, we can instead optimize the lower bound

$$Q^{(\pi^*)}(s, a) - Q^{(\pi^*)}(s, \hat{\pi}(s)) \leq \left(Q^{(\pi^*)}(s, a) - \arg\min_{a' \in A} Q^{(\pi^*)}(s, a') \right) \cdot \mathbb{1}(\hat{\pi}(s) = a),$$

which yields the optimization problem

$$\hat{\pi} = \arg\min_{\pi \in \Pi} \sum_{(s,a) \in Z} \left(Q^{(\pi^*)}(s, a) - \arg\min_{a' \in A} Q^{(\pi^*)}(s, a') \right) \cdot \mathbb{1}(\hat{\pi}(s) = a).$$

This strategy is proposed in [9] and shown to learn significantly more compact policies compared to the original DAGGER algorithm.

4.3 Updating the DNN Policy

Another shortcoming of Algorithm 1 is that it does not adjust the DNN policy π^* to account for limitations on the capabilities of the programmatic policy $\hat{\pi}$. Intuitively, if $\hat{\pi}$ cannot accurately approximate π^*, then π^* may suggest actions that lead to states where $\hat{\pi}$ cannot perform well, even if π^* performs well in these states. There has been work on addressing this issue. For example, *coaching* can be used to select actions that are more suitable for $\hat{\pi}$ [22]. Alternatively, π^* can be iteratively updated using gradient descent to better reflect $\hat{\pi}$ [49].

A related strategy is *adaptive teaching*, where rather than choosing π^* to be a DNN, it is instead a policy whose structure mirrors that of $\hat{\pi}$ [26]. In this case, we can directly update π^* on each training iteration to reflect the structure of $\hat{\pi}$. As an example, in the case of state machine policies, π^* can be chosen to be a "loop-free" policy, which consists of a linear sequence of modes. These modes can then be mapped to the modes of $\hat{\pi}$, and regularized so that their local policies and mode transitions mirror that of $\hat{\pi}$. Adaptive teaching has been shown to be an effective strategy for learning state machine policies [26].

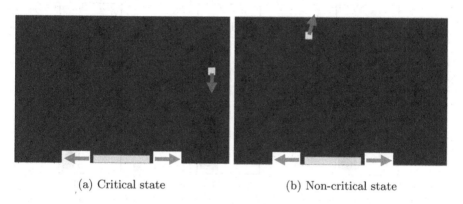

(a) Critical state   (b) Non-critical state

Fig. 5. A toy game of Pong. The paddle is the gray bar at the bottom, and the ball is the gray square. The red arrow shows the direction the ball is traveling, and the blue arrows show the possible actions (move paddle left vs. right). We show examples where the action taken in this state (a) does, and (b) does not significantly impact the cumulative reward accrued from this state. (Color figure online)

4.4 Program Synthesis for Supervised Learning

Recall that imitation learning reduces the reinforcement learning problem for programmatic policies to a supervised learning problem. We briefly discuss algorithms for solving this supervised learning problem. In general, this problem is an instance of *programming by example* [19,20], which is a special case of program synthesis [21] where the task is specified by a set of input-output examples. In our setting, the input-output examples are the state-action pairs in the dataset Z used to train the programmatic policy at each iteration of Algorithm 1.

An added challenge applying program synthesis in machine learning settings is that traditional programming by example algorithms are designed to compute a program that correctly fits *all* of the training examples. In contrast, in machine learning, there typically does not exist a single program that fits all of the training examples. Instead, we need to solve a *quantitative* synthesis problem where the goal is to minimize the number of errors on the training data.

One standard approach to solving such program synthesis problems is to simply enumerate over all possible programmatic policies $\pi \in \Pi$. In many cases, Π is specified as a context-free grammar, in which case standard algorithms can be used to enumerate programs in that grammar (typically up to a bounded depth) [5]. In addition, domain-specific techniques can be used to prune provably suboptimal portions of the search space to speed up enumeration [12]. For particularly large search spaces, an alternative strategy is to use a stochastic search algorithm that heuristically optimizes the objective; for example, Metropolis Hastings can be used to adaptively sample programs (e.g., with the unnormalized probability density function taken to be the objective value) [27,40].

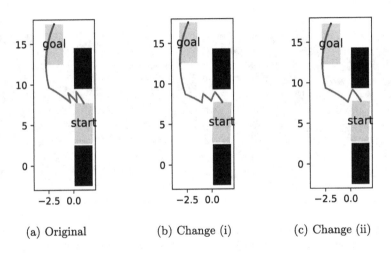

(a) Original   (b) Change (i)   (c) Change (ii)

Fig. 6. A human expert can modify our state machine policy to improve performance. (a) A trajectory using the original state machine policy shown in Fig. 2(b). (b) The human expert sets the steering angle to the maximum value 0.5. (c) The human expert sets the thresholds in the mode transitions so the blue car drives as close to the black cars as possible. Adapted from [26]. (Color figure online)

5   Case Studies

In this section, we describe a number of case studies that demonstrate the value of programmatic policies, demonstrating their interpretability (Sect. 5.1), verifiability (Sect. 5.2), and robustness (Sect. 5.3).

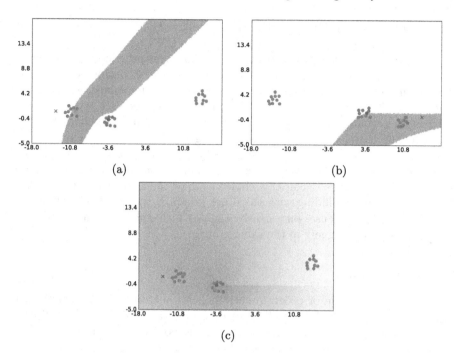

Fig. 7. Visualization of the programmatic attention layer in Fig. 4c, which has two rules R_1 and R_2. In this task, there are three groups of agents. The red circle denotes the agent currently choosing an action, the red cross denotes its goal, and the green circle denotes the agent selected by the rule. (a, b) Visualization of rule R_1 for two different states; orange denotes the region where the filter condition is satisfied—i.e., R_1 chooses a random agent in this region. (c) Visualization of rule R_2, showing the score output by the map operator; darker values are higher—i.e., the rule chooses the agent with the darkest value. Adapted from [27]. (Color figure online)

5.1   Interpretability

A key advantage of programmatic policies is that they are interpretable [26, 27, 50]. One consequence of their interpretability is that human experts can examine programmatic policies and modify them to improve performance. As an example, consider the state machine policy shown in Fig. 2b in Sect. 3. We have manually made the following changes to this policy: (i) increase the steering angle in mode m_1 to its maximum value 0.5 (so the car steers as much as possible when exiting the parking spot), and (ii) decrease the gap maintained between the agent and the black cars by changing the condition for transitioning from mode m_1 to mode m_2 to $d_f \leq 0.1$, and from mode m_2 to mode m_1 to $d_b \leq 0.1$ (so the blue car drives as far as possible without colliding with a black car before changing directions). Figure 6 visualizes the effects changes; in particular, it shows trajectories obtained using the original policy, the policy with change (i), and the policy with change (ii). As can be seen, the second modified policy exits the parking spot more quickly than the original policy. There is no straightforward way to make these kinds of changes to improve a DNN policy.

Similarly, we describe how it is possible to interpret programmatic attention layers in neurosymbolic transformers. In particular, Fig. 7 visualizes the synthesized programmatic attention policy described in Sect. 3 for a multi-agent control problem; in this example, there are three groups of agents, each trying to move towards their goals. Figures 7a & 7b visualize rule R_1 in two different states. In particular, R_1 selects a random far-away agent in the orange region to focus on. Note that in both states, the orange region is in the direction of the goal of the agent. Intuitively, the agent is focusing on an agent in the other group that is between itself and the goal; this choice enables the agent to plan a path to its goal that avoids colliding with the other group. Next, Fig. 7c visualizes rule R_2; this rule simply focuses on a nearby agent, which enables the agent to avoid collisions with other agents in the same group.

5.2   Verification

Another key advantage of programmatic policies is that they are significantly easier to formally verify. Intuitively, because they make significant use of discrete control flow structures, it is easier for formal methods to prune branches of the search space corresponding to unreachable program paths.

Verification is useful when there is an additional safety constraint that must be satisfied by the policy in addition to maximizing cumulative reward. A common assumption is that the agent should remain in a safe subset of the state space $S_{\text{safe}} \subseteq S$ during the entire rollout. Furthermore, in these settings, it is often assumed that the transitions are deterministic—i.e., the next state is $s' = f(s, a)$ for some deterministic transition function $f : S \times A \to S$. Finally, rather than considering a single initial state, we instead consider a subset of initial states $S_1 \subseteq S_{\text{safe}}$. Then, we consider the safety constraint that for any rollout ζ starting from $s_1 \in S_1$, we have $s_t \in S_{\text{safe}}$ for all $t \in [H]$; we use $\phi(\pi) \in \{\text{true}, \text{false}\}$ to indicate whether a given policy π satisfies this constraint. Our goal is to solve

$$\pi^* = \arg\max_{\pi \in \Pi_{\text{safe}}} J(\pi) \qquad \text{where} \qquad \Pi_{\text{safe}} = \{\pi \in \Pi \mid \phi(\pi)\}.$$

A standard strategy for verifying safety is to devise a logical formula that encodes a safe rollout; in particular, we can encode our safety constraint as follows:

$$\phi(\pi) \equiv \forall \vec{s} \, . \, \left[(s_1 \in S_1) \wedge \bigwedge_{t=1}^{H} (a_t = \pi(s_t) \wedge s_{t+1} = f(s_t, a_t)) \right] \Rightarrow \bigwedge_{t=1}^{H} (s_t \in S_{\text{safe}}),$$

where $\vec{s} = (s_1, ..., s_H)$ are the free variables, and we use $\equiv$ to distinguish equality of logical formulas from equality of variables within a formula. Intuitively, this formula says that if (i) s_1 is an initial state, and (ii) the actions are chosen by π and the transitions by f, then all states are safe.

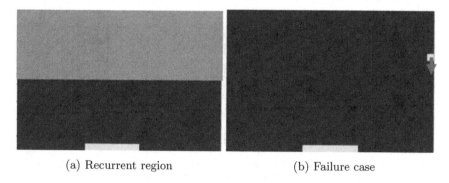

(a) Recurrent region   (b) Failure case

Fig. 8. (a) A recurrent region; proving that the ball always returns to this region implies that the policy plays correctly for an infinite horizon. (b) A failure case found by verification, where the paddle fails to keep the ball from exiting. (Color figure online)

With this expression for $\phi(\pi)$, to prove safety, it suffices to prove that $\neg\phi(\pi) \equiv$ false. The latter equivalence is an instance of Satisfiability Modulo Theory (SMT), and can automatically be checked by an SMT solver [15] as long as predicates of the form $s \in S_{\text{safe}}$, $a = \pi(s)$, and $s' = f(s, a)$ can be expressed in a theory that is supported by the SMT solver. A standard setting is where S_{safe} is a polytope, and π and f are piecewise affine; in these cases, each of these predicates can be expressed as conjunctions and disjunctions of linear inequalities, which are typically supported (e.g., the problem can be reduced to an integer program).

As an example, this strategy has been used to verify that the decision tree policy for a toy game of pong shown in Fig. 1 in Sect. 3 is correct—i.e., that it successfully blocks the ball from exiting. In this case, we can actually prove correctness over an infinite horizon. Rather than prove that the ball does not exit in H steps, we instead prove that for any state s_1 where the ball is in the top half of the screen (depicted in blue in Fig. 8a), the ball returns to this region after H steps. If this property is true, then the ball never exits the screen.

For this property, the SMT solver initially identified a failure case where the ball exits the screen, which is shown in Fig. 8b; in this corner case, the is at the very edge of the screen, and the paddle fails to keep the ball from exiting. This problem can be fixed by manually examining the decision tree and modifying it to correctly handle the failure case; the modified decision tree has been successfully proven to be correct—i.e., it always keeps the ball in the screen.

In another example, we used bounded verification to verify that the state machine policy in Fig. 2b does not result in any collisions for parallel parking task in Fig. 2a. We used dReach [29], an SMT solver designed to verify safety for hybrid systems, which are dynamical systems that include both continuous transitions (modeled using differential equations) and discrete, discontinuous ones (modeled using a finite-state machine). In particular, dReach performs bounded reachability analysis, where it unrolls the state machine modes up to

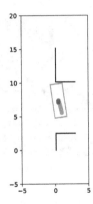

Fig. 9. A failure case found using our verification algorithm with tolerance parameter $\delta = 0.24$ for the state machine policy in Fig. 2b on the parallel parking task in Fig. 2a. Here, the car collides with the car in the front.

some bound. Furthermore, dReach is *sound* and δ-*complete*—i.e., if it says the system is safe, then it is guaranteed to be safe, and if it says the system is unsafe, then there exists some δ-bounded perturbation that renders the system unsafe. Thus, we can vary δ to quantify the robustness of the system to perturbations.

With $\delta = 0.1$, dReach proved that the policy in Fig. 2b is indeed safe for up to an unrolling of 7 modes of the state machine, which was enough for the controller to complete the task from a significant fraction of the initial state space. However, with $\delta = 0.24$, dReach identified a failure case where the car would collide with the car in the front (under some perturbations of the original model); this failure case is shown in Fig. 9. We manually fixed this problem by inspecting the state machine policy in Fig. 2b and modifying the switching conditions $G_{m_1}^{m_2}$ and $G_{m_2}^{m_1}$ to $d_f \leq 0.5$ and $d_b \leq 0.5$, respectively. With these changes, dReach proved that the policy is safe for $\delta = 0.24$.

More generally, similar strategies can be used to verify robustness and stability of programmatic controllers [9,50]. It can also be extended to compute regions of attraction—for instance, to show that a decision tree policy provably stabilizes a pendulum to the origin [39]. To improve performance, one strategy is to compose a provably safe programmatic policy with a higher performing but potentially unsafe DNN policy using *shielding* [1,7,8,32,51]; intuitively, this strategy uses the DNN policy as long as the programmatic policy can ensure safety. Finally, the techniques so far have focused on safety after training the policy; in some settings, it can be desirable to continue running reinforcement learning after deploying the policy to adapt to changing environments. To enable safety during learning, one strategy is to prove safety while accounting for uncertainty in the current model of the environment [3].

5.3   Robustness

Another advantage of programmatic policies is that they tend to be more robust than DNN policies—i.e., they generalize well to states outside of the distribution on which the policy was trained. For example, it has been shown that a programmatic policy trained to drive a car along one race track can generalize to other race tracks not seen during training, while DNN policies trained in the same way do not generalize as well [50]. We can formalize this notion by considering separate training and test distributions over tasks—e.g., the training distribution over tasks might include driving on just a single race track, whereas the test distribution includes driving on a number of additional race tracks. Then, a policy is robust if it performs well on the test distribution over tasks even when it is trained on the training distribution of tasks.

A special case is *inductive* generalization, where the tasks are indexed by natural numbers $i \in \mathbb{N}$, the training distribution is over small i, and the test distribution is over large i [26]. As a simple example, i may indicate the horizon over which the task is trained; then, a robust policy is one that is trained on short horizon tasks but generalizes to long horizon tasks.

Going back to the parallel parking task from Fig. 2 in Sect. 3; for this task, we can consider inductively generalization of a policy in terms of the number of back-and-forth motions needed to solve the task [26]. In particular, Figs. 10a, 10b, and 10c depict training tasks with relatively few back-and-forth motions, and Fig. 10d depicts a test task with a much larger number of back-and-forth motions. As shown in Fig. 10e, a DNN policy trained using deep reinforcement learning can solve additional tasks from the training distribution; however, Fig. 10f shows that this policy does not generalize to tasks from the test distribution. In contrast, a state machine policy performs well on both additional tasks from the training distribution (Fig. 10g) as well as tasks from the test distribution (Fig. 10h). Intuitively, the state machine policy is learning to the correct back-and-forth motion needed to solve the parallel parking problem. It can do so since (i) it is sufficiently expressive to represent the "correct" solution, yet (ii) it is sufficiently constrained that it learns a systematic policy. In contrast, the DNN policy can likely represent the correct solution, but because it is highly underconstrained, it finds an alternative solution that works on the training tasks, but does not generalize well to the test tasks. Thus, programmatic policies provide a promising balance between expressiveness and structure needed to solve challenging control tasks in a generalizable way.

For an illustration of these distinctions, we show the sequence of actions taken as a function of time by a programmatic policy compared to a DNN policy in Fig. 11. Here, the task is to fly a 2D quadcopter through an obstacle course by controlling its vertical acceleration. As can be seen, the state machine policy produces a smooth repeating pattern of actions; in contrast, the DNN policy acts highly erratically. This example further illustrates how programmatic policies are both complex (evidenced by the complexity of the red curve) yet structured (evidenced by the smoothness of the red curve and its repeating pattern). In contrast, DNN policies are expressive (as evidenced by the complexity of the red curve), but lack the structure needed to generalize robustly.

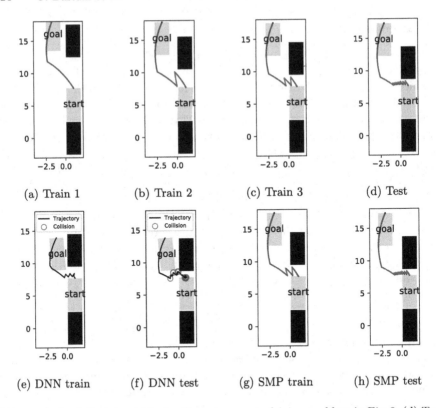

(a) Train 1   (b) Train 2   (c) Train 3   (d) Test

(e) DNN train   (f) DNN test   (g) SMP train   (h) SMP test

Fig. 10. (a, b, c) Training tasks for the autonomous driving problem in Fig. 2. (d) Test task, which is harder due to the increased number of back-and-forth motions required. (a) The trajectory taken by the DNN policy on a training task. (b) The trajectory taken by the DNN policy on a test task; as can be seen, it has several unsafe collisions. (c) The trajectory taken by the state machine policy (SMP) on a training task. (d) The trajectory taken by the SMP on a test task; as can be seen, it generalizes well to this task. Adapted from [26].

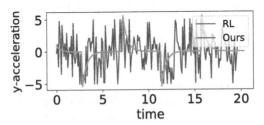

Fig. 11. The vertical acceleration (i.e., action) selected by the policy as a function of time, for each our programmatic policy (red) and a DNN policy (blue), for a 2D quadcopter task. Adapted from [26]. (Color figure online)

6   Conclusions and Future Work

In this chapter, we have describe an approach to reinforcement learning where we train programmatic policies such as decision trees, state machine policies, and list processing programs, instead of DNN policies. These policies can be trained using algorithms based on imitation learning, which first train a DNN policy using deep reinforcement learning and then train a programmatic policy to imitate the DNN policy. This strategy reduces the reinforcement learning problem to a supervised learning problem, that can be solved by existing algorithms such as program synthesis. Through a number of case studies, we have demonstrated that compared to DNN policies, programmatic policies are highly interpretable, are easier to formally verify, and generalized more robustly.

We leave a number of directions for future work. One important challenge is that synthesizing programmatic policies remains costly. Many state-of-the-art program synthesis algorithms rely heavily on domain-specific pruning strategies to improve performance, including strategies targeted at machine learning applications [12]. Leveraging these strategies can significantly increase the complexity of programmatic policies that can be learned in a tractable way.

Another interesting challenge is scaling verification algorithms to more realistic problems. The key limitation of existing approaches is that even if the programmatic policy has a compact representation, the model of the environment often does not. A natural question in this direction is whether we can learn programmatic models of the environment that are similarly easy to formally verify, while being a good approximation of the true environment.

Finally, we have described one strategy for constructing neurosymbolic policies that combine programs and DNNs—i.e., the neurosymbolic transformer. We believe a number of additional kinds of model compositions may be feasible—for example, leveraging a neural network to detect objects and then using a program to reason about them, or using programs to perform high-level reasoning such as path planning while letting a DNN policy take care of low-level control.

References

1. Alshiekh, M., Bloem, R., Ehlers, R., Könighofer, B., Niekum, S., Topcu, U.: Safe reinforcement learning via shielding. In: Thirty-Second AAAI Conference on Artificial Intelligence (2018)
2. Alur, R., Courcoubetis, C., Henzinger, T.A., Ho, P.-H.: Hybrid automata: an algorithmic approach to the specification and verification of hybrid systems. In: Grossman, R.L., Nerode, A., Ravn, A.P., Rischel, H. (eds.) HS 1991-1992. LNCS, vol. 736, pp. 209–229. Springer, Heidelberg (1993). https://doi.org/10.1007/3-540-57318-6_30
3. Anderson, G., Verma, A., Dillig, I., Chaudhuri, S.: Neurosymbolic reinforcement learning with formally verified exploration. In: Neural Information Processing Systems (2020)
4. Bain, M., Sammut, C.: A framework for behavioural cloning. In: Machine Intelligence 15, pp. 103–129 (1995)

5. Balog, M., Gaunt, A.L., Brockschmidt, M., Nowozin, S., Tarlow, D.: DeepCoder: learning to write programs. In: International Conference on Learning Representations (2017)

6. Bastani, H., et al.: Deploying an artificial intelligence system for COVID-19 testing at the greek border. Available at SSRN (2021)

7. Bastani, O.: Safe reinforcement learning with nonlinear dynamics via model predictive shielding. In: 2021 American Control Conference (ACC), pp. 3488–3494. IEEE (2021)

8. Bastani, O., Li, S., Xu, A.: Safe reinforcement learning via statistical model predictive shielding. In: Robotics: Science and Systems (2021)

9. Bastani, O., Pu, Y., Solar-Lezama, A.: Verifiable reinforcement learning via policy extraction. arXiv preprint arXiv:1805.08328 (2018)

10. Breiman, L., Friedman, J.H., Olshen, R.A., Stone, C.J.: Classification and Regression Trees. Routledge (2017)

11. Brockman, G., et al.: OpenAI gym. arXiv preprint arXiv:1606.01540 (2016)

12. Chen, Q., Lamoreaux, A., Wang, X., Durrett, G., Bastani, O., Dillig, I.: Web question answering with neurosymbolic program synthesis. In: Proceedings of the 42nd ACM SIGPLAN International Conference on Programming Language Design and Implementation, pp. 328–343 (2021)

13. Chen, Y., Wang, C., Bastani, O., Dillig, I., Feng, Yu.: Program synthesis using deduction-guided reinforcement learning. In: Lahiri, S.K., Wang, C. (eds.) CAV 2020. LNCS, vol. 12225, pp. 587–610. Springer, Cham (2020). https://doi.org/10.1007/978-3-030-53291-8_30

14. Collins, S., Ruina, A., Tedrake, R., Wisse, M.: Efficient bipedal robots based on passive-dynamic walkers. Science **307**(5712), 1082–1085 (2005)

15. de Moura, L., Bjørner, N.: Z3: an efficient SMT solver. In: Ramakrishnan, C.R., Rehof, J. (eds.) TACAS 2008. LNCS, vol. 4963, pp. 337–340. Springer, Heidelberg (2008). https://doi.org/10.1007/978-3-540-78800-3_24

16. Ellis, K., Ritchie, D., Solar-Lezama, A., Tenenbaum, J.B.: Learning to infer graphics programs from hand-drawn images. arXiv preprint arXiv:1707.09627 (2017)

17. Ellis, K., Solar-Lezama, A., Tenenbaum, J.: Unsupervised learning by program synthesis (2015)

18. Feser, J.K., Chaudhuri, S., Dillig, I.: Synthesizing data structure transformations from input-output examples. ACM SIGPLAN Not. **50**(6), 229–239 (2015)

19. Gulwani, S.: Automating string processing in spreadsheets using input-output examples. ACM Sigplan Not. **46**(1), 317–330 (2011)

20. Gulwani, S.: Programming by examples. Dependable Softw. Syst. Eng. **45**(137), 3–15 (2016)

21. Gulwani, S., Polozov, O., Singh, R., et al.: Program synthesis. Found. Trends® Program. Lang. **4**(1–2), 1–119 (2017)

22. He, H., Eisner, J., Daume, H.: Imitation learning by coaching. Adv. Neural. Inf. Process. Syst. **25**, 3149–3157 (2012)

23. Heess, N., Hunt, J.J., Lillicrap, T.P., Silver, D.: Memory-based control with recurrent neural networks. arXiv preprint arXiv:1512.04455 (2015)

24. Henzinger, T.A.: The theory of hybrid automata. In: Inan, M.K., Kurshan, R.P. (eds.) Verification of Digital and Hybrid Systems. NATO ASI Series, vol. 170, pp. 265–292. Springer, Berlin (2000). https://doi.org/10.1007/978-3-642-59615-5_13

25. Huang, J., Smith, C., Bastani, O., Singh, R., Albarghouthi, A., Naik, M.: Generating programmatic referring expressions via program synthesis. In: International Conference on Machine Learning, pp. 4495–4506. PMLR (2020)

26. Inala, J.P., Bastani, O., Tavares, Z., Solar-Lezama, A.: Synthesizing programmatic policies that inductively generalize. In: International Conference on Learning Representations (2020)

27. Inala, J.P., et al.: Neurosymbolic transformers for multi-agent communication. In: Neural Information Processing Systems (2020)

28. Kaelbling, L.P., Littman, M.L., Cassandra, A.R.: Planning and acting in partially observable stochastic domains. Artif. Intell. $101(1-2)$, 99–134 (1998)

29. Kong, S., Gao, S., Chen, W., Clarke, E.: dReach: δ-reachability analysis for hybrid systems. In: Baier, C., Tinelli, C. (eds.) TACAS 2015. LNCS, vol. 9035, pp. 200–205. Springer, Heidelberg (2015). https://doi.org/10.1007/978-3-662-46681-0_15

30. Kraska, T., et al.: SageDB: a learned database system. In: CIDR (2019)

31. Levine, S., Finn, C., Darrell, T., Abbeel, P.: End-to-end training of deep visuomotor policies. J. Mach. Learn. Res. $17(1)$, 1334–1373 (2016)

32. Li, S., Bastani, O.: Robust model predictive shielding for safe reinforcement learning with stochastic dynamics. In: 2020 IEEE International Conference on Robotics and Automation (ICRA), pp. 7166–7172. IEEE (2020)

33. Lillicrap, T.P., et al.: Continuous control with deep reinforcement learning. arXiv preprint arXiv:1509.02971 (2015)

34. Mnih, V., et al.: Human-level control through deep reinforcement learning. Nature $518(7540)$, 529–533 (2015)

35. Pepy, R., Lambert, A., Mounier, H.: Path planning using a dynamic vehicle model. In: 2006 2nd International Conference on Information & Communication Technologies, vol. 1, pp. 781–786. IEEE (2006)

36. Puterman, M.L.: Markov decision processes. Handb. Oper. Res. Manage. Sci. 2, 331–434 (1990)

37. Raghu, A., Komorowski, M., Celi, L.A., Szolovits, P., Ghassemi, M.: Continuous state-space models for optimal sepsis treatment: a deep reinforcement learning approach. In: Machine Learning for Healthcare Conference, pp. 147–163. PMLR (2017)

38. Ross, S., Gordon, G., Bagnell, D.: A reduction of imitation learning and structured prediction to no-regret online learning. In: Proceedings of the Fourteenth International Conference on Artificial Intelligence and Statistics, pp. 627–635. JMLR Workshop and Conference Proceedings (2011)

39. Sadraddini, S., Shen, S., Bastani, O.: Polytopic trees for verification of learning-based controllers. In: Zamani, M., Zufferey, D. (eds.) NSV 2019. LNCS, vol. 11652, pp. 110–127. Springer, Cham (2019). https://doi.org/10.1007/978-3-030-28423-7_8

40. Schkufza, E., Sharma, R., Aiken, A.: Stochastic superoptimization. ACM SIGARCH Comput. Archit. News $41(1)$, 305–316 (2013)

41. Schulman, J., Levine, S., Abbeel, P., Jordan, M., Moritz, P.: Trust region policy optimization. In: International Conference on Machine Learning, pp. 1889–1897. PMLR (2015)

42. Shah, A., Zhan, E., Sun, J.J., Verma, A., Yue, Y., Chaudhuri, S.: Learning differentiable programs with admissible neural heuristics. In: NeurIPS (2020)

43. Silver, D., et al.: Mastering the game of go with deep neural networks and tree search. Nature $529(7587)$, 484–489 (2016)

44. Sutton, R.S., Barto, A.G.: Reinforcement Learning: An Introduction. MIT Press, Cambridge (2018)

45. Sutton, R.S., McAllester, D.A., Singh, S.P., Mansour, Y.: Policy gradient methods for reinforcement learning with function approximation. In: Advances in Neural Information Processing Systems, pp. 1057–1063 (2000)

46. Tian, Y., et al.: Learning to infer and execute 3D shape programs. In: International Conference on Learning Representations (2018)
47. Valkov, L., Chaudhari, D., Srivastava, A., Sutton, C., Chaudhuri, S.: HOUDINI: lifelong learning as program synthesis. In: Proceedings of the 32nd International Conference on Neural Information Processing Systems, pp. 8701–8712 (2018)
48. Vaswani, A., et al.: Attention is all you need. In: Advances in Neural Information Processing Systems, pp. 5998–6008 (2017)
49. Verma, A., Le, H.M., Yue, Y., Chaudhuri, S.: Imitation-projected programmatic reinforcement learning. In: Neural Information Processing Systems (2019)
50. Verma, A., Murali, V., Singh, R., Kohli, P., Chaudhuri, S.: Programmatically interpretable reinforcement learning. In: International Conference on Machine Learning, pp. 5045–5054. PMLR (2018)
51. Wabersich, K.P., Zeilinger, M.N.: Linear model predictive safety certification for learning-based control. In: 2018 IEEE Conference on Decision and Control (CDC), pp. 7130–7135. IEEE (2018)
52. Wang, F., Rudin, C.: Falling rule lists. In: Artificial Intelligence and Statistics, pp. 1013–1022. PMLR (2015)
53. Young, H., Bastani, O., Naik, M.: Learning neurosymbolic generative models via program synthesis. In: International Conference on Machine Learning, pp. 7144–7153. PMLR (2019)

Interpreting and Improving Deep-Learning Models with Reality Checks

Chandan Singh, Wooseok Ha, and Bin Yu[✉]

University of California, Berkeley, Berkeley CA, USA
{cs1,haywse,binyu}@berkeley.edu

Abstract. Recent deep-learning models have achieved impressive predictive performance by learning complex functions of many variables, often at the cost of interpretability. This chapter covers recent work aiming to interpret models by attributing importance to features and feature groups for a single prediction. Importantly, the proposed attributions assign importance to interactions between features, in addition to features in isolation. These attributions are shown to yield insights across real-world domains, including bio-imaging, cosmology image and natural-language processing. We then show how these attributions can be used to directly improve the generalization of a neural network or to distill it into a simple model. Throughout the chapter, we emphasize the use of reality checks to scrutinize the proposed interpretation techniques. (Code for all methods in this chapter is available at ⟲github.com/csinva and ⟲github.com/Yu-Group, implemented in PyTorch [54]).

Keywords: Interpretability · Interactions · Feature importance · Neural network · Distillation

1 Interpretability: For What and For Whom?

Deep neural networks (DNNs) have recently received considerable attention for their ability to accurately predict a wide variety of complex phenomena. However, there is a growing realization that, in addition to predictions, DNNs are capable of producing useful information (i.e. interpretations) about domain

---

C. Singh and W. Ha—Equal contribution.

We gratefully acknowledge partial support from NSF TRIPODS Grant 1740855, DMS-1613002, 1953191, 2015341, IIS 1741340, ONR grant N00014-17-1-2176, the Center for Science of Information (CSoI), an NSF Science and Technology Center, under grant agreement CCF-0939370, NSF grant 2023505 on Collaborative Research: Foundations of Data Science Institute (FODSI), the NSF and the Simons Foundation for the Collaboration on the Theoretical Foundations of Deep Learning through awards DMS-2031883 and 814639, and a grant from the Weill Neurohub.

A. Holzinger et al. (Eds.): xxAI 2020, LNAI 13200, pp. 229–254, 2022.
https://doi.org/10.1007/978-3-031-04083-2_12

relationships contained in data. More precisely, interpretable machine learning can be defined as "the extraction of *relevant* knowledge from a machine-learning model concerning relationships either contained in data or learned by the model" [50].[1]

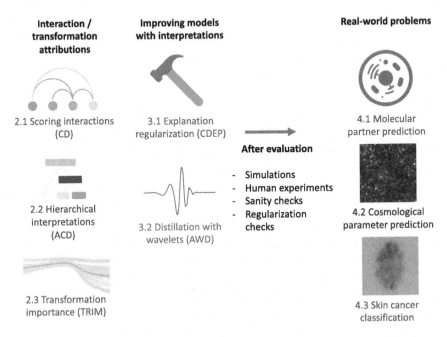

Interaction / transformation attributions	Improving models with interpretations	Real-world problems

2.1 Scoring interactions (CD)

3.1 Explanation regularization (CDEP)

4.1 Molecular partner prediction

After evaluation

2.2 Hierarchical interpretations (ACD)

3.2 Distillation with wavelets (AWD)

- Simulations
- Human experiments
- Sanity checks
- Regularization checks

4.2 Cosmological parameter prediction

2.3 Transformation importance (TRIM)

4.3 Skin cancer classification

Fig. 1. Chapter overview. We begin by defining interpretability and some of its desiderata, following [50] (Sect. 1). We proceed to overview different methods for computing interpretations for interactions/transformations (Sect. 2), including for scoring interactions [49], generating hierarchical interpretations [68], and calculating importances for transformations of features [67]. Next, we show how these interpretations can be used to improve models (Sect. 3), including by directly regularizing interpretations [60] and distilling a model through interpretations [31]. Finally, we show how these interpretations can be adapted to real-world applications (Sect. 4), including molecular partner prediction, cosmological parameter prediction, and skin-cancer classification.

Here, we view knowledge as being *relevant* if it provides insight for a particular audience into a chosen problem. This definition highlights that interpretability is poorly specified without the context of a particular audience and problem, and should be evaluated with the context in mind. This definition also implies that interpretable ML provides correct information (i.e. *knowledge*), and we use the term interpretation, assuming that the interpretation technique at

---

[1] We include different headings such as explainable AI (XAI), intelligible ML and transparent ML under this definition.

hand has passed some form of *reality check* (i.e. it faithfully captures some notion of reality).

Interpretations have found uses both in their own right, e.g. medicine [41], policy-making [11], and science [5,77], as well as in auditing predictions themselves in response to issues such as regulatory pressure [29] and fairness [22]. In these domains, interpretations have been shown to help with evaluating a learned model, providing information to repair a model (if needed), and building trust with domain experts [13]. However, this increasing role, along with the explosion in proposed interpretation techniques [4,27,31,50,53,75,81,84] has raised considerable concerns about the use of interpretation methods in practice [2,30]. Furthermore, it is unclear how interpretation techniques should be evaluated in the real-world context to advance our understanding of a particular problem. To do so, we first review some of the desiderata of interpretability, following [50] among many definitions [19,40,63], then discuss some methods for critically evaluating interpretations.

The PDR Desiderata for Interpretations. In general, it is unclear how to select and evaluate interpretation methods for a particular problem and audience. To help guide this process, we cover the PDR framework [50], consisting of three desiderata that should be used to select interpretation methods for a particular problem: predictive accuracy, descriptive accuracy, and relevancy. *Predictive accuracy* measures the ability of a model to capture underlying relationships in the data (and generally includes different measures of a model's quality of fit)—this can be seen as the most common form of reality check. In contrast, *descriptive accuracy* measures how well one can approximate what the model has learned using an interpretation method. Descriptive accuracy measures errors during the post-hoc analysis stage of modeling, when interpretations methods are used to analyze a fitted model. For an interpretation to be trustworthy, one should try to maximize both of the accuracies. In cases where either accuracy is not very high, the resulting interpretations may still be useful. However, it is especially important to check their trustworthiness through external validation, such as running an additional experiment. *Relevancy* guides which interpretation to select based on the context of the problem, often playing a key role in determining the trade-off between predictive and descriptive accuracy; however, predictive accuracy and relevancy are not always a trade-off and the examples are shown in Sect. 4.

Evaluating Interpretations and Additional Reality Checks. Techniques striving for interpretations can provide a large amount of fine-grained information, often not just for individual features but also for feature groups [49,68]. As such, it is important to ensure that this added information correctly reflects a model (i.e. has high descriptive accuracy), and can be useful in practice. This is challenging in general, but there are some promising directions. One direction, often used in statistical research including causal inference, uses simulation studies to evaluate interpretations. In this setting, a researcher defines a simple generative process, generates a large amount of data from that process, and trains their statistical or ML model on that data. Assuming a proper simulation setup, a sufficiently relevant and powerful model to recover the generative process, and sufficiently large training data, the trained model should achieve near-perfect generalization accuracy. The practitioner then measures whether their interpretations recover aspects of the original generative process. If the simulation captures the reality well, then it can be viewed as a weaker form of reality check.

Going a step further, interpretations can be tested by gathering new data in followup experiments or observations for retrospective validation. Another direction, which this chapter also focuses on, is to demonstrate the interpretations through domain knowledge which is relevant to a particular domain/audience. To do so, we closely collaborate with domain experts and showcase how interpretations can inform relevant knowledge in fundamental problems in cosmology and molecular-partner prediction. We highlight the use of reality checks to evaluate each proposed method in the chapter.

Chapter Overview. A vast line of prior work has focused on assigning importance to individual features, such as pixels in an image or words in a document. Several methods yield feature-level importance for different architectures. They can be categorized as gradient-based [7,65,71,73], decomposition-based [6,51,66] and others [15,26,57,85], with many similarities among the methods [3,43]. While many methods have been developed to attribute importance to individual features of a model's input, relatively little work has been devoted to understanding interactions between key features. These interactions are a crucial part of interpreting modern deep-learning models, as they are what enable strong predictive performance on structured data.

Here, we cover a line of work that aims to identify, attribute importance, and utilize interactions in neural networks for interpretation. We then explore how these attributions can be used to help improve the performance of DNNs. Despite their strong predictive performance, DNNs sometimes latch onto spurious correlations caused by dataset bias or overfitting [79]. As a result, DNNs often exploit bias regarding gender, race, and other sensitive attributes present in training datasets [20,28,52]. Moreover, DNNs are extremely computationally intensive and difficult to audit.

Figure 1 shows an overview of this chapter. We first overview different methods for computing interpretations (Sect. 2), including for scoring interactions [49], generating hierarchical interpretations [68], and calculating importances for transformations of features [67]. Next, we show how these interpretations can be used to improve models (Sect. 3), including by directly regularizing interpretations [60] and distilling a model through interpretations [31]. Finally, we show how these interpretations can be adapted to real-world problems (Sect. 4), including molecular partner prediction, cosmological parameter prediction, and skin-cancer classification.

2 Computing Interpretations for Feature Interactions and Transformations

This section reviews three recent methods developed to extract the interactions between features that an (already trained) DNN has learned. First, Sect. 2.1 shows how to compute importance scores for groups of features via contextual decomposition (CD), a method which works with LSTMs [49] and arbitrary DNNs, such as CNNs [68]. Next, Sect. 2.2 covers agglomerative contextual decomposition (ACD), where a group-level importance measure, in this case CD, is used as a joining metric in an agglomerative clustering procedure. Finally, Sect. 2.3 covers transformation importance (TRIM), which allows for computing scores for interactions on transformations of a model's input. Other methods have been recently developed for understanding model interactions with varying degrees of computational cost and faithfulness to the trained model [17,18,75,76,78,83].

2.1 Contextual Decomposition (CD) Importance Scores for General DNNs

Contextual decomposition breaks up the forward pass of a neural network in order to find an importance score of some subset of the inputs for a particular prediction. For a given DNN $f(x)$, its output is represented as a SoftMax operation applied to logits $g(x)$. These logits, in turn, are the composition of L layers g_i, $i = 1, \ldots, L$, such as convolutional operations or ReLU non-linearities:

$$f(x) = \text{SoftMax}(g(x)) = \text{SoftMax}(g_L(g_{L-1}(\ldots(g_2(g_1(x)))))). \tag{1}$$

Given a group of features $\{x_j\}_{j \in S}$, the CD algorithm, $g^{CD}(x)$, decomposes the logits $g(x)$ into a sum of two terms, $\beta(x)$ and $\gamma(x)$. $\beta(x)$ is the importance measure of the feature group $\{x_j\}_{j \in S}$, and $\gamma(x)$ captures contributions to $g(x)$ not included in $\beta(x)$.

$$g^{CD}(x) = (\beta(x), \gamma(x)), \tag{2}$$
$$\beta(x) + \gamma(x) = g(x). \tag{3}$$

Computing the CD decomposition for $g(x)$, requires layer-wise CD decompositions $g_i^{CD}(x) = (\beta_i, \gamma_i)$ for each layer $g_i(x)$, where $g_i(x)$ represents the vector of neural activations at the i-th layer. Here, β_i corresponds to the importance measure of $\{x_j\}_{j \in S}$ to layer i, and γ_i corresponds to the contribution of the rest of the input to layer i. Maintaining the decomposition requires $\beta_i + \gamma_i = g_i(x)$ for each i, the CD scores for the full network are computed by composing these decompositions.

$$g^{CD}(x) = g_L^{CD}(g_{L-1}^{CD}(\ldots(g_2^{CD}(g_1^{CD}(x)))))). \tag{4}$$

Note that the above equation shows the CD algorithm g^{CD} takes as input a vector x and for each layer it outputs the pair of vector scores $g_i^{CD}(x) = (\beta_i, \gamma_i)$; and the final output is given by a pair of numbers $g^{CD}(x) = (\beta(x), \gamma(x))$ such that the sum $\beta(x) + \gamma(x)$ equals the logits $g(x)$.

The initial CD work [49] introduced decompositions g_i^{CD} for layers used in LSTMs and the followup work [68] for layers used in CNNs and more generic deep architectures. Below, we give example decompositions for some commonly used layers, such as convolutional layer, linear layer, or ReLU activation.

When g_i is a convolutional or fully connected layer, the layer operation consists of a weight matrix W and a bias vector b. The weight matrix can be multiplied with β_{i-1} and γ_{i-1} individually, but the bias must be partitioned between the two. The bias is partitioned proportionally based on the absolute value of the layer activations. For the convolutional layer, this equation yields only one activation of the output; it must be repeated for each activation.

$$\beta_i = W\beta_{i-1} + \frac{|W\beta_{i-1}|}{|W\beta_{i-1}| + |W\gamma_{i-1}|} \cdot b; \tag{5}$$

$$\gamma_i = W\gamma_{i-1} + \frac{|W\gamma_{i-1}|}{|W\beta_{i-1}| + |W\gamma_{i-1}|} \cdot b. \tag{6}$$

Next, for the ReLU activation function,[2] importance score β_i is computed as the activation of β_{i-1} alone and then update γ_i by subtracting this from the total activation.

$$\beta_i = \text{ReLU}(\beta_{i-1}); \tag{7}$$

$$\gamma_i = \text{ReLU}(\beta_{i-1} + \gamma_{i-1}) - \text{ReLU}(\beta_{i-1}). \tag{8}$$

For a dropout layer, dropout is simply applied to β_{i-1} and γ_{i-1} individually. Computationally, a CD call is comparable to a forward pass through the network f.

Reality Check: Identifying Top-Scoring Phrases. When feasible, a common means of scrutinizing what a model has learned is to inspect its most important features and interactions. Table 1 shows the ACD-top-scoring phrases of different lengths for an LSTM trained on SST (here the phrases are considered from all sentences in the SST's validation set). These phrases were extracted by running ACD separately on each sample in validation set. The score of each phrase was then computed by averaging over the score it received in each occurrence in an ACD hierarchy. The extracted phrases are clearly reflective of the corresponding sentiment, providing additional evidence that ACD is able to capture meaningful positive and negative phrases. The paper [49] also shows that CD properly captures negation interactions for phrases.

---

[2] See [49, Sect. 3.2.2] for other activation functions such as sigmoid or hyperbolic tangent.

Table 1. Top-scoring phrases of different lengths extracted by CD on SST's validation set. The positive/negative phrases identified by CD are all indeed positive/negative.

Length	Positive	Negative
1	Pleasurable, glorious	Nowhere, grotesque, sleep
3	Amazing accomplishment, great fun	Bleak and desperate, conspicuously lacks
5	A pretty amazing accomplishment	Ultimately a pointless endeavour

2.2 Agglomerative Contextual Decomposition (ACD)

Next, we cover agglomerative contextual decomposition (ACD), a general technique that can be applied to a wide range of DNN architectures and data types. Given a prediction from a trained DNN, ACD produces a hierarchical clustering of the input features, along with the contribution of each cluster to the final prediction. This hierarchy is designed to identify clusters of features that the DNN learned are predictive. Throughout this subsection, we use the term CD interaction score between two groups of features to mean the difference between the scores of the combined group and the original groups.

Given the generalized CD scores introduced above, we now introduce the clustering procedure used to produce ACD interpretations. At a high level, this method is equivalent to agglomerative hierarchical clustering, where the CD interaction score is used as the joining metric to determine which clusters to join at each step. This procedure builds the hierarchy by starting with individual features and iteratively combining them based on the highest interaction scores provided by CD. The displayed ACD interpretation is the hierarchy, along with the CD importance score at each node.

The clustering procedure proceeds as follows. After initializing by computing the CD scores of each feature individually, the algorithm iteratively selects all groups of features within k% of the highest-scoring group (where k is a hyperparameter) and adds them to the hierarchy. Each time a new group is added to the hierarchy, a corresponding set of candidate groups is generated by adding individual contiguous features to the original group. For text, the candidate groups correspond to adding one adjacent word onto the current phrase, and for images adding any adjacent pixel onto the current image patch. Candidate groups are ranked according to the CD interaction score, which is the difference between the score of the candidate and the original groups.

Reality Check: Human Experiment. Human experiments show that ACD allows users to better reason about the accuracy of DNNs. Each subject was asked to fill out a survey asking whether, using ACD, they could identify the more accurate of two models across three datasets (SST [70], MNIST [36] and ImageNet [16]), and ACD was compared against three baselines: CD [49], Integrated Gradients (IG) [73], and occlusion [38,82]. Each model uses a standard

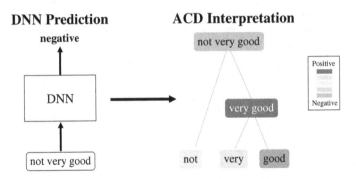

Fig. 2. ACD illustrated through the toy example of predicting the phrase "not very good" as negative. Given the network and prediction, ACD constructs a hierarchy of meaningful phrases and provides importance scores for each identified phrase. In this example, ACD identifies that "very" modifies "good" to become the very positive phrase "very good", which is subsequently negated by "not" to produce the negative phrase "not very good".

architecture that achieves high classification accuracy, and has an analogous model with substantially poorer performance obtained by randomizing some fraction of its weights while keeping the same predicted label. The objective of this experiment was to determine if subjects could use a small number of interpretations produced by ACD to identify the more accurate of the two models (Fig. 2).

For each question, 11 subjects were given interpretations from two different models (one high-performing and one with randomized weights), and asked to identify which of the two models had a higher generalization accuracy. To prevent subjects from simply selecting the model that predicts more accurately for the given example, for each question a subject is shown two sets of examples: one where only the first model predicts correctly and one where only the second model predicts correctly (although one model generalizes to *new* examples much better).

Figure 3 shows the results of the survey. For SST, humans were better able to identify the strongly predictive model using ACD compared to other baselines, with only ACD and CD outperforming random selection (50%). Based on a one-sided two-sample t-test, the gaps between ACD and IG/Occlusion are significant, but not the gap between ACD and CD. In the simple setting of MNIST, ACD performs similarly to other methods. When applied to ImageNet, a more complex dataset, ACD substantially outperforms prior, non-hierarchical methods, and is the only method to outperform random chance. The paper [68] also contains results showing that the ACD hierarchy is robust to adversarial perturbations.

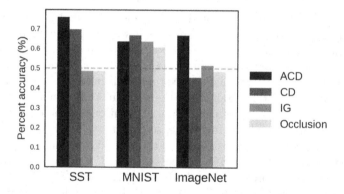

Fig. 3. Results for human studies. Binary accuracy for whether a subject correctly selected the more accurate model using different interpretation techniques.

2.3 Transformation Importance with Applications to Cosmology (TRIM)

Both CD and ACD show how to attribute importance to interactions between features. However, in many cases, raw features such as pixels in an image or words in a document may not be the most meaningful spaces to perform interpretation. When features are highly correlated or features in isolation are not semantically meaningful, the resulting attributions need to be improved.

To meet this challenge, TRIM (Transformation Importance) attributes importance to transformations of the input features (see Fig. 4). This is critical for making interpretations relevant to a particular audience/problem, as attributions in a domain-specific feature space (e.g. frequencies or principal components) can often be far more interpretable than attributions in the raw feature space (e.g. pixels or biological readings). Moreover, features after transformation can be more independent, semantically meaningful, and comparable across data points. The work here focuses on combining TRIM with CD, although TRIM can be combined with any local interpretation method.

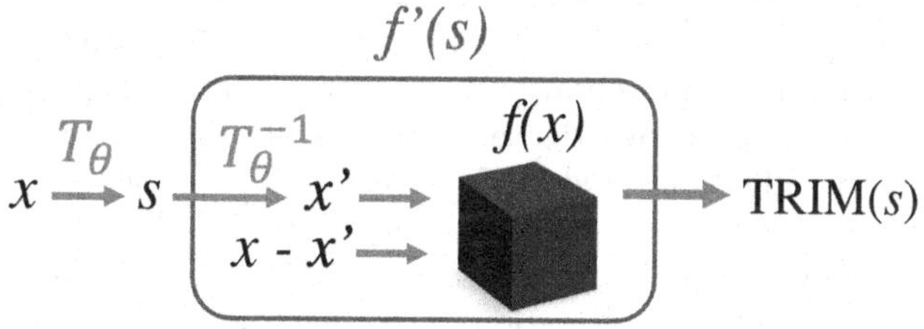

$$f'(s)$$

$$x \xrightarrow{T_\theta} s \overbrace{\begin{array}{l} T_\theta^{-1} \\ x \text{ - } x' \end{array} \xrightarrow{} x' \xrightarrow{}}^{f(x)} \quad \blacksquare \longrightarrow \text{TRIM}(s)$$

Fig. 4. TRIM: attributing importance to a transformation of an input $T_\theta(x)$ given a model $f(x)$.

TRIM aims to interpret the prediction made by a model f given a single input x. The input x is in some domain $\mathcal{X}$, but we desire an explanation for its representation s in a different domain $\mathcal{S}$, defined by a mapping $T : \mathcal{X} \to \mathcal{S}$, such that $s = T(x)$. For example, if x is an image, s may be its Fourier representation, and T would be the Fourier transform. Notably, this process is *entirely post-hoc*: the model f is already fully trained on the domain $\mathcal{X}$. By reparametrizing the network as shown in Fig. 4, we can obtain attributions in the domain $\mathcal{S}$. If we require that the mapping T be invertible, so that $x = T^{-1}(s)$, we can represent each data point x with its counterpart s in the desired domain, and the function to interpret becomes $f' = f \circ T^{-1}$; the function f' can be interpreted with any existing local interpretation method *attr* (e.g. LIME [57] or CD [49,68])). Note that if the transformation T is not perfectly invertible (i.e. $x \neq x'$), then the residuals $x - x'$ may also be required for local interpretation. For example, they are required for any gradient-based attribution method to aid in computing $\partial f' / \partial s$.[3] Once we have the reparameterized function $f'(s)$, we need only specify which part of the input to interpret, before calculating the TRIM score:

Definition 1. *Given a model f, an input x, a mask M, a transformation T, and an attribution method attr,*

$$TRIM(s) = attr\,(f'; s)$$
$$where\ f' = f \circ T^{-1}, s = M \odot T(x)$$

Here M is a mask used to specify which parts of the transformed space to interpret and $\odot$ denotes elementwise multiplication.

In the work here, the choice of attribution method *attr* is CD, and *attr* $(f; x', x)$ represents the CD score for the features x' as part of the input x. This formulation does not require that x' simply be a binary masked version of x; rather, the selection of the mask M allows a human/domain scientist to decide which transformed features to score. In the case of image classification, rather than simply scoring a pixel, one may score the contribution of a frequency band to the prediction $f(x)$. This general setup allows for attributing importance to a wide array of transformations. For example, T could be any invertible transform (e.g. a wavelet transform), or a linear projection (e.g. onto a sparse dictionary). Moreover, we can parameterize the transformation T_θ and learn the parameters θ to produce a desirable representation (e.g. sparse or disentangled).

As a simple example, we investigate a text-classification setting using TRIM. We train a 3-layer fully connected DNN with ReLU activations on the Kaggle Fake News dataset,[4] achieving a test accuracy of 94.8%. The model is trained directly on a bag-of words representation, but TRIM can provide a more succinct space via a topic model transformation. The topic model is learned via latent dirichlet allocation [10], which provides an invertible linear mapping between a

---

[3] If the residual is not added, the gradient of $f' = f \circ T^{-1}$ requires $\partial f / \partial x|_{x'}$, which can potentially cause evaluation of f at the out-of-distribution examples $x' \neq x$.

[4] https://www.kaggle.com/c/fake-news/overview.

document's bag-of-words representation and its topic-representation, where each topic assigns different linear weights to each word. Figure 5 shows the mean attributions for different topics when the model predicts *Fake*. Interestingly, the topic with the highest mean attribution contains recognizable words such as *clinton* and *emails*.

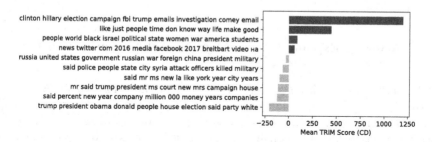

Fig. 5. TRIM attributions for a fake-news classifier based on a topic model transformation. Each row shows one topic, labeled with the top ten words in that topic. Higher attributions correspond to higher contribution to the class *fake*. Calculated over all points which were accurately classified as *fake* in the test set (4,160 points).

Simulation. In the case of a perfectly invertible transformation, such as the Fourier transform, TRIM simply measures the ability of the underlying attribution method (in this case CD) to correctly attribute importance in the transformed space. We run synthetic simulations showing the ability of TRIM with CD to recover known groundtruth feature importances. Features are generated i.i.d. from a standard normal distribution. Then, a binary classification outcome is defined by selecting a random frequency and testing whether that frequency is greater than its median value. Finally, we train a 3-layer fully connected DNN with ReLU activations on this task and then test the ability of different methods to assign this frequency the highest importance. Table 2 shows the percentage of errors made by different methods in such a setup. CD has the lowest error on average, compared to popular baselines.

Table 2. Error (%) in recovering a groundtruth important frequency in simulated data using different attribution methods with TRIM, averaged over 500 simulated datasets.

CD	DeepLift [66]	SHAP [43]	Integrated gradients [73]
0.4 ± 0.282	3.6 ± 0.833	4.0 ± 0.897	4.2 ± 0.876

3 Using Attributions to Improve Models

This section shows two methods for using the attributions introduced in Sect. 2 to directly improve DNNs. Section 3.1 shows how CD scores can be penalized during training to improve generalization in interesting ways and Sect. 3.2 shows how attribution scores can be used to distill a DNN into a simple data-driven wavelet model.

3.1 Penalizing Explanations to Align Neural Networks with Prior Knowledge (CDEP)

While much work has been put into developing methods for explaining DNNs, relatively little work has explored the potential to use these explanations to help build a better model. Some recent work proposes forcing models to attend to certain regions [12,21,48], penalizing the gradients or expected gradients of a neural network [8,21,23,42,61,62], or using layer-wise relevance propagation to prune/improve models [72,80]. A newly emerging line of work investigates how domain experts can use explanations during the training loop to improve their models (e.g. [64]).

Here, we cover contextual decomposition explanation penalization (CDEP), a method which leverages CD to enable the insertion of domain knowledge into a model [60]. Given prior knowledge in the form of importance scores, CDEP works by allowing the user to directly penalize importances of certain features or feature interactions. This forces the DNN to not only produce the correct prediction, but also the correct explanation for that prediction. CDEP can be applied to arbitrary DNN architectures and is often orders of magnitude faster and more memory efficient than recent gradient-based methods [23,62]; CDEP offers significant computational improvements, since, unlike gradient-based attributions, the CD score is computed along the forward pass, only first derivatives are required for optimization, early layers can be frozen, and all activations of a DNN do not need to be cached to perform backpropagation; furthermore, with gradient-based methods the training requires the storage of activations and gradients for all layers of the network as well as the gradient with respect to the input, whereas penalizing CD requires only a small constant amount of memory more than standard training.

CDEP works by augmenting the traditional objective function used to train a neural network, as displayed in Eq. (9) with an additional component. In addition to the standard prediction loss $\mathcal{L}$, which teaches the model to produce the correct predictions by penalizing wrong predictions, we add an explanation error $\mathcal{L}_{\text{expl}}$, which teaches the model to produce the correct explanations for its predictions by penalizing wrong explanations. In place of the prediction and labels $f_\theta(X), y$, used in the prediction error $\mathcal{L}$, the explanation error $\mathcal{L}_{\text{expl}}$ uses the explanations produced by an interpretation method $\text{expl}_\theta(X)$, along with targets provided by the user expl_X. The two losses are weighted by a hyperparameter $\lambda \in \mathbb{R}$:

$$\hat{\theta} = \underset{\theta}{\mathrm{argmin}}\ \overbrace{\mathcal{L}\left(f_\theta(X), y\right)}^{\text{Prediction error}} + \lambda \overbrace{\mathcal{L}_{\mathrm{expl}}\left(\mathrm{expl}_\theta(X), \mathrm{expl}_X\right)}^{\text{Explanation error}} \tag{9}$$

CDEP uses CD as the explanation function used to compute $\mathrm{expl}_\theta(X)$, allowing the penalization of interactions between features. We now substitute the above CD scores into the generic equation in Eq. (9) to arrive at CDEP as it is used in this chapter. We collect from the user, for each input x_i, a collection of feature groups $x_{i,S}$, $x_i \in \mathbb{R}^d$, $S \subseteq \{1, ..., d\}$, along with explanation target values $\mathrm{expl}_{x_{i,S}}$, and use the $\|\cdot\|_1$ loss for $\mathcal{L}_{\mathrm{expl}}$. This yields a vector $\beta(x_j)$ for any subset of features in an input x_j which we would like to penalize. We can then collect prior knowledge label explanations for this subset of features, expl_{x_j} and use it to regularize the explanation:

$$\hat{\theta} = \underset{\theta}{\mathrm{argmin}}\ \overbrace{\sum_i \sum_c - y_{i,c} \log f_\theta(x_i)_c}^{\text{Prediction error}} + \lambda \overbrace{\sum_i \sum_S \|\beta(x_{i,S}) - \mathrm{expl}_{x_{i,S}}\|_1}^{\text{Explanation error}} \tag{10}$$

In the above, i indexes each individual example in the dataset, S indexes a subset of the features for which we penalize their explanations, and c sums over each class.

The choice of prior knowledge explanations expl_X is dependent on the application and the existing domain knowledge. CDEP allows for penalizing arbitrary interactions between features, allowing the incorporation of a very broad set of domain knowledge. In the simplest setting, practitioners may precisely provide prior knowledge human explanations for each data point. To avoid assigning human labels, one may utilize programmatic rules to identify and assign prior knowledge importance to regions, which are then used to help the model identify important/unimportant regions. In a more general case, one may specify importances of different feature interactions.

Towards Reality Check: ColorMNIST Task. Here, we highlight CDEP's ability to alter which features a DNN uses to perform digit classification. Similar to one previous study [39], we alter the MNIST dataset to include three color channels and assign each class a distinct color, as shown in Fig. 6. An unpenalized DNN trained on this biased data will completely misclassify a test set with inverted colors, dropping to 0% accuracy (see Table 3), suggesting that it learns to classify using the colors of the digits rather than their shape.

Interestingly, this task can be approached by minimizing the contribution of pixels in isolation (which only represent color) while maximizing the importance of groups of pixels (which can represent shapes). To do this, CDEP penalizes the CD contribution of sampled single-pixel values, following Eq. (10). Minimizing the contribution of single pixels encourages the DNN to focus instead on groups of pixels. Table 3 shows that CDEP can partially divert the network's focus on color to also focus on digit shape. The table includes 2 baselines: penalization of the squared gradients (RRR) [62] and Expected Gradients (EG) [23]. The

Fig. 6. ColorMNIST: the shapes remain the same between the training set and the test set, but the colors are inverted. (Color figure online)

baselines do not improve the test accuracy of the model on this task above the random baseline, while CDEP significantly improves the accuracy to 31.0%.

Table 3. Test Accuracy on ColorMNIST. CDEP is the only method that captures and removes color bias. All values averaged over thirty runs. Predicting at random yields a test accuracy of 10%.

	Vanilla	CDEP	RRR	Expected gradients
ColorMNIST	0.2 ± 0.2	**31.0 $\pm$ 2.3**	0.2 ± 0.1	10.0 ± 0.1

The paper [60] further shows how CDEP can be applied to diverse applications, such as notions of fairness in the COMPAS dataset [35] and in natural-language processing.

3.2 Distilling Adaptive Wavelets from Neural Networks with Interpretations

One promising approach to acquiring highly predictive interpretable models is model distillation. Model distillation is a technique which distills the knowledge in one model into another model. Here, we focus on the case where we distill a DNN into a simple, wavelet model. Wavelets have many useful properties, including fast computation, an orthonormal basis, and interpretation in both spatial and frequency domains [44]. Here, we cover adaptive wavelet distillation (AWD), a method to learn a valid wavelet by distilling information from a trained DNN [31].

Equation (11) shows the three terms in the formulation of the method. x_i represents the i-th input signal, $\hat{x}_i$ represents the reconstruction of x_i, h and g represent the lowpass and highpass wavelet filters, and Ψx_i denotes the wavelet coefficients of x_i. λ is a hyperparameter penalizing the sparsity of the wavelet coefficients, which can help to learn a compact representation of the input signal and γ is a hyperparameter controlling the strength of the interpretation loss, which controls how much to use the information coming from a trained model f:

$$\underset{h,g}{\text{minimize}} \ \mathcal{L}(h,g) = \underbrace{\frac{1}{m}\sum_i \|x_i - \widehat{x}_i\|_2^2}_{\text{Reconstruction loss}} + \underbrace{\frac{1}{m}\sum_i W(h,g,x_i;\lambda)}_{\text{Wavelet loss}} + \underbrace{\gamma \sum_i \|\text{TRIM}_f(\Psi x_i)\|_1}_{\text{Interpretation loss}},$$

$$(11)$$

Here the reconstruction loss ensures that the wavelet transform is invertible, allowing for reconstruction of the original data. Hence the transform does not lose any information in the input data.

The wavelet loss ensures that the learned filters yield a valid wavelet transform. Specifically, [45,47] characterize the sufficient and necessary conditions on h and g to build an orthogonal wavelet basis. Roughly speaking, these conditions state that in the frequency domain the mass of the lowpass filer h is concentrated on the range of low frequencies while the highpass filter g contains more mass in the high frequencies. We also desire the learned wavelet to provide sparse representations so we add the ℓ_1 norm penalty on the wavelet coefficients. Combining all these conditions via regularization terms, we define the wavelet loss at the data point x_i as

$$W(h,g,x_i;\lambda) = \lambda \|\Psi x_i\|_1 + (\sum_n h[n] - \sqrt{2})^2 + (\sum_n g[n])^2 + (\|h\|_2^2 - 1)^2$$
$$+ \sum_w (|\widehat{h}(w)|^2 + |\widehat{h}(w+\pi)|^2 - 2)^2 + \sum_k (\sum_n h[n]h[n-2k] - \mathbf{1}_{k=0})^2,$$

where g is set as $g[n] = (-1)^n h[N-1-n]$ and where N is the support size of h (see [31] for further details on the formulations of wavelet loss).

Finally, the interpretation loss enables the distillation of knowledge from the pre-trained model f into the wavelet model. It ensures that attributions in the space of wavelet coefficients Ψx_i are sparse, where the attributions of wavelet coefficients is calculated by TRIM, as described in Sect. 2.3. This forces the wavelet transform to produce representations that concisely explain the model's predictions at different scales and locations.

A key difference between AWD and existing adaptive wavelet techniques (e.g. [55,56]) is that they use *interpretations from a trained model* to learn the wavelets; this incorporates information not just about the signal but also an outcome of interest and the inductive biases learned by a DNN. This can help learn an interpretable representation that is well-suited to efficient computation and effective prediction.

Reality Check: Molecular Partner Prediction. For evaluation, see Sect. 4.1, which shows an example of how a distilled AWD model can provide a simpler, more interpretable model while improving prediction accuracy.

4 Real-Data Problems Showcasing Interpretations

In this section, we focus on three real-data problems where the methods introduced in Sect. 2 and Sect. 3 are able to provide useful interpretations in context. Sect. 4.1 describes how AWD can distill DNNs used in cell biology, Sect. 4.2 describes how TRIM + CD yield insights in a cosmological context, and Sect. 4.3 describes how CDEP can be used to ignore spurious correlations in a medical imaging task.

4.1 Molecular Partner Prediction

We now turn our attention to a crucial question in cell biology: understanding clathrin-mediated endocytosis (CME) [32,34]. It is the primary pathway by which things are transported into the cell, making it essential functions of higher eukaryotic life [46]. Many questions about this process remain unanswered, prompting a line of studies aiming to better understand this process [33]. One major challenge with analysis of CME, is the ability to readily distinguish between abortive coats (ACs) and successful clathrin-coated pits (CCPs). Doing so enables an understanding of what mechanisms allow for successful endocytosis. This is a challenging problem where DNNs have recently been shown to outperform classical statistical and ML methods.

Figure 7 shows the pipeline for this challenging problem. Tracking algorithms run on videos of cells identify time-series traces of endocytic events. An LSTM model learns to classify which endocytic events are successful and CD scores identify which parts of the traces the model uses. Using these CD scores, domain experts are able to validate that the model does, in fact use reasonable features such as the max value of the time-series traces and the length of the trace.

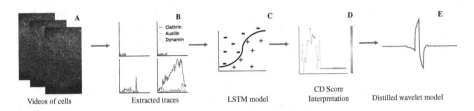

Fig. 7. Molecular partner prediction pipeline. (**A**) Tracking algorithms run on videos of cells identify (**B**) time-series traces of endocytic events. (**C**) An LSTM model learns to classify which endocytic events are successful and (**D**) CD scores identify which parts of the traces the model uses. (**E**) AWD distills the LSTM model into a simple wavelet model which is able to obtain strong predictive performance.

However, the LSTM model is still relatively difficult to understand and computationally intensive. To create an extremely transparent model, we extract only the maximum 6 wavelet coefficients at each scale. By taking the maximum coefficients, these features are expected to be invariant to the specific locations where a CME event occurs in the input data. This results in a final model with 30 coefficients (6 wavelet coefficients at 5 scales). These wavelet coefficients are used to train a linear model, and the best hyperparameters are selected via cross-validation on the training set. Figure 7 shows the best learned wavelet (for one particular run) extracted by AWD corresponding to the setting of hyperparameters $\lambda = 0.005$ and $\gamma = 0.043$. Table 4 compares the results for AWD to the original LSTM and the initialized, non-adaptive DB5 wavelet model, where the performance is measured via a standard R^2 score, a proportion of variance in the response that is explained by the model. The AWD model not only closes the gap between the standard wavelet model (DB5) and the neural network, it considerably improves the LSTM's performance (a 10% increase of R^2 score). Moreover, we calculate the compression rates of the AWD wavelet and DB5—these rates measure the proportion of wavelet coefficients in the test set, in which the magnitude and the attributions are both above 10^{-3}. The AWD wavelet exhibits much better compression than DB5 (an 18% reduction), showing the ability of AWD to simultaneously provide sparse representations and explain the LSTM's predictions concisely. The AWD model also dramatically decreases the computation time at test time, a more than 200-fold reduction when compared to LSTM.

In addition to improving prediction accuracy, AWD enables domain experts to vet their experimental pipelines by making them more transparent. By inspecting the learned wavelet, AWD allows for checking what clathrin signatures signal a successful CME event; it indicates that the distilled wavelet aims to identify a large buildup in clathrin fluorescence (corresponding to the building of a clathrin-coated pit) followed by a sharp drop in clathrin fluorescence (corresponding to the rapid deconstruction of the pit). This domain knowledge is extracted from the pre-trained LSTM model by AWD using only the saliency interpretations in the wavelet space.

Table 4. Performance comparisons for different models in molecular-partner prediction. AWD substantially improves predictive accuracy, compression rate, and computation time on the test set. A higher R^2 score, and lower compression factor, and lower computation time indicate better results. For AWD, values are averaged over 5 different random seeds.

	AWD (Ours)	Standard wavelet (DB5)	LSTM
Regression (R^2 score)	**0.262 (0.001)**	0.197	0.237
Compression factor	**0.574 (0.010)**	0.704	N/A
Computation time	**0.0002 s**	**0.0002 s**	0.0449 s

To see the effect of interpretation loss on learning the wavelet transforms and increased performance, we also learn the wavelet transform while setting the interpreration loss to be zero. In this case, the best regression R^2 score selected via cross-validation is 0.231, and the adaptive wavelets without the interpretation loss still outperforms the baseline wavelet but fail to outperform the neural network models.

4.2 Cosmological Parameter Prediction

We now turn to a cosmology example, where attributing importance to transformations helps understand cosmological models in a more meaningful feature space. Specifically, we consider weak gravitational lensing convergence maps, i.e. maps of the mass distribution in the Universe integrated up to a certain distance from the observer. In a cosmological experiment (e.g. a galaxy survey), these mass maps are obtained by measuring the distortion of distant galaxies caused by the deflection of light by the mass between the galaxy and the observer [9]. These maps contain a wealth of physical information of interest to cosmologists, such as the total matter density in the universe, Ω_m. Current research aims at identifying the most informative features in these maps for inferring the true cosmological parameters, with DNN-based inference methods often obtaining state-of-the-art results [25,58,59].

In this context, it is important to not only have a DNN that predicts well, but also understand what it learns. Knowing which features are important provides deeper understanding and can be used to design optimal experiments or analysis methods. Moreover, because this DNN is trained on numerical simulations (realizations of the Universe with different cosmological parameters), it is important to validate that it uses physical features rather than latching on to numerical artifacts in the simulations. TRIM can help understand and validate that the DNN learns appropriate physical features by analyzing attributing importance in the spectral domain.

A DNN is trained to accurately predict Ω_m from simulated weak gravitational lensing convergence maps (full details in [67]). To understand what features the model is using, we desire an interpretation in the space of the power spectrum. The images in Fig. 8 show how different information is contained within different frequency bands in the mass maps. The plot in Fig. 8 shows the TRIM attributions with CD (normalized by the predicted value) for different frequency bands when predicting the parameter Ω_m. Interestingly, the most important frequency band for the predictions seems to peak at scales around $\ell = 10^4$ and then decay for higher frequencies.[5] A physical interpretation of this result is that the DNN concentrates on the most discriminative part of the Power Spectrum, i.e. at scales large enough not to be dominated by sample variance, and smaller than the frequency cutoff at which the simulations lose power due to resolution effects.

Figure 9 shows some of the curves from Fig. 8 separated based on their cosmology, to show how the curves vary with the value of Ω_m. Increasing the value of Ω_m increases the contribution of scales close to $\ell = 10^4$, making other

---

[5] Here the unit of frequency used is angular multipole ℓ.

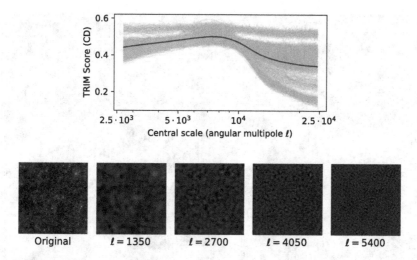

Fig. 8. Different scales (i.e. frequency bands) contribute differently to the prediction of Ω_m. Each blue line corresponds to one testing image and the red line shows the mean. Images show the features present at different scales. The bandwidth is $\Delta_\ell = 2,700$. (Color figure online)

frequencies relatively unimportant. This seems to correspond to known cosmological knowledge, as these scales seem to correspond to galaxy clusters in the mass maps, which are structures very sensitive to the value of Ω_m. The fact that the importance of these features varies with Ω_m would seem to indicate that at lower Ω_m the model is using a different source of information, not located at any single scale, for making its prediction.

4.3 Improving Skin Cancer Classification via CDEP

In recent years, deep learning has achieved impressive results in diagnosing skin cancer [24]. However, the datasets used to train these models often include spurious features which make it possible to attain high test accuracy without learning the underlying phenomena [79]. In particular, a popular dataset from ISIC (International Skin Imaging Collaboration) has colorful patches present in approximately 50% of the non-cancerous images but not in the cancerous images as can be seen in Fig. 10 [14]. We use CDEP to remedy this problem by penalizing the DNN placing importance on the patches during training.

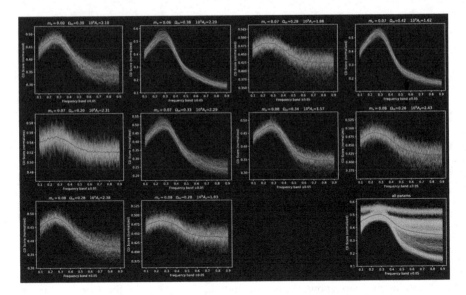

Fig. 9. TRIM attributions vary with the value of Ω_m.

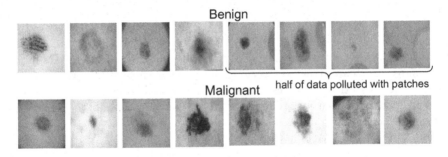

Fig. 10. Example images from the ISIC dataset. Half of the benign lesion images include a patch in the image. Training on this data results in the neural network overly relying on the patches to classify images; CDEP avoids this.

The task in this section is to classify whether an image of a skin lesion contains (1) benign melanoma or (2) malignant melanoma. In a real-life task, this would for example be done to determine whether a biopsy should be taken. In order to identify the spurious patches, binary maps of the patches for the skin cancer task are segmented using SLIC, a common image-segmentation algorithm [1]. After the spurious patches were identified, they are penalized using to have zero importance.

Table 5 shows results comparing the performance of a DNN trained with and without CDEP. We report results on two variants of the test set. The first, which we refer to as "no patches" only contains images of the test set that do not include patches. The second also includes images with those patches. Training with CDEP improves the AUC and F1-score for both test sets, compared to both a Vanilla DNN and using the RRR method introduced in [62]. Further visual inspection shows that the DNN attributes low importance to regions in the images with patches.

Table 5. Results from training a DNN on ISIC to recognize skin cancer (averaged over three runs). Results shown for the entire test set and for only the test-set images that do not include patches ("no patches"). The network trained with CDEP generalizes better, getting higher AUC and F1 on both.

	AUC (no patches)	F1 (no patches)	AUC (all)	F1 (all)
Vanilla	0.93	0.67	0.96	0.67
RRR	0.76	0.45	0.87	0.45
CDEP	**0.95**	**0.73**	**0.97**	**0.73**

5   Discussion

Overall, the interpretation methods here are shown to (1) accurately recover known importances for features/feature interactions [49], (2) correctly inform human decision-making and be robust to adversarial perturbations [68], and (3) reliably alter a neural network's predictions when regularized appropriately [60]. For each case, we demonstrated the use of reality checks through predictive accuracy (the most common form of reality check) or through domain knowledge which is relevant to a particular domain/audience.

There is considerable future work to do in developing and evaluating attributions, particularly in distilling/building interpretable models for real-world domains and understanding how to better make useful interpretation methods. Below we discuss them in turn.

5.1   Building/Distilling Accurate and Interpretable Models

In the ideal case, a practitioner can develop a simple model to make their predictions, ensuring interpretability by obviating the need for post-hoc interpretation. Interpretable models tend to be faster, more computationally efficient, and smaller than their DNN counterparts. Moreover, interpretable models allow for easier inspection of knowledge extracted from the learned models and make reality checks more transparent. AWD [31] represents one effort to use attributions to distill DNNs into an interpretable wavelet model, but the general idea can go much further. There are a variety of interpretable models, such as rule-based models [37, 69, 74] or additive models [13] whose fitting process could benefit from accurate attributions. Moreover, AWD and related techniques could be extended beyond the current setting to unsupervised/reinforcement learning settings or to incorporate multiple layers. Alternatively, attributions can be used as feature engineering tools, to help build simpler, more interpretable models. More useful features can help enable better exploratory data analysis, unsupervised learning, or reality checks.

5.2 Making Interpretations Useful

Furthermore, there is much work remaining to improve the relevancy of interpretations for a particular audience/problem. Given the abundance of possible interpretations, it is particularly easy for researchers to propose novel methods which do not truly solve any real-world problems or fail to faithfully capture some aspects of reality. A strong technique to avoid this is to directly test newly introduced methods in solving a domain problem. Here, we discussed several real-data problems that have benefited from improved interpretations Sect. 4, spanning from cosmology to cell biology. In instances like this, where interpretations are used directly to solve a domain problem, their relevancy is indisputable and reality checks can be validated through domain knowledge. A second, less direct, approach is the use of human studies where humans are asked to perform tasks, such as evaluating how much they trust a model's predictions [68]. While challenging to properly construct and perform, these studies are vital to demonstrating that new interpretation methods are, in fact, relevant to any potential practitioners. We hope the plethora of open problems in various domains such as science, medicine, and public policy can help guide and benefit from improved interpretability going forward.

References

1. Achanta, R., Shaji, A., Smith, K., Lucchi, A., Fua, P., Süsstrunk, S.: SLIC superpixels compared to state-of-the-art superpixel methods. IEEE Trans. Pattern Anal. Mach. Intell. **34**(11), 2274–2282 (2012)
2. Adebayo, J., Gilmer, J., Muelly, M., Goodfellow, I., Hardt, M., Kim, B.: Sanity checks for saliency maps. In: Advances in Neural Information Processing Systems, pp. 9505–9515 (2018)
3. Ancona, M., Ceolini, E., Oztireli, C., Gross, M.: Towards better understanding of gradient-based attribution methods for deep neural networks. In: 6th International Conference on Learning Representations (ICLR 2018) (2018)
4. Andreas, J., Rohrbach, M., Darrell, T., Klein, D.: Neural module networks. In: Proceedings of the IEEE Conference on Computer Vision and Pattern Recognition, pp. 39–48 (2016)
5. Angermueller, C., Pärnamaa, T., Parts, L., Stegle, O.: Deep learning for computational biology. Mol. Syst. Biol. **12**(7), 878 (2016)
6. Bach, S., Binder, A., Montavon, G., Klauschen, F., Müller, K.R., Samek, W.: On pixel-wise explanations for non-linear classifier decisions by layer-wise relevance propagation. PLoS One **10**(7), e0130140 (2015)
7. Baehrens, D., Schroeter, T., Harmeling, S., Kawanabe, M., Hansen, K., MÃžller, K.R.: How to explain individual classification decisions. J. Mach. Learn. Res. **11**(Jun), 1803–1831 (2010)

8. Bao, Y., Chang, S., Yu, M., Barzilay, R.: Deriving machine attention from human rationales. arXiv preprint arXiv:1808.09367 (2018)
9. Bartelmann, M., Schneider, P.: Weak gravitational lensing. Phys. Rep. **340**(4–5), 291–472 (2001)
10. Blei, D.M., Ng, A.Y., Jordan, M.I.: Latent Dirichlet allocation. J. Mach. Learn. Res. **3**(Jan), 993–1022 (2003)
11. Brennan, T., Oliver, W.L.: The emergence of machine learning techniques in criminology. Criminol. Public Policy **12**(3), 551–562 (2013)
12. Burns, K., Hendricks, L.A., Saenko, K., Darrell, T., Rohrbach, A.: Women also snowboard: overcoming bias in captioning models. arXiv preprint arXiv:1803.09797 (2018)
13. Caruana, R., Lou, Y., Gehrke, J., Koch, P., Sturm, M., Elhadad, N.: Intelligible models for healthcare: predicting pneumonia risk and hospital 30-day readmission. In: Proceedings of the 21th ACM SIGKDD International Conference on Knowledge Discovery and Data Mining, pp. 1721–1730. ACM (2015)
14. Codella, N., et al.: Skin lesion analysis toward melanoma detection 2018: a challenge hosted by the international skin imaging collaboration (ISIC). arXiv preprint arXiv:1902.03368 (2019)
15. Dabkowski, P., Gal, Y.: Real time image saliency for black box classifiers. arXiv preprint arXiv:1705.07857 (2017)
16. Deng, J., Dong, W., Socher, R., Li, L.J., Li, K., Fei-Fei, L.: ImageNet: a large-scale hierarchical image database. In: CVPR 2009 (2009)
17. Devlin, S., Singh, C., Murdoch, W.J., Yu, B.: Disentangled attribution curves for interpreting random forests and boosted trees. arXiv preprint arXiv:1905.07631 (2019)
18. Dhamdhere, K., Agarwal, A., Sundararajan, M.: The shapley taylor interaction index. arXiv preprint arXiv:1902.05622 (2019)
19. Doshi-Velez, F., Kim, B.: A roadmap for a rigorous science of interpretability. arXiv preprint arXiv:1702.08608 (2017)
20. Dressel, J., Farid, H.: The accuracy, fairness, and limits of predicting recidivism. Sci. Adv. **4**(1), eaao5580 (2018)
21. Du, M., Liu, N., Yang, F., Hu, X.: Learning credible deep neural networks with rationale regularization. arXiv preprint arXiv:1908.05601 (2019)
22. Dwork, C., Hardt, M., Pitassi, T., Reingold, O., Zemel, R.: Fairness through awareness. In: Proceedings of the 3rd Innovations in Theoretical Computer Science Conference, pp. 214–226. ACM (2012)
23. Erion, G., Janizek, J.D., Sturmfels, P., Lundberg, S., Lee, S.I.: Learning explainable models using attribution priors. arXiv preprint arXiv:1906.10670 (2019)
24. Esteva, A., et al.: Dermatologist-level classification of skin cancer with deep neural networks. Nature **542**(7639), 115 (2017)
25. Fluri, J., Kacprzak, T., Lucchi, A., Refregier, A., Amara, A., Hofmann, T., Schneider, A.: Cosmological constraints with deep learning from KiDS-450 weak lensing maps. Phys. Rev. D **100**(6), 063514 (2019)
26. Fong, R.C., Vedaldi, A.: Interpretable explanations of black boxes by meaningful perturbation. arXiv preprint arXiv:1704.03296 (2017)
27. Frosst, N., Hinton, G.: Distilling a neural network into a soft decision tree. arXiv preprint arXiv:1711.09784 (2017)
28. Garg, N., Schiebinger, L., Jurafsky, D., Zou, J.: Word embeddings quantify 100 years of gender and ethnic stereotypes. Proc. Natil. Acad. Sci. **115**(16), E3635–E3644 (2018)

29. Goodman, B., Flaxman, S.: European union regulations on algorithmic decision-making and a "right to explanation". arXiv preprint arXiv:1606.08813 (2016)
30. Gupta, A., Arora, S.: A simple saliency method that passes the sanity checks. arXiv preprint arXiv:1905.12152 (2019)
31. Ha, W., Singh, C., Lanusse, F., Upadhyayula, S., Yu, B.: Adaptive wavelet distillation from neural networks through interpretations. Adv. Neural Inf. Process. Syst. **34** (2021)
32. He, K., et al.: Dynamics of Auxilin 1 and GAK in clathrin-mediated traffic. J. Cell Biol. **219**(3) (2020)
33. Kaksonen, M., Roux, A.: Mechanisms of clathrin-mediated endocytosis. Nat. Rev. Mol. Cell Biol. **19**(5), 313 (2018)
34. Kirchhausen, T., Owen, D., Harrison, S.C.: Molecular structure, function, and dynamics of clathrin-mediated membrane traffic. Cold Spring Harb. Perspect. Biol. **6**(5), a016725 (2014)
35. Larson, J., Mattu, S., Kirchner, L., Angwin, J.: How we analyzed the COMPAS recidivism algorithm. ProPublica **9** (2016)
36. LeCun, Y.: The MNIST database of handwritten digits (1998). http://yann.com/exdb/mnist/
37. Letham, B., Rudin, C., McCormick, T.H., Madigan, D., et al.: Interpretable classifiers using rules and Bayesian analysis: building a better stroke prediction model. Ann. Appl. Stat. **9**(3), 1350–1371 (2015)
38. Li, J., Monroe, W., Jurafsky, D.: Understanding neural networks through representation erasure. arXiv preprint arXiv:1612.08220 (2016)
39. Li, Y., Vasconcelos, N.: REPAIR: removing representation bias by dataset resampling. arXiv preprint arXiv:1904.07911 (2019)
40. Lipton, Z.C.: The mythos of model interpretability. arXiv preprint arXiv:1606.03490 (2016)
41. Litjens, G., et al.: A survey on deep learning in medical image analysis. Med. Image Anal. **42**, 60–88 (2017)
42. Liu, F., Avci, B.: Incorporating priors with feature attribution on text classification. arXiv preprint arXiv:1906.08286 (2019)
43. Lundberg, S.M., Lee, S.I.: A unified approach to interpreting model predictions. In: Advances in Neural Information Processing Systems, pp. 4768–4777 (2017)
44. Mallat, S.: A Wavelet Tour of Signal Processing, Third Edition: The Sparse Way. Academic Press (2008)
45. Mallat, S.G.: A theory for multiresolution signal decomposition: the wavelet representation. IEEE Trans. Pattern Anal. Mach. Intell. **11**(7), 674–693 (1989)
46. McMahon, H.T., Boucrot, E.: Molecular mechanism and physiological functions of clathrin-mediated endocytosis. Nat. Rev. Mol. Cell Biol. **12**(8), 517 (2011)
47. Meyer, Y.: Wavelets and Operators: Volume 1. No. 37, Cambridge University Press (1992)
48. Mitsuhara, M., et al.: Embedding human knowledge in deep neural network via attention map. arXiv preprint arXiv:1905.03540 (2019)
49. Murdoch, W.J., Liu, P.J., Yu, B.: Beyond word importance: contextual decomposition to extract interactions from LSTMs. In: ICLR (2018)
50. Murdoch, W.J., Singh, C., Kumbier, K., Abbasi-Asl, R., Yu, B.: Definitions, methods, and applications in interpretable machine learning. Proc. Natl. Acad. Sci. **116**(44), 22071–22080 (2019)
51. Murdoch, W.J., Szlam, A.: Automatic rule extraction from long short term memory networks (2017)

52. Obermeyer, Z., Powers, B., Vogeli, C., Mullainathan, S.: Dissecting racial bias in an algorithm used to manage the health of populations. Science **366**(6464), 447–453 (2019)
53. Olah, C., Mordvintsev, A., Schubert, L.: Feature visualization. Distill **2**(11), e7 (2017)
54. Paszke, A., et al.: Automatic differentiation in Pytorch (2017)
55. Recoskie, D.: Learning sparse orthogonal wavelet filters (2018)
56. Recoskie, D., Mann, R.: Learning sparse wavelet representations. arXiv preprint arXiv:1802.02961 (2018)
57. Ribeiro, M.T., Singh, S., Guestrin, C.: Why should i trust you?: explaining the predictions of any classifier. In: Proceedings of the 22nd ACM SIGKDD International Conference on Knowledge Discovery and Data Mining, pp. 1135–1144. ACM (2016)
58. Ribli, D., Pataki, B.Á., Csabai, I.: An improved cosmological parameter inference scheme motivated by deep learning. Nat. Astron. **3**(1), 93 (2019)
59. Ribli, D., Pataki, B.Á., Zorrilla Matilla, J.M., Hsu, D., Haiman, Z., Csabai, I.: Weak lensing cosmology with convolutional neural networks on noisy data. Mon. Not. R. Astron. Soc. **490**(2), 1843–1860 (2019)
60. Rieger, L., Singh, C., Murdoch, W., Yu, B.: Interpretations are useful: penalizing explanations to align neural networks with prior knowledge. In: International Conference on Machine Learning, pp. 8116–8126. PMLR (2020)
61. Ross, A.S., Doshi-Velez, F.: Improving the adversarial robustness and interpretability of deep neural networks by regularizing their input gradients. In: Thirty-Second AAAI Conference on Artificial Intelligence (2018)
62. Ross, A.S., Hughes, M.C., Doshi-Velez, F.: Right for the right reasons: training differentiable models by constraining their explanations. arXiv preprint arXiv:1703.03717 (2017)
63. Rudin, C.: Please stop explaining black box models for high stakes decisions. arXiv preprint arXiv:1811.10154 (2018)
64. Schramowski, P., et al.: Making deep neural networks right for the right scientific reasons by interacting with their explanations. Nat. Mach. Intell. **2**(8), 476–486 (2020)
65. Selvaraju, R.R., Cogswell, M., Das, A., Vedantam, R., Parikh, D., Batra, D.: Grad-CAM: visual explanations from deep networks via gradient-based localization. https://arxiv.org/abs/1610.02391 v3 **7**(8) (2016)
66. Shrikumar, A., Greenside, P., Shcherbina, A., Kundaje, A.: Not just a black box: learning important features through propagating activation differences. arXiv preprint arXiv:1605.01713 (2016)
67. Singh, C., Ha, W., Lanusse, F., Boehm, V., Liu, J., Yu, B.: Transformation importance with applications to cosmology. arXiv preprint arXiv:2003.01926 (2020)
68. Singh, C., Murdoch, W.J., Yu, B.: Hierarchical interpretations for neural network predictions. In: International Conference on Learning Representations (2019). https://openreview.net/forum?id=SkEqro0ctQ
69. Singh, C., Nasseri, K., Tan, Y.S., Tang, T., Yu, B.: imodels: a python package for fitting interpretable models. J. Open Sour. Softw. **6**(61), 3192 (2021). https://doi.org/10.21105/joss.03192
70. Socher, R., et al.: Recursive deep models for semantic compositionality over a sentiment treebank. In: Proceedings of the 2013 Conference on Empirical Methods in Natural Language Processing, pp. 1631–1642 (2013)
71. Springenberg, J.T., Dosovitskiy, A., Brox, T., Riedmiller, M.: Striving for simplicity: the all convolutional net. arXiv preprint arXiv:1412.6806 (2014)

72. Sun, J., Lapuschkin, S., Samek, W., Binder, A.: Explain and improve: LRP-inference fine-tuning for image captioning models. Inf. Fusion **77**, 233–246 (2022)
73. Sundararajan, M., Taly, A., Yan, Q.: Axiomatic attribution for deep networks. In: ICML (2017)
74. Tan, Y.S., Singh, C., Nasseri, K., Agarwal, A., Yu, B.: Fast interpretable greedy-tree sums (FIGS). arXiv preprint arXiv:2201.11931 (2022)
75. Tsang, M., Cheng, D., Liu, Y.: Detecting statistical interactions from neural network weights. arXiv preprint arXiv:1705.04977 (2017)
76. Tsang, M., Sun, Y., Ren, D., Liu, Y.: Can i trust you more? Model-agnostic hierarchical explanations. arXiv preprint arXiv:1812.04801 (2018)
77. Vu, M.A.T., et al.: A shared vision for machine learning in neuroscience. J. Neurosci. 0508–17 (2018)
78. Wang, R., Wang, X., Inouye, D.I.: Shapley explanation networks. arXiv preprint arXiv:2104.02297 (2021)
79. Winkler, J.K., et al.: Association between surgical skin markings in dermoscopic images and diagnostic performance of a deep learning convolutional neural network for melanoma RecognitionSurgical skin markings in dermoscopic images and deep learning convolutional neural network recognition of MelanomaSurgical skin markings in dermoscopic images and deep learning convolutional neural network recognition of melanoma. JAMA Dermatol. (2019). https://doi.org/10.1001/jamadermatol.2019.1735
80. Yeom, S.K., et al.: Pruning by explaining: a novel criterion for deep neural network pruning. Pattern Recogn. **115**, 107899 (2021)
81. Yosinski, J., Clune, J., Nguyen, A., Fuchs, T., Lipson, H.: Understanding neural networks through deep visualization. arXiv preprint arXiv:1506.06579 (2015)
82. Zeiler, M.D., Fergus, R.: Visualizing and understanding convolutional networks. In: Fleet, D., Pajdla, T., Schiele, B., Tuytelaars, T. (eds.) ECCV 2014. LNCS, vol. 8689, pp. 818–833. Springer, Cham (2014). https://doi.org/10.1007/978-3-319-10590-1_53
83. Zhang, H., Cheng, X., Chen, Y., Zhang, Q.: Game-theoretic interactions of different orders. arXiv preprint arXiv:2010.14978 (2020)
84. Zhang, Q., Cao, R., Shi, F., Wu, Y.N., Zhu, S.C.: Interpreting CNN knowledge via an explanatory graph. arXiv preprint arXiv:1708.01785 (2017)
85. Zintgraf, L.M., Cohen, T.S., Adel, T., Welling, M.: Visualizing deep neural network decisions: prediction difference analysis. arXiv preprint arXiv:1702.04595 (2017)

Beyond the Visual Analysis of Deep Model Saliency

Sarah Adel Bargal[1]([⊠])(ⓘ), Andrea Zunino[2](ⓘ), Vitali Petsiuk[1](ⓘ), Jianming Zhang[3](ⓘ), Vittorio Murino[4](ⓘ), Stan Sclaroff[1](ⓘ), and Kate Saenko[1](ⓘ)

[1] Boston University, Boston, MA 02215, USA
{sbargal,vpetsiuk,sclaroff,saenko}@bu.edu
[2] Leonardo Labs, 16149 Genova, Italy
andrea.zunino.ext@leonardo.com
[3] Adobe Research, San Jose, CA 95110, USA
jianmzha@adobe.com
[4] University of Verona, 37134 Verona, Italy
vittorio.murino@univr.it

Abstract. Increased explainability in machine learning is traditionally associated with lower performance, *e.g.* a decision tree is more explainable, but less accurate than a deep neural network. We argue that, in fact, increasing the explainability of a deep classifier can improve its generalization. In this chapter, we survey a line of our published work that demonstrates how spatial and spatiotemporal visual explainability can be obtained, and how such explainability can be used to train models that generalize better on unseen in-domain and out-of-domain samples, refine fine-grained classification predictions, better utilize network capacity, and are more robust to network compression.

Keywords: Explainability · Interpretability · Deep learning · Saliency

1 Introduction

Deep learning is now widely used in state-of-the-art Artificial Intelligence (AI) technology. A Deep Neural Network (DNN) model however is, thus far, a "black box." AI applications in finance, medicine, and autonomous vehicles demand justifiable predictions, barring most deep learning methods from use. Understanding what is going on inside the "black box" of a DNN, what the model has learned, and how the training data influenced that learning are all instrumental as AI serves humans and should be accountable to humans and society.

In response, Explainable AI (XAI) popularizes a series of visual explanations called *saliency methods*, that highlight pixels that are "important" for a model's final prediction to which we contribute multiple works that target understanding deep model behavior through the analysis of saliency maps that highlight regions of evidence used by the model. We then contribute works that utilize such saliency to obtain models that have

---

S. A. Bargal and A. Zunino—Equal contribution.

© The Author(s) 2022
A. Holzinger et al. (Eds.): xxAI 2020, LNAI 13200, pp. 255–269, 2022.
https://doi.org/10.1007/978-3-031-04083-2_13

improved accuracy, network utilization, robustness, and domain generalization. In this work, we provide an overview of our contributions in this field.

XAI in Visual Data. Grounding model decisions in visual data has the benefit of being clearly interpretable by humans. The evidence upon which a deep convolutional model participates in the class conditional probability for a specific class is highlighted in the form of a saliency map. In our work [36], we present applications of spatial grounding in model interpretation, data annotation assistance for facial expression analysis and medical imaging tasks, and as a diagnostic tool for model misclassifications. We do so in a discriminative way that highlights evidence for every possible outcome given the same input for any deep *convolutional* neural network classifier.

We also propose a black-box grounding techniques RISE [22] and D-RISE [23]. Unlike the majority of previous approaches RISE can produce saliency maps without the access to the internal states of the base model, such as weights, gradients or feature maps. The advantages of such a black-box approach are that RISE does not assume any specifics about the base model architecture, it can be used to test proprietary models that do not allow full access, the implementation is very easily adapted to a new base model. The saliency is computed by perturbing the input image using a set of randomized masks while keeping track of the changes in the output. Major changes in the output are reflected in increased saliency of the perturbed region of the input, see Fig. 2.

Deep *recurrent* models are state-of-the-art for many vision tasks including video action recognition and video captioning. Models are trained to caption or classify activity in videos, but little is known about the evidence used to make such decisions. Our work was the first to formulate top-down saliency in deep recurrent models for space-time grounding of videos [1]. We do so using a *single contrastive* backward pass of an already trained model. This enables the visualization of spatiotemporal cues that contribute to a deep model's classification/captioning output and localization of segments within a video that correspond with a specific action, or phrase from a caption, without explicitly optimizing/training for these tasks.

XAI for Improved Models. We propose three frameworks that utilize explanations to improve model accuracy. The first proposes a guided dropout regularizer for deep networks [39] based on the explanation of a network prediction defined as the firing of neurons in specific paths. The explanation at each neuron is utilized to determine the probability of dropout, rather than dropping out neurons uniformly at random as in standard dropout. This results in dropping out with higher probability neurons that contribute more to decision making at training time, forcing the network to learn alternative paths in order to maintain loss minimization, resulting in a plasticity-like behavior, a characteristic of human brains. This demonstrates better generalization ability, an increased utilization of network neurons, and a higher resilience to network compression for image/video recognition.

Our second training strategy not only leads to a more explainable AI system for object classification, but as a consequence, suffers no perceptible accuracy degradation [40]. Our training strategy enforces a periodic saliency-based feedback to encourage the model to focus on the image regions that directly correspond to the ground-truth object. We propose explainability as a means for bridging the visual-semantic gap between different domains where model explanations are used as a means of disentagling domain specific information from otherwise relevant features. We demonstrate that this leads to improved generalization to new domains without hindering performance on the original domain.

Our third strategy is applied at test time and improves model accuracy by zooming in on the evidence, and ensuring the model has "the right reasons" for a prediction, being defined as reasons that are coherent with those used to make similar correct decisions at training time [2, 3]. The reason/evidence upon which a deep neural network makes a prediction is defined to be the spatial grounding, in the pixel space, for a specific class conditional probability in the model output. We use evidence grounding as the signal to a module that assesses how much one can trust a Convolutional Neural Network (CNN) prediction over another.

The rest of this chapter is organized as follows. Section 2 presents saliency approaches that target explaining how deep neural network models associate input regions to output predictions. Sections 3, 4, and 5 present approaches that utilize explainability in the form of saliency (Sect. 2) to obtain models that possess state-of-the-art in-domain and out-of-domain accuracy, have improved neuron utilization, and are more robust to network compression. Section 6 concludes the presented line of works.

2 Saliency-Based XAI in Vision

In this section we propose sample white- and black-box methods for saliency-based explainability for vision models.

2.1 White-Box Models

We first present sample white-box grounding techniques developed for the purpose of explainability of deep vision models. Formulation of white-box techniques assumes knowledge of model architectures and parameters.

Spatial. In a standard spatial CNN, the forward activation of neuron a_j is computed by $\widehat{a}_j = \phi(\sum_i w_{ij}\widehat{a}_i + b_i)$, where $\widehat{a}_i$ is the activation coming from the previous layer, ϕ is a nonlinear activation function, w_{ij} and b_i are the weight from neuron i to neuron j and the added bias at layer i, respectively. Excitation Backprop (EB) was proposed in [37] to identify the task-relevant neurons in any intermediate layer of a pre-trained CNN network. EB devises a backpropagation formulation that is able to reconstruct the evidence used by a deep model to make decisions. It computes the probability of each neuron recursively using conditional probabilities $P(a_i|a_j)$ in a top-down order starting from a probability distribution over the output units, as follows:

$$P(a_i) = \sum_{a_j \in \mathcal{P}_i} P(a_i|a_j)P(a_j) \tag{1}$$

where $\mathcal{P}_i$ is the parent node set of a_i. EB passes top-down signals through excitatory connections having non-negative activations, excluding from the competition inhibitory ones. EB is designed with an assumption of non-negative activations that are positively correlated with the detection of specific visual features. Most modern CNNs use ReLU activation functions, which satisfy this assumption. Therefore, negative weights can be assumed to not positively contribute to the final prediction. Assuming C_j the child node set of a_j, for each $a_i \in C_j$, the conditional winning probability $P(a_i|a_j)$ is defined as

$$P(a_i|a_j) = \begin{cases} Z_j \hat{a}_i w_{ij}, & \text{if } w_{ij} \geq 0, \\ 0, & \text{otherwise} \end{cases} \qquad (2)$$

where Z_j is a normalization factor such that a probability distribution is maintained, i.e. $\sum_{a_i \in C_j} P(a_i|a_j) = 1$. Recursively propagating the top-down signal and preserving the sum of backpropagated probabilities, it is possible to highlight the salient neurons in each layer using Eq. 1, i.e. neurons that mostly contribute to a specific task. This has been shown to accurately localize spatial objects in images (corresponding to object classes) in a weakly-supervised way.

Spatiotemporal. Spatiotemporal explainability is instrumental for applications like action detection and image/video captioning [32]. We extend EB to become spatiotemporal [1]. This work is the first to formulate top-down saliency in deep recurrent models for space-time grounding of videos. In this section we explain the details of our spatiotemporal grounding framework: cEB-R. As illustrated in Fig. 1, we have three main modules: RNN Backward, Temporal normalization, and CNN Backward.

The *RNN Backward* module implements an excitation backprop formulation for RNNs. Recurrent models such as LSTMs are well-suited for top-down temporal saliency as they explicitly propagate information over time. The extension of EB for Recurrent Networks, EB-R, is not straightforward since EB must be implemented through the unrolled time steps of the RNN and since the original RNN formulation contains $tanh$ non-linearities which do not satisfy the EB assumption. [6,10] have conducted an analysis over variations of the standard RNN formulation, and discovered that different non-linearities performed similarly for a variety of tasks. Based on this, we use *ReLU* nonlinearities and corresponding derivatives, instead of $tanh$. This satisfies the EB assumption, and results in similar performance on both tasks.

Working backwards from the RNN's output layer, we compute the conditional winning probabilities from the set of output nodes O, and the set of dual output nodes $\overline{O}$:

$$P^t(a_i|a_j) = \begin{cases} Z_j \hat{a}_i^t w_{ij}, & \text{if } w_{ij} \geq 0, \\ 0, & \text{otherwise.} \end{cases} \qquad (3)$$

$$\overline{P}^t(a_i|a_j) = \begin{cases} Z_j \hat{a}_i^t \overline{w}_{ij}, & \text{if } \overline{w}_{ij} \geq 0, \\ 0, & \text{otherwise.} \end{cases} \qquad (4)$$

$Z_j = 1/\sum_{i:w_{ij} \geq 0} \hat{a}_i^t w_{ij}$ is a normalization factor such that the sum of all conditional probabilities of the children of a_j (Eqs. 3, 4) sum to 1; $w_{ij} \in W$ where W is the set of model weights and w_{ij} is the weight between child neuron a_i and parent neuron a_j; $\overline{w}_{ij} \in \overline{W}$ where $\overline{W}$ obtained by negating the model weights at the classification layer only. $\overline{P}^t(a_i|a_j)$ is only needed for *contrastive* attention.

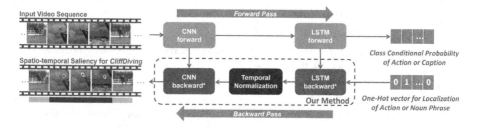

Fig. 1. Our proposed framework spatiotemporally highlights/grounds the evidence that an RNN model used in producing a class label or caption for a given input video. In this example, by using our proposed back-propagation method, the evidence for the activity class *CliffDiving* is highlighted in a video that contains *CliffDiving* and *HorseRiding*. Our model employs a single backward pass to produce saliency maps that highlight the evidence that a given RNN used in generating its outputs.

We compute the neuron winning probabilities starting from the prior distribution encoding a given action/caption as follows:

$$P^t(a_i) = \sum_{a_j \in \mathcal{P}_i} P^t(a_i|a_j)P^t(a_j) \tag{5}$$

$$\overline{P}^t(a_i) = \sum_{a_j \in \mathcal{P}_i} \overline{P}^t(a_i|a_j)\overline{P}^t(a_j) \tag{6}$$

where $\mathcal{P}_i$ is the set of parent neurons of a_i.

Replacing $tanh$ non-linearities with *ReLU* non-linearities to extend EB in time does not suffice for temporal saliency. EB performs normalization at every layer to maintain a probability distribution. For spatiotemporal localization, the *Temporal Normalization* module normalizes signals from the desired n^{th} time-step of a T-frame clip in both time and space (assuming S neurons in current layer) before being further backpropagated into the CNN:

$$P_N^t(a_i) = P^t(a_i)/\sum_{t=1}^{T}\sum_{i=1}^{S} P^t(a_i). \tag{7}$$

$$\overline{P}_N^t(a_i) = \overline{P}^t(a_i)/\sum_{t=1}^{T}\sum_{i=1}^{S} \overline{P}^t(a_i). \tag{8}$$

cEB-R computes the difference between the normalized saliency maps obtained by EB-R starting from O, and EB-R starting from $\overline{O}$ using negated weights of the classification layer. cEB-R is more discriminative as it grounds the evidence that is unique to a selected class/word and not common to other classes used at training time. This is conducted as follows:

$$Map^t(a_i) = P_N^t(a_i) - \overline{P}_N^t(a_i). \tag{9}$$

For every video frame f_t at time step t, we use the backprop of [37] for all CNN layers in the *CNN Backward* module:

$$P^t(a_i|a_j) = \begin{cases} Z_j \widehat{a}_i^t w_{ij}, & \text{if } w_{ij} \geq 0, \\ 0, & \text{otherwise} \end{cases} \tag{10}$$

$$Map^t(a_i) = \sum_{a_j \in P_i} P^t(a_i|a_j)Map^t(a_j) \tag{11}$$

where $\hat{a}_i^t$ is the activation when frame f_t is passed through the CNN. Map^t at the desired CNN layer is the cEB-R saliency map for f_t. Computationally, the complexity of cEB-R is on the order of a *single* backward pass. Note that for EB-R, $P_N^t(a_j)$ is used instead of $Map^t(a_j)$ in Eq. 11.

The general framework has been applied to action localization. We ground the evidence of a specific action using a model trained on this task. The input is a video sequence and the action to be localized, and the output is spatiotemporal saliency maps for this action in the video. Performing cEB-R results in a sequence of saliency maps Map^t for $t = 1, ..., T$. These maps can then be used for localizing the action by finding temporal regions of highest aggregate saliency. This has also been applied to other spatiotemporal applications such as image and video captioning.

2.2 Black-Box Models

Black-box methods operate under the assumption that no internal information about the model is available. Thus we can only observe the final output of the model for each input that we provide. In this paradigm to explain the black-box model one has to come up with a way to query the model in such a way, that the outputs would reveal some of the underlying behaviour of the model. This methods are typically slower than white-box approaches since information is obtained at the cost of additional queries to the model.

One way to construct the queries is to run the model on similar versions of the input and analyze the differences in the output. For example, to compute how important different regions of the inputs are, *i.e.* compute saliency, one can mask out certain parts of the image. Significant changes in the output would mean the importance of the masked region.

Our method RISE [22] builds on this idea. We probe the base model by perturbing the input image using random masks and record its responses to each of the masked images. The saliency map S is computed as a weighted sum of the used masks, where the weights come from the probabilities predicted by the base model (see Fig. 2):

$$S_{I,f} = \frac{1}{\sum_{M \in \mathcal{M}} M} \sum_{M \in \mathcal{M}} f(I \odot M) \cdot M, \tag{12}$$

where f is the base model, I is the input image and $\mathcal{M}$ is the set of generated masks. The mask M has large weight $f(I \odot M)$ in the sum only if the score of the base model is high on the masked image, *i.e.* the mask preserves important regions. We generate masks as a uniformly random binary grid (bilinearly upsampled) to refrain from imposing any priors on the resulting saliency maps.

RISE can be applied to explain models that predict a distribution over labels given an image such as classification and captioning models. Classification saliency methods fail when directly applied to the object detection models. To generate such saliency maps for object detectors we propose D-RISE method [23]. It accounts for the differences in object detection model's structure and output format. To measure the effect of

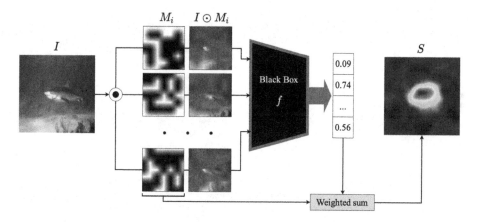

Fig. 2. RISE overview

the masks on the model output we propose a similarity metric between detection two proposals d_t and d_j:

$$s(d_t, d_j) = s_L(d_t, d_j) \cdot s_P(d_t, d_j) \cdot s_O(d_t, d_j), \tag{13}$$

This metrics computes similarity values for the three components of the detection proposals: localization (bounding box L), classification (class probabilities P), and objectness score (O).

$$s_L(d_t, d_j) = \text{IoU}(L_t, L_j), \tag{14}$$

$$s_P(d_t, d_j) = \frac{P_t \cdot P_j}{\|P_t\|\|P_j\|}, \tag{15}$$

$$s_O(d_t, d_j) = O_j. \tag{16}$$

Using the masking technique and the similarity metric D-RISE can compute saliency maps for object detectors in the similar querying manner. We use D-RISE to gain insights into the use of context by the detector. We demonstrate how to use saliency to better understand the use of correlations in the data by the model, *e.g.* ski poles are used when detecting the ski class. We also demonstrate the utility of saliency maps for detecting accidental or adversarial biases in the data.

3   XAI for Improved Models: Excitation Dropout

Dropout avoids overfitting on training data, allowing for better generalization on unseen test data. In this work, we target at determining how the dropped neurons are selected, answering the question *Which neurons to drop out?*

Our approach [39] is inspired by brain plasticity [8,17,18,29]. We deliberately, and temporarily, paralyze/injure neurons to enforce learning alternative paths in a deep network. At training time, neurons that are more relevant to the correct prediction,

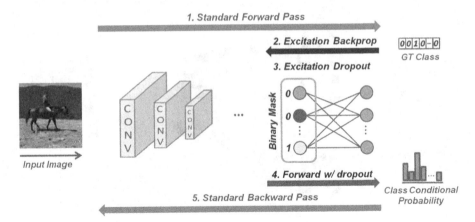

Fig. 3. Training pipeline of Excitation Dropout. *Step 1:* A minibatch goes through the standard forward pass. *Step 2:* Backward EB is performed until the specified dropout layer; this gives a neuron saliency map at the dropout layer in the form of a probability distribution. *Step 3:* The probability distribution is used to generate a binary mask for each image of the batch based on a Bernoulli distribution determining whether each neuron will be dropped out or not. *Step 4:* A forward pass is performed from the specified dropout layer to the end of the network, zeroing the activations of the dropped out neurons. *Step 5:* The standard backward pass is performed to update model weights.

i.e. neurons having a high saliency, are given a higher dropout probability. The relevance of a neuron for making a certain prediction is quantified using Excitation Backprop [37]. Excitation Backprop conveniently yields a probability distribution at each layer that reflects neuron saliency, or neuron contribution to the prediction being made. This is utilized in the training pipeline of our approach, named *Excitation Dropout*, which is summarized in Fig. 3.

Method. In the standard formulation of dropout [9,31], the suppression of a neuron in a given layer is modeled by a Bernoulli random variable p which is defined as the probability of retaining a neuron, $0 < p \leq 1$. Given a specific layer where dropout is applied, during the training phase, each neuron is turned off with a probability $1 - p$.

We argue for a different approach that is *guided* in the way it selects neurons to be dropped. In a training iteration, certain paths have high excitation contributing to the resulting classification, while other regions of the network have low responses. We encourage the learning of alternative paths (plasticity) through the temporary damaging of the currently highly excited path. We re-define the probability of retaining a neuron as a function of its contribution in the currently highly excited path

$$p = 1 - \frac{(1 - P) * (N - 1) * p_{EB}}{((1 - P) * N - 1) * p_{EB} + P} \tag{17}$$

where p_{EB} is the probability backpropagated through the EB formulation (Eq. 1) in layer l, P is the *base* probability of retaining a neuron when all neurons are equally contributing to the prediction and N is the number of neurons in a fully-connected layer

l or the number of filters in a convolutional layer l. The retaining probability defined in Eq. 17 drops the neurons that contribute the most to the recognition of a specific class, with higher probability. Dropping out highly relevant neurons, we retain less relevant ones and thus encourage them to awaken.

Results. We evaluate the effectiveness of Excitation Dropout on popular network architectures that employ dropout layers including AlexNet [14], VGG16 [28], VGG19 [28], and CNN-2 [19]. We perform dropout in the first fully-connected layer of the networks and find that it results in a 1%–5% accuracy improvement in comparison to Standard Dropout and other proposed dropout variants in the literature including Adaptive, Information, Standard, and Curriculum Dropout. These results have been validated on image and video datasets including UCF101 [30], Cifar10 [13], Cifar100 [13], and Caltech256 [7].

Excitation Dropout shows a higher number of active neurons, a higher entropy over activations, and a probability distribution p_{EB} that is more spread (higher entropy over p_{EB}) among the neurons of the layer, leading to a lower peak probability of p_{EB} and therefore less specialized neurons. These results are observed to have consistent trends over all training iterations for examined image and video recognition datasets. Excitation Dropout also enables networks to have a higher robustness against network compression for all examined datasets. It is capable of maintaining a much less steep decline of GT probability as more neurons are pruned. Explainability has also been recently used to prune networks for transfer learning from large corpora to more specialized tasks [35].

4   XAI for Improved Models: Domain Generalization

While Sect. 3 focuses on dropping neurons 'relevant' to a prediction as a means of network regularization within a particular domain, we now propose using such relevance to focus on domain agnostic features that can aid domain generalization.

We develop a training strategy [40] for deep neural network models that increases explainability, suffers no perceptible accuracy degradation on the training domain, and improves performance on unseen domains.

We posit that the design of algorithms that better mimic the way humans reason, or "explain", can help mitigate domain bias. Our approach utilizes explainability as a means for bridging the visual-semantic gap between different domains as presented in Fig. 4. Specifically, our training strategy is guided by model explanations and available human-labeled explanations, mimicking interactive human feedback [26]. Explanations are defined as regions of visual evidence upon which a network makes a decision. This is represented in the form of a saliency map conveying how much each pixel contributed to the network's decision.

Our training strategy periodically guides the forward activations of spatial layer(s) of a Convolutional Neural Network (CNN) trained for object classification. The activations are guided to focus on regions in the image that directly correspond to the ground-truth (GT) class label, as opposed to context that may more likely be domain dependent. The proposed strategy aims to reinforce explanations that are non-domain specific, and alleviate explanations that are domain specific. Classification models are

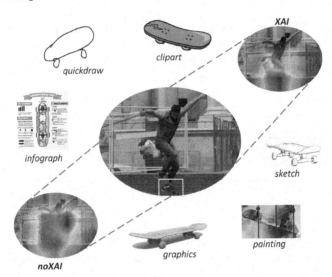

Fig. 4. In this figure we demonstrate how explainability (XAI) can be used to achieve domain generalization from a *single source*. Training a deep neural network model to enforce explainability, *e.g.* focusing on the skateboard region (red is most salient, and blue is least salient) for the ground-truth class skateboard in the central training image, enables improved generalization to other domains where the background is not necessarily class-informative. (Color figure online)

compact and fast in comparison to more complex semantic segmentation models. This allows the compact classification model to possess some properties of a segmentation model without increasing model complexity or test-time overhead.

Method. We enforce focusing on objects in an image by scaling the forward activations of a particular spatial layer l in the network at certain epochs. We generate a multiplicative binary *mask* for guiding the focus of the network in the layer in which we are enforcing XAI. For an explainable image x^i, the binary mask is a binarization of the achieved saliency map, *i.e.* $mask_{j,k}^i = \mathbb{1}(s_{j,k}^i > 0) \ \forall j \ \forall k, \ j = 1, \ldots, W$ and $k = 1, \ldots, H$, where W and H are the spatial dimension of a layers' output neuron activations; The mask is active at locations of non-zero saliency. This re-inforces the activations corresponding to the active saliency regions that have been classified as being explainable. For images that need an improved explanation, the binary mask is assigned to be the GT spatial annotation $mask_{j,k}^i = g_{j,k}^i \ \forall j \ \forall k, \ j = 1, \ldots, W$ and $k = 1, \ldots, H$; The mask is active at GT locations. This increases the frequency at which the network reinforces activations at locations that are likely to be non-domain specific and suppresses activations at locations that are likely to be domain specific. We then perform element-wise multiplication of our computed *mask* with the forward activations of layer l; *i.e.* $a_{j,k}^{l,i} = mask_{j,k}^i * a_{j,k}^{l,i} \ \forall j \ \forall k, \ j = 1, \ldots, W$ and $k = 1, \ldots, H$.

Results. The identification of evidence within a visual input using top-down neural attention formulations [27] can be a powerful tool for domain analysis. We demonstrate that more explainable deep classification models could be trained without hindering their performance.

We train ResNet architectures for the single-label classification task for the popular MSCOCO [15] and PASCAL VOC [4] datasets. The XAI model resulted in a 25% increase in the number of correctly classified images that result in better localization/explainability using the popular pointing game metric. The XAI model has learnt to rely less on context information, without hurting the performance.

Thus far, evaluation assumed that a saliency map whose peak overlaps with the GT spatial annotation of the object is a better explanation. We then conduct a human study to confirm our intuitive quantification of an "explainable" model. The study asks users what they think is a better explanation for the presence of an object. XAI evidence won for 67% of the whole image population and 80% of the images with a winner choice.

Finally, we demonstrate how the explainable model better generalizes from real images of MSCOCO/PASCAL VOC to six unseen target domains from the Domain-Net [20] and Syn2Real [21] datasets (clipart, quickdraw, infograph, painting, sketch, and graphics).

5   XAI for Improved Models: Guided Zoom

In state-of-the-art deep single-label classification models, the top-k $(k = 2, 3, 4, \ldots)$ accuracy is usually significantly higher than the top-1 accuracy. This is more evident in fine-grained datasets, where differences between classes are quite subtle. Exploiting the information provided in the top k predicted classes boosts the final prediction of a model. We propose Guided Zoom [3], a novel way in which explainability could be used to improve model performance. We do so by making sure the model has "the right reasons" for a prediction. The reason/evidence upon which a deep neural network makes a prediction is defined to be the grounding, in the pixel space, for a specific class conditional probability in the model output. Guided Zoom examines how reasonable the evidence used to make each of the top-k predictions is. In contrast to work that implements reasonableness in the loss function *e.g.* [24, 25], test time evidence is deemed reasonable in Guided Zoom if it is coherent with evidence used to make similar correct decisions at training time. This leads to better informed predictions.

Method. We now describe how Guided Zoom utilizes multiple discriminative evidence, does not require part annotations, and implicitly enforces part correlations. This is done through explanations of the main modules depicted in Fig. 5.

Conventional CNNs trained for image classification output class conditional probabilities upon which predictions are made. The class conditional probabilities are the result of some corresponding evidence in the input image. From correctly classified training examples, we generate a reference pool $\mathcal{P}$ of (evidence, prediction) pairs over which the *Evidence CNN* will be trained for the same classification task. We recover/ground such evidence using several grounding techniques [1, 22, 27]. We extract the image patch corresponding to the peak saliency region. This patch highlights the most discriminative evidence. However, the next most discriminative patches may also be good additional evidence for differentiating fine-grained categories.

Also, grounding techniques only highlight part(s) of an object. However, a more inclusive segmentation map can be extracted from the already trained model at test time using an iterative adversarial erasing of patches [33]. We augment our reference pool with patches resulting from performing iterative adversarial erasing of the most

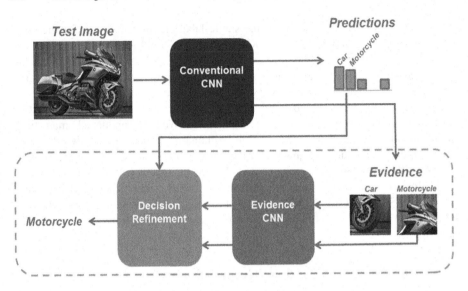

Fig. 5. Pipeline of Guided Zoom. A conventional CNN outputs class conditional probabilities for an input image. Salient patches could reveal that evidence is weak. We refine the class prediction of the conventional CNN by introducing two modules: 1) *Evidence CNN* determines the consistency between the evidence of a test image prediction and that of correctly classified training examples of the same class. 2) *Decision Refinement* uses the output of *Evidence CNN* to refine the prediction of the conventional CNN.

discriminative evidence from an image. We notice that adversarial erasing results in implicit part localization from most to least discriminative parts. All patches extracted from this process inherit the ground-truth label of the original image. By labeling different parts with the same image ground-truth label, we are implicitly forcing part-label correlations in *Evidence CNN*.

Including such additional evidence in our reference pool gives a richer description of the examined classes compared to models that recursively zoom into one location while ignoring other discriminative cues [5]. We note that we add an evidence patch to the reference pool only if the removal of the previous salient patch does not affect the correct classification of the sample image. Erasing is performed by adding a black-filled square on the previous most salient evidence to encourage a highlight of the next salient evidence. We then train a CNN model, *Evidence CNN*, on the generated evidence pool.

At test time, we analyze whether the evidence upon which a prediction is made is reasonable. We do so by examining the consistency of a test (evidence, prediction) with our reference pool that is used to train *Evidence CNN*. We exploit the visual evidence used for each of the top-k predictions for *Decision Refinement*. The refined prediction will be inclined toward each of the top-k classes by an amount proportional to how coherent its evidence is with the reference pool. For example, if the (evidence, prediction) of the second-top predicted class is more coherent with the reference pool of this class, then the refined prediction will be more inclined toward the second-top class.

Assuming test image s^j, where $j \in 1, \ldots, m$ and m is the number of testing examples, s^j is passed through the conventional CNN resulting in $v^{j,0}$, a vector of class conditional probabilities having some top-k classes $c_1, \ldots, c_k$ to be considered for the prediction refinement. We obtain the evidence for each of the top-k predicted classes $e_0^{j,c_1}, \ldots, e_0^{j,c_k}$, and pass each one through the *Evidence CNN* to get the output class conditional probability vectors $v_0^{j,c_1}, \ldots, v_0^{j,c_k}$. We then perform adversarial erasing to get the next most salient evidence $e_l^{j,c_1}, \ldots, e_l^{j,c_k}$ and their corresponding class conditional probability vectors $v_l^{j,c_1}, \ldots, v_l^{j,c_k}$, for $l \in 1, \ldots, L$. Finally, we compute a weighted combination of all class conditional probability vectors proportional to their saliency (a lower l has more discriminative evidence and is therefore assigned a higher weight w_l). The estimated, refined class c_{ref}^j is determined as the class having the maximum aggregate prediction in the weighted combination.

Results. We show that Guided Zoom results in an improvement of a model's classification accuracy on four fine-grained classification datasets: CUB-200-2011 Birds [34], Stanford Dogs [11], FGVC-Aircraft [16], and Stanford Cars [12] of various bird species, dog species, aircraft models, and car models.

Guided Zoom is a generic framework that can be directly applied to any deep convolutional model for decision refinement within the top-k predictions. Guided zoom demonstrates that multi-zooming is more beneficial than a single recursive zoom [5]. We also demonstrate that Guided Zoom further improves the performance of existing multi-zoom approaches [38]. Choosing random patches to be used with original images, as opposed to Guided Zoom patches results in comparable results to using the original images on their own. Therefore, Guided Zoom presents performance gains that are complementary to data augmentation.

6 Conclusion

This chapter presents sample white- and black-box approaches to providing visual grounding as a form of explainable AI. It also presents a human judgement verification that such visual explainability techniques mostly agree with evidence humans use for the presence of visual cues. This chapter then demonstrates three strategies on how this preliminary form of explainable AI (also widely known as saliency maps) can be integrated into automated algorithms, that do not require human feedback, to improve fine-grained accuracy, in-domain and out-of-domain generalization, network utilization, and robustness to network compression.

References

1. Bargal, S.A., Zunino, A., Kim, D., Zhang, J., Murino, V., Sclaroff, S.: Excitation backprop for RNNs. In: Proceedings of the IEEE Conference on Computer Vision and Pattern Recognition (CVPR) (2018)
2. Bargal, S.A., et al.: Guided zoom: questioning network evidence for fine-grained classification. In: Proceedings of the British Machine Vision Conference (BMVC) (2019)
3. Bargal, S.A., et al.: Guided zoom: zooming into network evidence to refine fine-grained model decisions. IEEE Trans. Pattern Anal. Mach. Intell. (PAMI) **01**, 1 (2021)
4. Everingham, M., Eslami, S.A., Van Gool, L., Williams, C.K., Winn, J., Zisserman, A.: The pascal visual object classes challenge: a retrospective. Int. J. Comput. Vis. (IJCV) **111**(1), 98–136 (2015)

5. Fu, J., Zheng, H., Mei, T.: Look closer to see better: recurrent attention convolutional neural network for fine-grained image recognition. In: Proceedings of the IEEE Conference on Computer Vision and Pattern Recognition (CVPR) (2017)

6. Greff, K., Srivastava, R.K., Koutník, J., Steunebrink, B.R., Schmidhuber, J.: LSTM: a search space odyssey. IEEE Trans. Neural Netw. Learn. Syst. (2016)

7. Griffin, G., Holub, A., Perona, P.: Caltech-256 object category dataset. Technical report 7694, California Institute of Technology (2007). http://authors.library.caltech.edu/7694

8. Hebb, D.O.: The Organization of Behavior: A Neuropsychological Theory. Psychology Press, Hove (2005)

9. Hinton, G.E., Srivastava, N., Krizhevsky, A., Sutskever, I., Salakhutdinov, R.: Improving neural networks by preventing co-adaptation of feature detectors. CoRR abs/1207.0580 (2012)

10. Jozefowicz, R., Zaremba, W., Sutskever, I.: An empirical exploration of recurrent network architectures. In: Proceedings of the International Conference on Machine Learning (ICML) (2015)

11. Khosla, A., Jayadevaprakash, N., Yao, B., Li, F.F.: Novel dataset for fine-grained image categorization: Stanford dogs. In: Proceedings of the IEEE Conference on Computer Vision and Pattern Recognition Workshop (CVPRw) (2011)

12. Krause, J., Stark, M., Deng, J., Fei-Fei, L.: 3D object representations for fine-grained categorization. In: Proceedings of the IEEE Conference on Computer Vision and Pattern Recognition Workshop (CVPRw) (2013)

13. Krizhevsky, A., Hinton, G., et al.: Learning multiple layers of features from tiny images. Citeseer (2009)

14. Krizhevsky, A., Sutskever, I., Hinton, G.E.: Imagenet classification with deep convolutional neural networks. In: Advances in Neural Information Processing Systems (NIPS) (2012)

15. Lin, T.Y., et al.: Microsoft COCO: common objects in context. In: Proceedings of the European Conference on Computer Vision (ECCV) (2014)

16. Maji, S., Rahtu, E., Kannala, J., Blaschko, M., Vedaldi, A.: Fine-grained visual classification of aircraft. arXiv preprint arXiv:1306.5151 (2013)

17. Miconi, T., Clune, J., Stanley, K.O.: Differentiable plasticity: training plastic neural networks with backpropagation. arXiv preprint arXiv:1804.02464 (2018)

18. Mittal, D., Bhardwaj, S., Khapra, M.M., Ravindran, B.: Recovering from random pruning: on the plasticity of deep convolutional neural networks. In: Winter Conference on Applications of Computer Vision (2018)

19. Morerio, P., Cavazza, J., Volpi, R., Vidal, R., Murino, V.: Curriculum dropout. In: Proceedings of the IEEE International Conference on Computer Vision (ICCV) (2017)

20. Peng, X., Bai, Q., Xia, X., Huang, Z., Saenko, K., Wang, B.: Moment matching for multi-source domain adaptation. In: Proceedings of the IEEE International Conference on Computer Vision (ICCV) (2019)

21. Peng, X., Usman, B., Kaushik, N., Hoffman, J., Wang, D., Saenko, K.: Visda: The visual domain adaptation challenge. arXiv preprint arXiv:1710.06924 (2017)

22. Petsiuk, V., Das, A., Saenko, K.: RISE: randomized input sampling for explanation of black-box models. In: Proceedings of the British Machine Vision Conference (BMVC) (2018)

23. Petsiuk, V., et al.: Black-box explanation of object detectors via saliency maps. In: Proceedings of the IEEE Conference on Computer Vision and Pattern Recognition (CVPR) (2021)

24. Rieger, L., Singh, C., Murdoch, W., Yu, B.: Interpretations are useful: penalizing explanations to align neural networks with prior knowledge. In: International Conference on Machine Learning, pp. 8116–8126. PMLR (2020)

25. Ross, A.S., Hughes, M.C., Doshi-Velez, F.: Right for the right reasons: Training differentiable models by constraining their explanations. arXiv preprint arXiv:1703.03717 (2017)

26. Schramowski, P., et al.: Making deep neural networks right for the right scientific reasons by interacting with their explanations. Nat. Mach. Intell. **2**(8), 476–486 (2020)
27. Selvaraju, R.R., Cogswell, M., Das, A., Vedantam, R., Parikh, D., Batra, D.: Grad-cam: visual explanations from deep networks via gradient-based localization. In: Proceedings of the IEEE International Conference on Computer Vision (ICCV) (2017)
28. Simonyan, K., Zisserman, A.: Very deep convolutional networks for large-scale image recognition. arXiv preprint arXiv:1409.1556 (2014)
29. Song, S., Miller, K.D., Abbott, L.F.: Competitive hebbian learning through spike-timing-dependent synaptic plasticity. Nat. Neurosci. **3**(9), 919 (2000)
30. Soomro, K., Zamir, A.R., Shah, M.: UCF101: A dataset of 101 human actions classes from videos in the wild. arXiv preprint arXiv:1212.0402 (2012)
31. Srivastava, N., Hinton, G., Krizhevsky, A., Sutskever, I., Salakhutdinov, R.: Dropout: a simple way to prevent neural networks from overfitting. J. Mach. Learn. Res. (JMLR) **15**(1), 1929–1958 (2014)
32. Sun, J., Lapuschkin, S., Samek, W., Binder, A.: Explain and improve: LRP-inference fine-tuning for image captioning models. Inf. Fusion **77**, 233–246 (2022)
33. Wei, Y., Feng, J., Liang, X., Cheng, M.M., Zhao, Y., Yan, S.: Object region mining with adversarial erasing: a simple classification to semantic segmentation approach. In: Proceedings of the IEEE Conference on Computer Vision and Pattern Recognition (CVPR) (2017)
34. Welinder, P., et al.: Caltech-UCSD birds 200. In: Technical Report CNS-TR-2010-001, California Institute of Technology (2010)
35. Yeom, S.K., et al.: Pruning by explaining: a novel criterion for deep neural network pruning. Pattern Recogn. **115**, 107899 (2021)
36. Zhang, J., Bargal, S.A., Lin, Z., Brandt, J., Shen, X., Sclaroff, S.: Top-down neural attention by excitation backprop. Int. J. Comput. Vis. (IJCV) **126**(10), 1084–1102 (2018)
37. Zhang, J., Lin, Z., Brandt, J., Shen, X., Sclaroff, S.: Top-down neural attention by excitation backprop. In: Proceedings of the European Conference on Computer Vision (ECCV) (2016)
38. Zheng, H., Fu, J., Mei, T., Luo, J.: Learning multi-attention convolutional neural network for fine-grained image recognition. In: Proceedings of the IEEE International Conference on Computer Vision (ICCV) (2017)
39. Zunino, A., Bargal, S.A., Morerio, P., Zhang, J., Sclaroff, S., Murino, V.: Excitation Dropout: encouraging plasticity in deep neural networks. Int. J. Comput. Vis. (IJCV) (2021)
40. Zunino, A., et al.: Explainable deep classification models for domain generalization. In: Proceedings of the IEEE/CVF Conference on Computer Vision and Pattern Recognition, pp. 3233–3242 (2021)

ECQ^x: Explainability-Driven Quantization for Low-Bit and Sparse DNNs

Daniel Becking[1] , Maximilian Dreyer[1], Wojciech Samek[1,2](✉) ,
Karsten Müller[1](✉) , and Sebastian Lapuschkin[1](✉)

[1] Department of Artificial Intelligence, Fraunhofer Heinrich Hertz Institute,
Berlin, Germany
{daniel.becking,maximilian.dreyer,wojciech.samek,
karsten.mueller,sebastian.lapuschkin}@hhi.fraunhofer.de
[2] BIFOLD - Berlin Institute for the Foundations of Learning and Data,
Berlin, Germany

Abstract. The remarkable success of deep neural networks (DNNs) in various applications is accompanied by a significant increase in network parameters and arithmetic operations. Such increases in memory and computational demands make deep learning prohibitive for resource-constrained hardware platforms such as mobile devices. Recent efforts aim to reduce these overheads, while preserving model performance as much as possible, and include parameter reduction techniques, parameter quantization, and lossless compression techniques.

In this chapter, we develop and describe a novel quantization paradigm for DNNs: Our method leverages concepts of explainable AI (XAI) and concepts of information theory: Instead of assigning weight values based on their distances to the quantization clusters, the assignment function additionally considers weight relevances obtained from Layer-wise Relevance Propagation (LRP) and the information content of the clusters (entropy optimization). The ultimate goal is to preserve the most relevant weights in quantization clusters of highest information content.

Experimental results show that this novel Entropy-Constrained and XAI-adjusted Quantization (ECQ^x) method generates ultra low-precision (2–5 bit) and simultaneously sparse neural networks while maintaining or even improving model performance. Due to reduced parameter precision and high number of zero-elements, the rendered networks are highly compressible in terms of file size, up to $103\times$ compared to the full-precision unquantized DNN model. Our approach was evaluated on different types of models and datasets (including Google Speech Commands, CIFAR-10 and Pascal VOC) and compared with previous work.

Keywords: Neural Network Quantization · Layer-wise Relevance Propagation (LRP) · Explainable AI (XAI) · Neural Network Compression · Efficient Deep Learning

© The Author(s) 2022
A. Holzinger et al. (Eds.): xxAI 2020, LNAI 13200, pp. 271–296, 2022.
https://doi.org/10.1007/978-3-031-04083-2_14

1   Introduction

Solving increasingly complex real-world problems continuously contributes to the success of deep neural networks (DNNs) [37,38]. DNNs have long been established in numerous machine learning tasks and for this have been significantly improved in the past decade. This is often achieved by over-parameterizing models, i.e., their performance is attributed to their growing topology, adding more layers and parameters per layer [18,41]. Processing a very large number of parameters comes at the expense of memory and computational efficiency. The sheer size of state-of-the-art models makes it difficult to execute them on resource-constrained hardware platforms. In addition, an increasing number of parameters implies higher energy consumption and increasing run times.

Such immense storage and energy requirements however contradict the demand for efficient deep learning applications for an increasing number of hardware-constrained devices, e.g., mobile phones, wearable devices, Internet of Things, autonomous vehicles or robots. Specific restrictions of such devices include limited energy, memory, and computational budget. Beyond these, typical applications on such devices, e.g., healthcare monitoring, speech recognition, or autonomous driving, require low latency and/or data privacy. These latter requirements are addressed by executing and running the aforementioned applications directly on the respective devices (also known as "edge computing") instead of transferring data to third-party cloud providers prior to processing.

In order to tailor deep learning to resource-constrained hardware, a large research community has emerged in recent years [10,45]. By now, there exists a vast amount of tools to reduce the number of operations and model size, as well as tools to reduce the precision of operands and operations (bit width reduction, going from floating point to fixed point). Topics range from neural architecture search (NAS), knowledge distillation, pruning/sparsification, quantization and lossless compression to hardware design.

Beyond all, quantization and sparsification are very promising and show great improvements in terms of neural network efficiency optimization [21,43]. Sparsification sets less important neurons or weights to zero and quantization reduces parameters' bit widths from default 32 bit float to, e.g., 4 bit integer. These two techniques enable higher computational throughput, memory reduction and skipping of arithmetic operations for zero-valued elements, just to name a few benefits. However, combining both high sparsity and low precision is challenging, especially when relying only on the weight magnitudes as a criterion for the assignment of weights to quantization clusters.

In this work, we propose a novel neural network quantization scheme to render low-bit *and* sparse DNNs. More precisely, our contributions can be summarized as follows:

1. Extending the state-of-the-art concept of entropy-constrained quantization (ECQ) to utilize concepts of XAI in the clustering assignment function.
2. Use relevances observed from Layer-wise Relevance Propagation (LRP) at the granularity of per-weight decisions to correct the magnitude-based weight assignment.

3. Obtaining state-of-the-art or better results in terms of the trade-off between efficiency and performance compared to the previous work.

The chapter is organized as follows: First, an overview of related work is given. Second, in Sect. 3, basic concepts of neural network quantization are explained, followed by entropy-constrained quantization. Section 4 describes the ECQ extension towards ECQ^x as an explainability-driven approach. Here, LRP is introduced and the per-weight relevance derivation for the assignment function presented. Next, the ECQ^x algorithm is described in detail. Section 5 presents the experimental setup and obtained results, followed by the final conclusion in Sect. 6.

2 Related Work

A large body of literature exists that has focused on improving DNN model efficiency. Quantization is an approach that has shown great success [14]. While most research focuses on reducing the bit width for inference, [52] and others focus on quantizing weights, gradients and activations to also accelerate backward pass and training. Quantized models often require fine-tuning or re-training to adjust model parameters and compensate for quantization-induced accuracy degradation. This is especially true for precisions <8 bit (cf. Fig. 1 in Sect. 3). Trained quantization is often referred to as "quantization-aware training", for which additional trainable parameters may be introduced (e.g., scaling parameters [6] or directly trained quantization levels (centroids) [53]). A precision reduction to even 1 bit was introduced by BinaryConnect [8]. However, this kind of quantization usually results in severe accuracy drops. As an extension, ternary networks allow weights to be zero, i.e., constraining them to 0 in addition to w_- and w_+, which yields results that outperform the binary counterparts [28]. In DNN quantization, most clustering approaches are based on distance measurements between the unquantized weight distribution and the corresponding centroids. The works in [7] and [32] were pioneering in using Hessian-weighted and entropy-constrained clustering techniques. More recently the work of [34] use concepts from XAI for DNN quantization. They use DeepLIFT importance measures which are restricted to the granularity of convolutional channels, whereas our proposed ECQ^x computes LRP relevances per weight.

Another method for reducing the memory footprint and computational cost of DNNs is sparsification. In the scope of sparsification techniques, weights with small saliency (i.e., weights which minimally affect the model's loss function) are set to zero, resulting in a sparser computational graph and higher compressible matrices. Thus, it can be interpreted as a special form of quantization, having only one quantization cluster with centroid value 0 to which part of the parameter elements are assigned to. This sparsification can be carried out as unstructured sparsification [17], where any weight in the matrix with small saliency is set to zero, independently of its position. Alternatively, a structured sparsification is applied, where an entire regular subset of parameters is set to zero, e.g., entire convolutional filters, matrix rows or columns [19]. "Pruning" is conceptually related to sparsification but actually removes the respective weights rather than setting them to zero. This has the effect of changing the number of input and output shapes of layers and weight matrices[1]. Most pruning/sparsification approaches are magnitude-based, i.e., weight saliency is approximated by the weight values, which is straightforward. However, since the early 1990s methods that use, e.g., second-order Taylor information for weight saliency [27] have been used alongside other criteria ranging from random pruning to correlation and similarity measures (for the interested reader we recommend [21]). In [51], LRP relevances were first used for structured pruning.

Generating efficient neural network representations can also be a result of combining multiple techniques. In Deep Compression [16], a three-stage model compression pipeline is described. First, redundant connections are pruned iteratively. Next, the remaining weights are quantized. Finally, entropy coding is applied to further compress the weight matrices in a lossless manner. This three stage model is also used in the new international ISO/IEC standard on Neural Network compression and Representation (NNR) [24], where efficient data reduction, quantization and entropy coding methods are combined. For coding, the highly efficient universal entropy coder DeepCABAC [47] is used, which yields compression gains of up to 63×. Although the proposed method achieves high compression gains, the compressed representation of the DNN weights require decoding prior to performing inference. In contrast, compressed matrix formats like Compressed Sparse Row (CSR) derive a representation that enables inference directly in the compressed format [49].

Orthogonal to the previously described approaches is the research area of Neural Architecture Search (NAS) [12]. Both manual [36] and automated [44] search strategies have played an important role in optimizing DNN architectures in terms of latency, memory footprint, energy consumption, etc. Microstructural changes include, e.g., the replacement of standard convolutional layers by more efficient types like depth-wise or point-wise convolutions, layer decomposition or factorization, or kernel size reduction. The macro architecture specifies the type of modules (e.g., inverted residual), their number and connections.

Knowledge distillation (KD) [20] is another active branch of research that aims at generating efficient DNNs. The KD paradigm leverages a large teacher model that is used to train a smaller (more efficient) student model. Instead of using the "hard" class labels to train the student, the key idea of model distillation is to deploy the teacher's class probabilities, as they can contain more information about the input.

---

[1] In practice, pruning is often simulated by masking, instead of actually restructuring the model's architecture.

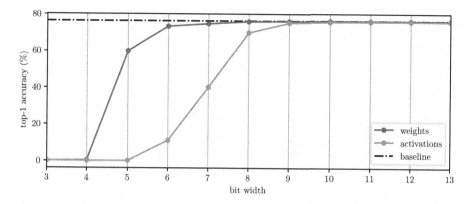

Fig. 1. Difference in sensitivity between activation and weight quantization of the EfficientNet-B0 model pre-trained on ImageNet. As a quantization scheme uniform quantization without re-training was used. Activations are more sensitive to quantization since model performance drops significantly faster. Going below 8 bit is challenging and often requires (quantization-aware) re-training of the model to compensate for the quantization error. Data originates from [50].

3   Neural Network Quantization

For neural network computing, the default precision used on general hardware like GPUs or CPUs is 32 bit floating-point ("single-precision"), which causes high computational costs, power consumption, arithmetic operation latency and memory requirements [43]. Here, quantization techniques can also reduce the number of bits required to represent weight parameters and/or activations of the full-precision neural network, as they map the respective data values to a finite set of discrete quantization levels (clusters). Providing n such clusters allows to represent each data point in only $\log_2 n$ bit. However, the continuous reduction of the number of clusters generally leads to an increasingly large error and degraded performances (see the EfficientNet-B0[2] example in Fig. 1).

This trade-off is a well-known problem in information theory and is addressed by rate-distortion optimization, a concept in lossy data compression. It aims to determine the minimal number of bits per data symbol (bitrate) at which the reconstruction of the compressed data does not exceed a certain level of distortion. Applying this to the domain of neural network quantization, the objective is to minimize the bitrate of the weight parameters while keeping model degradation caused by quantization below a certain threshold, i.e., the predictive performance of the model should not be affected by reduced parameter precisions. In contrast to multimedia compression approaches, e.g., for audio or video coding, the compression of DNNs has unique challenges and opportunities. Foremost, the neural network parameters to be compressed are not perceived directly by

---

[2] https://github.com/lukemelas/EfficientNet-PyTorch, Apache License, Version 2.0 - Copyright (c) 2019 Luke Melas-Kyriazi.

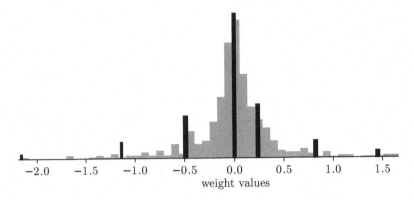

Fig. 2. Quantizing a neural network's layer weights (binned weight distribution shown as green bars) to 7 discrete cluster centers (centroids). The centroids (black bars) were generated by k-means clustering and the height of each bar represents the number of layer weights which are assigned to the respective centroid.

a user, as e.g., for video data. Therefore, the coding or compression error or distortion cannot be directly used as performance measure. Instead, such accuracy measurement needs to be deducted from a subsequent inference step. Then, current neural networks are highly over-parameterized [11] which allows for high errors/differences between the full-precision and the quantized parameters (while still maintaining model performance). Also, the various layer types and the location of a layer within the DNN have different impacts on the loss function, and thus different sensitivities to quantization.

Quantization can be further classified into uniform and non-uniform quantization. The most intuitive way to initialize centroids is by arranging them equidistantly over the range of parameter values (uniform). Other quantization schemes make use of non-uniform mapping functions, e.g., k-means clustering, which is determined by the distribution of weight values (see Fig. 2). As non-uniform quantization captures the underlying distribution of parameter values better, it may achieve less distortion compared to equidistantly arranged centroids. However, non-uniform schemes are typically more difficult to deploy on hardware, e.g., they require a codebook (look-up table), whereas uniform quantization can be implemented using a single scaling factor (step size) which allows a very efficient hardware implementation with fixed-point integer logic.

3.1   Entropy-Constrained Quantization

As discussed in [49], and experimentally shown in [50], lowering the entropy of DNN weights provides benefits in terms of memory as well as computational complexity. The Entropy-Constrained Quantization (ECQ) algorithm is a clustering algorithm that also takes the entropy of the weight distributions into account. More precisely, the first-order entropy $H = -\sum_c P_c \log_2 P_c$ is used, where P_c is the ratio of the number of parameter elements in the c-th cluster to the number of all parameter elements (i.e., the source distribution). To recall, the entropy H is the theoretical limit of the average number of bits required to represent any element of the distribution [39].

Thus, ECQ assigns weight values not only based on their distances to the centroids, but also based on the information content of the clusters. Similar to other rate-distortion-optimization methods, ECQ applies Lagrange optimization:

$$\mathbf{A}^{(l)} = \underset{c}{\operatorname{argmin}}\ d(\mathbf{W}^{(l)}, w_c^{(l)}) - \lambda^{(l)} \log_2(P_c^{(l)}). \tag{1}$$

Per network layer l, the assignment matrix $\mathbf{A}^{(l)}$ maps a centroid to each weight based on a minimization problem consisting of two terms: Given the full-precision weight matrix $\mathbf{W}^{(l)}$ and the centroid values $w_c^{(l)}$, the first term in Eq. (1) measures the squared distance between all weight elements and the centroids, indexed by c. The second term in Eq. (1) is weighted by the scalar Lagrange parameter $\lambda^{(l)}$ and describes the entropy constraint. More precisely, the information content I is considered, i.e., $I = -\log_2(P_c^{(l)})$, where the probability $P_c^{(l)} \in [0, 1]$ defines how likely a weight element $w_{ij}^{(l)} \in \mathbf{W}^{(l)}$ is going to be assigned to centroid $w_c^{(l)}$. Data elements with a high occurrence frequency, or a high probability, contain a low information content, and vice versa. P is calculated layer-wise as $P_c^{(l)} = N_{w_c}^{(l)}/N_{\mathbf{W}}^{(l)}$, with $N_{w_c}^{(l)}$ being the number of full-precision weight elements assigned to the cluster with centroid value $w_c^{(l)}$ (based on the squared distance), and $N_{\mathbf{W}}^{(l)}$ being the total number of parameters in $\mathbf{W}^{(l)}$. Note that $\lambda^{(l)}$ is scaled with a factor based on the number of parameters a layer has in proportion to other layers in the network to mitigate the constraint for smaller layers.

The entropy regularization term motivates sparsity and low-bit weight quantization in order to achieve smaller coded neural network representations. Based on the specific neural network coding optimization, we developed ECQ. This algorithm is based on previous work in Entropy-Constrained Trained Ternarization (EC2T) [28]. EC2T trains sparse and ternary DNNs to state-of-the-art accuracies.

In our developed ECQ, we generalize the EC2T method, such that DNNs of variable bit width can be rendered. Also, ECQ does not train centroid values to facilitate integer arithmetic on general hardware. The proposed quantization-aware training algorithm includes the following steps:

1. Quantize weight parameters by applying ECQ (but keep a copy of the full-precision weights).
2. Apply Straight-Through Estimator (STE) [5]:
   (a) Compute forward and backward pass through quantized model version.
   (b) Update full-precision weights with scaled gradients obtained from quantized model.

4   Explainability-Driven Quantization

Explainable AI techniques can be applied to find relevant features in input as well as latent space. Covering large sets of data, identification of relevant and functional model substructures is thus possible. Assuming over-parameterization of DNNs, the authors of [51] exploit this for pruning (of irrelevant filters) to great effect. Their successful implementation shows the potential of applying XAI for the purpose of quantization as well, as sparsification is part of quantization, e.g., by assigning weights to the zero-cluster. Here, XAI opens up the possibility to go beyond regarding model weights as static quantities and to consider the interaction of the model with given (reference) data. This work aims to combine the two orthogonal approaches of ECQ and XAI in order to further improve sparsity and efficiency of DNNs. In the following, the LRP method is introduced, which can be applied to extract relevances of individual neurons, as well as weights.

4.1   Layer-Wise Relevance Propagation

Layer-wise Relevance Propagation (LRP) [3] is an attribution method based on the conservation of flows and proportional decomposition. It explicitly is aligned to the layered structure of machine learning models. Regarding a model with n layers

$$f(x) = f_n \circ \cdots \circ f_1(x), \tag{2}$$

LRP first calculates all activations during the forward pass starting with f_1 until the output layer f_n is reached. Thereafter, the prediction score $f(x)$ of any chosen model output is redistributed layer-wise as an initial quantity of relevance R_n back towards the input. During this backward pass, the redistribution process follows a conservation principle analogous to Kirchhoff's laws in electrical circuits. Specifically, all relevance that flows into a neuron is redistributed towards neurons of the layer below. In the context of neural network predictors, the whole LRP procedure can be efficiently implemented as a forward-backward pass with modified gradient computation, as demonstrated in, e.g., [35].

Considering a layer's output neuron j, the distribution of its assigned relevance score R_j towards its lower layer input neurons i can be, in general, achieved by applying the basic decomposition rule

$$R_{i \leftarrow j} = \frac{z_{ij}}{z_j} R_j, \tag{3}$$

where z_{ij} describes the contribution of neuron i to the activation of neuron j [3,29] and z_j is the aggregation of the pre-activations z_{ij} at output neuron j, i.e., $z_j = \sum_i z_{ij}$. Here, the denominator enforces the conservation principle over all i contributing to j, meaning $\sum_i R_{i \leftarrow j} = R_j$. This is achieved by ensuring the decomposition of R_j is in proportion to the relative flow of activations z_{ij}/z_j in the forward pass. The relevance of a neuron i is then simply an aggregation of all incoming relevance quantities

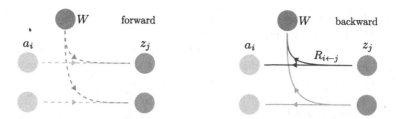

Fig. 3. LRP can be utilized to calculate relevance scores for weight parameters W, which contribute to the activation of output neurons z_j during the forward pass in interaction with data-dependent inputs a_i. In the backward pass, relevance messages $R_{i \leftarrow j}$ can be aggregated at neurons/input activations a_i, but also at weights W.

$$R_i = \sum_j R_{i \leftarrow j}. \tag{4}$$

Given the conservation of relevance in the decomposition step of Eq. (3), this means that $\sum_i R_i = \sum_j R_j$ holds for consecutive neural network layers. Next to component-wise non-linearities, linearly transforming layers (e.g., dense or convolutional) are by far the most common and basic building blocks of neural networks such as VGG-16 [41] or ResNet [18]. While LRP treats the former via identity backward passes, relevance decomposition formulas can be given for the latter explicitly in terms of weights w_{ij} and input activations a_i. Let the output of a linear neuron be given as $z_j = \sum_{i,0} z_{ij} = \sum_{i,0} a_i w_{ij}$ with bias "weight" w_{0j} and respective activation $a_0 = 1$. In accordance to Eq. (3), relevance is then propagated as

$$R_{i \leftarrow j} = \underbrace{a_i w_{ij}}_{z_{ij}} \frac{R_j}{z_j} = a_i \underbrace{w_{ij}}_{\frac{\partial z_j}{\partial a_i}} \frac{R_j}{z_j} = w_{ij} \underbrace{a_i}_{\frac{\partial z_j}{\partial w_{ij}}} \frac{R_j}{z_j} . \tag{5}$$

Equation (5) exemplifies, that the explicit computation of the backward directed relevances $R_{i \leftarrow j}$ in linear layers can be replaced equivalently by a (modified) "gradient × input" approach. Therefore, the activation a_i or weight w_{ij} can act as the input and target wrt. which the partial derivative regarding output z_j is computed. The scaled relevance term R_j/z_j takes the role of the upstream gradient to be propagated.

At this point, LRP offers the possibility to calculate relevances not only of neurons, but also of individual weights, depending on the aggregation strategy, as illustrated in Fig. 3. This can be achieved by aggregating relevances at the corresponding (gradient) targets, i.e., plugging Eq. (5) into Eq. (4). For a dense layer, this yields

$$R_{w_{ij}} = R_{i \leftarrow j} \tag{6}$$

with an individual weight as the aggregation target contributing (exactly) once to an output. A weight of a convolutional filter however is applied multiple times within a neural network layer. Here, we introduce a variable k signifying one such application context, e.g., one specific step in the application of a filter w in a (strided) convolution, mapping the filter's inputs i to an output j. While the relevance decomposition formula within one such context k does not change from Eq. (3), we can uniquely identify its backwards distributed relevance messages as $R^k_{i \leftarrow j}$. With that, the aggregation of relevance at the convolutional filter w at a given layer is given with

$$R_{w_{ij}} = \sum_k R^k_{i \leftarrow j}, \tag{7}$$

where k iterates over all applications of this filter weight.

Note that in modern deep learning frameworks, derivatives wrt. activations or weights can be computed efficiently by leveraging the available automatic differentiation functionality (autograd) [33]. Specifying the gradient target, autograd then already merges the relevance decomposition and aggregation steps outlined above. Thus, computation of relevance scores for filter weights in convolutional layers is also appropriately supported, for Eq. (3), as well as any other relevance decomposition rule which can be formulated as a modified gradient backward pass, such as Eqs. (8) and (9). The ability to compute the relevance of individual weights is a critical ingredient for the eXplainability-driven Entropy-Constrained Quantization strategy introduced in Sect. 4.2.

In the following, we will briefly introduce further LRP decomposition rules used throughout our study. In order to increase numerical stability of the basic decomposition rule in Eq. (3), the LRP ε-rule introduces a small term ε in the denominator:

$$R_{i \leftarrow j} = \frac{z_{ij}}{z_j + \varepsilon \cdot \text{sign}(z_j)} R_j. \tag{8}$$

The term ε absorbs relevance for weak or contradictory contributions to the activation of neuron j. Note here, in order to avoid divisions by zero, the $\text{sign}(z)$ function is defined to return 1 if $z \geq 0$ and -1 otherwise. In the case of a deep rectifier network, it can be shown [1] that the application of this rule to the whole neural network results in an explanation that is similar to (simple) "gradient $\times$ input" [40]. A common problem within deep neural networks is, that the gradient becomes increasingly noisy with network depth [35], partly a result from gradient shattering [4]. The ε parameter is able to suppress the influence of that noise given sufficient magnitude. With the aim of achieving robust decompositions, several purposed rules next to Eqs. (3) and (8) have been proposed in literature (see [29] for an overview).

One particular rule choice, which reduces the problem of gradient shattering and which has been shown to work well in practice, is the $\alpha\beta$-rule [3,30]

$$R_{i \leftarrow j} = \left(\alpha \frac{(z_{ij})^+}{(z_j)^+} - \beta \frac{(z_{ij})^-}{(z_j)^-} \right) R_j, \tag{9}$$

where $(\cdot)^+$ and $(\cdot)^-$ denote the positive and negative parts of the variables z_{ij} and z_j, respectively. Further, the parameters α and β are chosen subject to the constraints $\alpha - \beta = 1$ and $\beta \geq 0$ (i.e., $\alpha \geq 1$) in order to propagate relevance conservatively throughout the network. Setting $\alpha = 1$, the relevance flow is computed only with respect to the positive contributions $(z_{ij})^+$ in the forward pass. When alternatively parameterizing with, e.g., $\alpha = 2$ and $\beta = 1$, which is a common choice in literature, negative contributions are included as well, while favoring positive contributions.

Recent works recommend a composite strategy of decomposition rule assignments mapping multiple rules purposely to different parts of the network [25, 29]. This leads to an increased quality of relevance attributions for the intention of explaining prediction outcomes. In the following, a composite strategy consisting of the ε-rule for dense layers and the $\alpha\beta$-rule with $\beta = 1$ for convolutional layers is used. Regarding LRP-based pruning, Yeom et al. [51] utilize the $\alpha\beta$-rule (9) with $\beta = 0$ for convolutional as well as dense layers. However, using $\beta = 0$, subparts of the network that contributed solely negatively, might receive no relevance. In our case of quantization, all individual weights have to be considered. Thus, the $\alpha\beta$-rule with $\beta = 1$ is used for convolutional layers, because it also includes negative contributions in the relevance distribution process and reduces gradient shattering. The LRP implementation is based on the software package Zennit [2], which offers a flexible integration of composite strategies and readily enables extensions required for the computation of relevance scores for weights.

4.2 eXplainability-Driven Entropy-Constrained Quantization

For our novel eXplainability-driven Entropy-Constrained Quantization (ECQˣ), we modify the ECQ assignment function to optimally re-assign the weight clustering based on LRP relevances in order to achieve higher performance measures and compression efficiency. The rationale behind using LRP to optimize the ECQ quantization algorithm is two-fold:

Assignment Correction: In the quantization process, the entropy regularization term encourages weight assignments to more populated clusters in order to minimize the overall entropy. Since weights are usually normally distributed around zero, the entropy term also strongly encourages sparsity. In practice, this quantization scheme works well rendering sparse and low-bit neural networks for various machine learning tasks and network architectures [28,48,50].

From a scientific point of view, however, one might wonder why the shift of numerous weights from their nearest-neighbor clusters to a more distant cluster does not lead to greater model degradation, especially when assigned to zero. The quantization-aware re-training and fine-tuning can, up to a certain extent, compensate for this shift. Here, the LRP-generated relevances show potential to further improve quantization in two ways: 1) by re-adding "highly relevant" weights (i.e., preventing their assignment to zero if they have a high relevance), and 2) by assigning additional, "irrelevant" weights to zero (i.e., preventing their distance- and entropy-based assignment to a non-zero centroid).

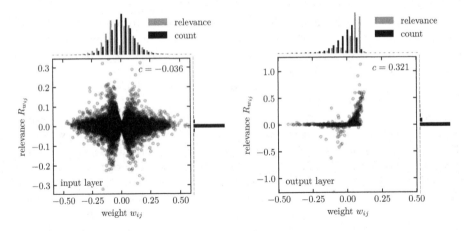

Fig. 4. Weight relevance $R_{w_{ij}}$ vs. weight value w_{ij} for the input layer (left) and output layer (right) of the full-precision MLP_GSC model (introduced in Sect. 5.1). The black histograms to the top and right of each panel display the distributions of weights (top) and relevances (right). The blue histograms further show the amount of relevance (blue) of each weight histogram bin. All relevances are collected over the validation set with equally weighted samples (i.e., by choosing $R_n = 1$). The value c measures the Pearsson correlation coefficient between weights and relevances.

We evaluated the discrepancy between weight relevance and magnitude in a correlation analysis depicted in Fig. 4. Here, all weight values w_{ij} are plotted against their associated relevance $R_{w_{ij}}$ for the input layer (left) and output layer (right) of the full-precision model MLP_GSC (which will be introduced in Sect. 5.1). In addition, histograms of both parameters are shown above and to the right of each relevance-weight-chart in Fig. 4 to better visualize the correlation between w_{ij} and $R_{w_{ij}}$. In particular, a weight of high magnitude is not necessarily also a relevant weight. And in contrast, there are also weights of small or medium magnitude that have a high relevance and thus should not be omitted in the quantization process. This phenomenon is especially true for layers closer to the input. The outcome of this analysis strongly motivates the use of LRP relevances for the weight assignment correction process of low-bit and sparse ECQ^x.

Regularizing Effect for Training: Since the previously described re-adding (which is also referred to as "regrowth" in literature) and removing of weights due to LRP depends on the propagated input data, weight relevances can change from data batch to data batch. In our quantization-aware training, we apply the STE, and thus the re-assignment of weights, after each forward-backward pass.

The regularizing effect which occurs due to dynamic re-adding and removing weights is probably related to the generalization effect which random Dropout [42] has on neural networks. However, as elaborated in the extensive survey by Hoefler et al. [21], in terms of dynamic sparsification, re-adding ("drop in") the best weights is as crucial as removing ("drop out") the right ones. Instead of randomly dropping weights, the work in [9] shows that re-adding weights based on largest gradients is related to Hebbian learning and biologically more plausible. LRP relevances go beyond the gradient criterion, which is why we consider it a suitable candidate.

In order to embed LRP relevances in the assignment function (1), we update the cost for the zero centroid ($c = 0$) by extending it as

$$\rho \, \mathbf{R}_{W^{(l)}} \cdot \left(d(\mathbf{W}^{(l)}, w^{(l)}_{c=0}) - \lambda^{(l)} \, \log_2(P^{(l)}_{c=0}) \right) \tag{10}$$

with relevance matrix $\mathbf{R}_{W^{(l)}}$ containing all weight relevances $R_{w_{ij}}$ of layer l with row/input index i and column/output index j, as specified in Eq. (7). The relevance-dependent assignment matrix $\mathbf{A}^{(l)}_{\text{x}}$ is thus described by:

$$\mathbf{A}^{(l)}_{\text{x}}(\mathbf{W}^{(l)}) = \underset{c}{\arg\min} \begin{cases} \rho \, \mathbf{R}_{W^{(l)}} \cdot \left(d(\mathbf{W}^{(l)}, w^{(l)}_{c=0}) - \lambda^{(l)} \, \log_2(P^{(l)}_{c=0}) \right), & \text{if } c = 0 \\ d(\mathbf{W}^{(l)}, w^{(l)}_c) - \lambda^{(l)} \, \log_2(P^{(l)}_c) & , \text{if } c \neq 0 \end{cases} \tag{11}$$

where ρ is a normalizing scaling factor, which also takes relevances of the previous data batches into account (momentum). The term $\rho \, \mathbf{R}_{W^{(l)}}$ increases the assignment cost of the zero cluster for relevant weights and decreases it for irrelevant weights.

Figure 5 shows an example of one ECQ^{x} iteration that includes the following steps: 1) ECQ^{x} computes a forward-backward pass through the quantized model, deriving its weight gradients. LRP relevances $\mathbf{R}_W$ are computed by redistributing modified gradients according to Eq. (7). 2) LRP relevances are then scaled by a normalizing scaling factor ρ, and 3) weight gradients are scaled by multiplying the non-zero centroid values (e.g., the upper left gradient of -0.03 is multiplied by the centroid value 1.36). 4) The scaled gradients are then applied to the full-precision (FP) background model which is a copy of the initial unquantized neural network and is used only for weight assignment, i.e. it is updated with the scaled gradients of the quantized network but does not perform inference itself, 5) The FP model is updated using the ADAM optimizer [23]. Then, weights are assigned to their nearest-neighbor cluster centroids. 6) Finally, the assignment $\mathbf{A}_{\text{x}}$ cost for each weight to each centroid is calculated using the λ-scaled information content of clusters (i.e., I_- (blue) ≈ 1.7, I_0 (green) $= 1.0$ and I_+ (purple) ≈ 2.4 in this example) and ρ-scaled relevances. Here, relevances above the exemplary threshold (i.e., mean $\bar{\mathbf{R}}_W \approx 0.3$) increase the cost for the zero cluster assignment, while relevances below (highlighted in red) decrease it. Each weight is assigned such that the cost function is minimized according to Eq. (11). 7) Depending on the intensity of the entropy and relevance constraints (controlled by λ and ρ), different assignment candidates can be rendered to fit a specific deep learning task. In the example shown in Fig. 5, an exemplary candidate grid was selected, which is depicted at the top left of the Figure. The weight at grid coordinate $D2$, for example, was assigned to the zero cluster due to its irrelevance and the weight at $C3$ due to the entropy constraint.

284   D. Becking et al.

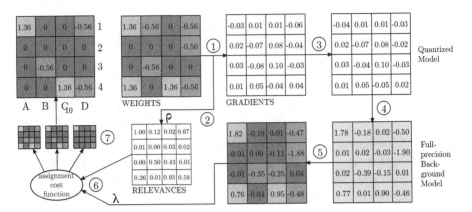

Fig. 5. Exemplary ECQ^x weight update. For simplicity, 3 centroids are used (i.e., symmetric 2 bit case). The process involves the following steps: 1) Derive gradients and LRP relevances from forward-backward pass. 2) LRP relevance scaling. 3) Gradients scaling. 4) Gradient attachment to full precision background model. 5) Background model update and nearest-neighbor clustering. 6) Computing of the assignment cost for each weight using the λ-scaled information content of clusters and the ρ-scaled relevances. Assign each weight by minimizing the cost. 7) Choosing an appropriate candidate (of various λ and ρ settings).

In the case of dense or convolutional layers, LRP relevances can be computed efficiently using the autograd functionality, as mentioned in Sect. 4.1. For a classification task, it is sensible to use the target class score as a starting point for the LRP backward pass. This way, the relevance of a neuron or weight describes its contribution to the target class prediction. Since the output is propagated throughout the network, all relevance is proportional to the output score. Consequently, relevances of each sample in a training batch are, in general, weighted differently according to their respective model output, or prediction confidence. However, with the aim of suppressing relevances for inaccurate predictions, it is sensible to weigh samples according to the model output, because a low output score usually corresponds to an unconfident decision of the model.

After the relevance calculation of a whole data batch, the relevance scores $\mathbf{R}_{W^{(l)}}$ are transformed to their absolute value and normalized, such that $\mathbf{R}_{W^{(l)}} \in [0, 1]$. Even though negative contributions work against an output, they might still be relevant to the network functionality, and their influence is thus considered instead of omitted. On one hand, they can lead to positive contributions for other classes. On the other, they can be relevant to balancing neuron activations throughout the network.

The relevance matrices $\mathbf{R}_{W^{(l)}}$ resulting from LRP are usually sparse, as can be seen in the weight histograms of Fig. 4. In order to control the effect of LRP in the assignment function, the relevances are exponentially transformed by β, applying a similar effect as for gamma correction in image processing:

$$\mathbf{R}'_{W^{(l)}} = \left(\mathbf{R}_{W^{(l)}}\right)^{\beta}$$

with $\beta \in [0, 1]$. Here, the parameter β is initially chosen such that the mean relevance $\hat{\mathbf{R}}_{W^{(l)}}$ does not change the assignment, e.g., $\rho \left(\hat{\mathbf{R}}_{W^{(l)}} \right)^{\beta} = 1$ or $\beta = -\frac{\ln \rho}{\ln \hat{\mathbf{R}}_{W^{(l)}}}$. In order to further control the sparsity of a layer, the target sparsity p is introduced. If the assignment increases a layer's sparsity by more than the target sparsity p, parameter β is accordingly minimized. Thus, in ECQˣ, LRP relevances are directly included in the assignment function and their effect can be controlled by parameter p. An experimental validation of the developed ECQˣ method, including state-of-the-art comparison and parameter variation tests, is given in the following section.

5   Experiments

In the experiments, we evaluate our novel quantization method ECQˣ using two widely used neural network architectures, namely a convolutional neural network (CNN) and a multilayer perceptron (MLP). More precisely, we deploy VGG16 for the task of small-scale image classification (CIFAR-10), ResNet18 for the Pascal Visual Object Classes Challenge (Pascal VOC) and an MLP with 5 hidden layers and ReLU non-linearities solving the task of keyword spotting in audio data (Google Speech Commands).

In the first subsection, the experimental setup and test conditions are described, while the results are shown and discussed in the second subsection. In particular, results for ECQˣ hyperparameter variation are shown, followed by a comparison against classical ECQ and results for bit width variation. Finally, overall results for ECQˣ for different accuracy and compression measurements are shown and discussed.

5.1   Experimental Setup

All experiments were conducted using the PyTorch deep learning framework, version 1.7.1 with torchvision 0.8.2 and torchaudio 0.7.2 extensions. As a hardware platform we used Tesla V100 GPUs with CUDA version 10.2. The quantization-aware training of ECQˣ was executed for 20 epochs in all experiments. As an optimizer we used ADAM with an initial learning rate of 0.0001. In the scope of the training procedure, we consider *all* convolutional and fully-connected layers of the neural networks for quantization, including the input and output layers. Note that numerous approaches in related works keep the input and/or output layers in full-precision (32 bit float), which may compensate for the model degradation caused by quantization, but is usually difficult to bring into application and incurs significant overhead in terms of energy consumption.

Google Speech Commands. The Google Speech Commands (GSC [46]) dataset consists of 105,829 utterances of 35 words recorded from 2,618 speakers. The standard is to discriminate ten words "Yes", "No", "Up", "Down", "Left", "Right", "On", "Off", "Stop", and "Go", and adding two additional labels, one for "Unknown Words", and another for "Silence" (no speech detected). Following the official Tensorflow example code for training[3], we implemented the corresponding data augmentation with PyTorch's torchaudio package. It includes randomly adding background noise with a probability of 80% and time shifting the audio by $[-100, 100]$ms with a probability of 50%. To generate features, the audio is transformed to MFCC fingerprints (Mel Frequency Cepstral Coefficients). We use 15 bins and a window length of 2000 ms. To solve GSC, we deploy an MLP (which we name *MLP_GSC* in the following) consisting of an input layer, five hidden layers and an output layer featuring 512, 512, 256, 256, 128, 128 and 12 output features, respectively. The MLP_GSC was pre-trained for 100 epochs using stochastic gradient descent (SGD) optimization with a momentum of 0.9, an initial learning rate of 0.01 and a cosine annealing learning rate schedule.

CIFAR-10. The CIFAR-10 [26] dataset consists of natural images with a resolution of 32×32 pixels. It contains 10 classes, with 6,000 images per class. Data is split to 50,000 training and 10,000 test images. We use standard data pre-processing, i.e., normalization, random horizontal flipping and cropping. To solve the task, we deploy a VGG16 from the torchvision model zoo[4]. The VGG16 classifier is adapted from 1,000 ImageNet classes to ten CIFAR classes by replacing its three fully-connected layers (with dimensions [25,088, 4,096], [4,096, 4,096], [4,096, 1,000]) by two ([512, 512], [512, 10]), as a consequence of CIFAR's smaller image size. We also implemented a VGG16 supporting batch normalization ("BatchNorm" in the following), i.e., VGG16_bn from torchvision. The VGGs were transfer-learned for 60 epochs using ADAM optimization and an initial learning rate of 0.0005.

Pascal VOC. The Pascal Visual Object Classes Challenge 2012 (VOC2012) [13] provides 11,540 images associated with 20 classes. The dataset has been split into 80% for training/validation and 20% for testing. We applied normalization, random horizontal flipping and center cropping to 224×224 pixels. As a neural network architecture, the pre-trained ResNet18 from the torchvision model zoo was deployed. Its classifier was adapted to predict 20 instead of 1,000 classes and the model was transfer-learned for 30 epochs using ADAM optimization with an initial learning rate of 0.0001.

---

[3] https://github.com/tensorflow/tensorflow/tree/master/tensorflow/examples/speech_commands.

[4] https://pytorch.org/vision/stable/models.html.

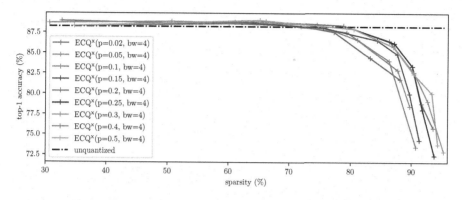

Fig. 6. Hyperparameter p controls the LRP-introduced sparsity.

5.2 ECQ^x Results

In this subsection, we compare ECQ^x to state-of-the-art ECQ quantization, analysing accuracy preservation vs. sparsity increase. Furthermore, we investigate ECQ^x compressibility, behavior on BatchNorm layers, and an appropriate choice of hyperparameters.

ECQ^x Hyperparameter Variation. In ECQ^x, two important hyperparameters, λ and p, influence the performance and thus are optimized for the comparative experiments described below. The parameter λ increases the intensity of the entropy constraint and thus distributes the working points of each trial over a range of sparsities (see Fig. 6). The p hyperparameter defines an upper bound for the per-layer percentage of zero values, allowing a maximum amount of p additional sparsity, on top of the λ-introduced sparsity. It thus implicitly controls the intensity of the LRP constraint.

Figure 6 shows results using several p values for the 4 bit ($bw = 4$) quantization of the MLP_GSC model. Note, that the variation of bit width bw is discussed below the comparative results. For smaller p, less sparse models are rendered with higher top-1 accuracies in the low-sparsity regime (e.g., $p = 0.02$ or $p = 0.05$ between 30–50% total network sparsity). In the regime of higher sparsity, larger values of p show a better sparsity-accuracy trade-off. Note, that larger p do not only set more weights to zero but also re-add relevant weights (regrowth). For $p = 0.4$ and $p = 0.5$, both lines are congruent since no layer is achieving more than 40% additional LRP-introduced sparsity with the initial β value (cf. Sect. 4.2).

ECQ^x vs. ECQ Analysis. As shown in Fig. 7, the LRP-driven ECQ^x approach renders models with higher performance and simultaneously higher efficiency. In this comparison, efficiency is determined in terms of sparsity, which can be exploited to compress the model more or to skip arithmetic operations with zero values. Both methods achieve a quantization to 4 bit integer without any performance degradation of the model. Performance is even slightly increased

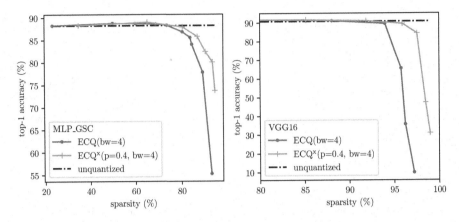

Fig. 7. Resulting model performances, when applying ECQ vs. ECQ^x 4 bit quantization on MLP_GSC (left) and VGG16 (right). Each point corresponds to a model rendered with a specific λ which is a regulator for the entropy constraint and thus incrementally enhances sparsity. Abbreviations in the legend labels refer to bit width (*bw*) and target sparsity (*p*), which is defined in Sect. 4.2.

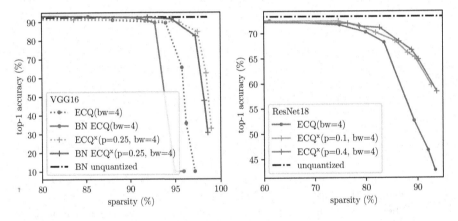

Fig. 8. Resulting model performances, when applying ECQ vs. ECQ^x 4 bit quantization on VGG16, VGG16 with BatchNorm (BN) modules (left) and ResNet18 (right).

due to quantization when compared to the unquantized baseline. In the regime of high sparsity, model accuracy of the previous state-of-the-art (ECQ) drops significantly faster compared to the LRP-adjusted quantization scheme.

Regarding the handling of BatchNorm modules for LRP, it is proposed in literature to merge the BatchNorm layer parameters with the preceding linear layer [15] into a single linear transformation. This canonization process is sensible, because it reduces the number of computational steps in the backward pass while maintaining functional equivalence between the original and the canonized model in the forward pass.

It has been further shown, that network canonization can increase explanation quality [15]. With the aim of computing weight relevance scores for a BatchNorm layer's adjacent linear layer in its original (trainable) state, keeping the layers separate is more favorable than merging. Therefore, the $\alpha\beta$-rule with $\beta = 1$ is also applied to BatchNorm layers. The quantization results of the VGG architecture with BatchNorm modules and ResNet18 are shown in Fig. 8.

In order to capture the computational overhead of LRP in terms of additional training time, we compared the average training times of the different model architectures per epoch. Relevance-dependent quantization (ECQ^x) requires approximately $1.2\times$, $2.4\times$, and $3.2\times$ more processing time than baseline quantization (ECQ) for the MLP_GSC, VGG16, and ResNet18 architectures, respectively. This extra effort can be explained with the additional forward-backward passes performed in Zennit for LRP computation. More concretely, using Zennit as a plug-in XAI module, it computes one additional forward pass layer-wise and redistributes the relevances to the preceding layers according to the decomposition and aggregation rules specified in Sect. 4.1. For redistribution, Zennit computes one additional backward pass for ε-rule associated layers and two additional backward passes for $\alpha\beta$-rule associated layers in order to derive positive α and negative β relevance contributions. To recap, in the applied composite strategy, the ε-rule is used for dense layers and the $\alpha\beta$-rule for convolutional layers and BatchNorm parameters, which results in the extra computational cost for VGG16 and ResNet18 compared to MLP_GSC, which consists solely of dense layers. In addition, aggregation of relevances for convolutional filters is not required for dense layers. Note that the above mentioned values for additional computational overhead of ECQ^x due to relevance computation can be interpreted as an upper-bound and that there are options to minimize the effort, e.g., by 1) not considering relevances for cluster assignments in each training iteration, 2) leveraging pre-computed outputs or even gradients from the quantized base model instead of separately computing forward-backward passes with a model copy in the Zennit module. Whereas 1) corresponds to a change in the quantization setup, 2) requires parallelization optimizations of the software framework.

Bit Width Variation. Bit width reduction has multiple benefits over full-precision in terms of memory, latency, power consumption, and chip area efficiency. For instance, a reduction from standard 32 bit precision to 8 bit or 4 bit directly leads to a memory reduction of almost $4\times$ and $8\times$. Arithmetic with lower bit width is exponentially faster if the hardware supports it. E.g., since the release of NVIDIA's Turing architecture, 4 bit integer is supported which increases the throughput of the RTX 6000 GPU to 522 TOPS (tera operations per second), when compared to 8 bit integer (261 TOPS) or 32 bit floating point (14.2 TFLOPS) [31]. Furthermore, Horowitz showed that, for a 45 nm technology, low-precision logic is significantly more efficient in terms of energy and area [22]. For example, performing 8 bit integer addition and multiplication is $30\times$ and $19\times$ more energy efficient compared to 32 bit floating point addition

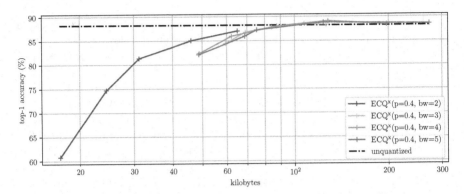

Fig. 9. Resulting MLP_GSC model performances vs. memory footprint, when applying ECQ^x with 2 bit to 5 bit quantization.

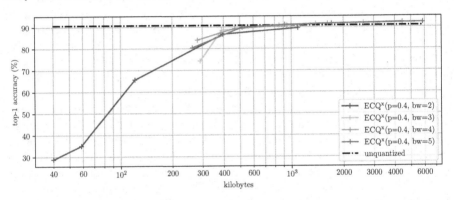

Fig. 10. Resulting VGG16 model performances vs. memory footprint, when applying ECQ^x with 2 bit to 5 bit quantization.

and multiplication. The respective chip area efficiency is increased by 116× and 27× as compared to 32 bit float. It is also shown that memory reads and writes have the highest energy cost, especially when reading data from external DRAM. This further motivates bit width reduction because it can reduce the number of overall RAM accesses since more data fits into the same caches/registers when having a reduced precision.

In order to investigate different bit widths in the regime of ultra low precision, we compare the compressibility and model performances of the MLP_GSC and VGG16 networks when quantized to 2 bit, 3 bit, 4 bit and 5 bit integer values (see Figs. 9 and 10). Here, we directly encoded the integer tensors with the DeepCABAC codec of the ISO/IEC MPEG NNR standard [24]. The least sparse working points of each trial, i.e., the rightmost data points of each line, show the expected behaviour, namely that compressibility is increased by continuously reducing the bit width from 5 bit to 2 bit. However, this effect decreases or even reverses when the bit width is in the range of 3 bit to 5 bit. In other words, reducing the number of centroids from $2^5 = 32$ to $2^3 = 8$ does not necessarily lead to a further significant reduction in the resulting bitstream size if sparsity is predominant. The 2 bit quantization still minimizes the size of the bit stream, even if, especially for the VGG model, more accuracy is sacrificed for this purpose. Note that compressibility is only one reason for reducing bit width besides, for example, speeding up model inference due to increased throughput.

ECQ^x Results Overview. In addition to the performance graphs in the previous subsections, all quantization results are summarized in Table 1. Here, ECQ^x and ECQ are compared specifically for a 2 and 4 bit quantization as these fit particularly well to power-of-two hardware registers. The ECQ^x 4 bit quantization achieves a compression ratio for VGG16 of 103× with a negligible drop in accuracy of −0.1%. In comparison, ECQ achieves the same compression ratio only with a model degradation of −1.23% top-1 accuracy. For the 4 bit quantization of MLP_GSC, ECQ^x achieves its highest accuracy ("drop", i.e., increase of +0.71% compared to the unquantized baseline model) with a compression ratio that is almost 10% larger compared to the highest achievable accuracy of ECQ (+0.47%). For sparsities beyond 70%, ECQ significantly reduces the model's predictive performance, e.g., at a sparsity of 80.39% ECQ shows a loss of −1.40% whereas ECQ^x only degrades by −0.34%. ResNet18 sacrifices performance at each quantization setting, but especially for ECQ^x the accuracy loss is negligible. The 2 bit representations of ResNet18 sacrifice more than −5% top-1 accuracy compared to the unquantized model, which may be compensated with more than 20 epochs of quantization-aware training, but is also due to the higher complexity of the Pascal VOC task.

And finally, the 2 bit results in Table 1 show two major findings: 1) With only a minor model degradation all weight layers of the MLP_GSC and VGG networks can also be quantized to only 4 discrete centroid values while still maintaining a high level of sparsity, 2) ECQ^x renders higher compressible models in comparison to ECQ, as indicated by the higher compression ratios CR.

Table 1. Quantization results for ECQ^x for 2 bit and 4 bit quantization: highest accuracy, highest compression gain without model degradation (if possible) and highest compression gain with negligible degradation. Underlined values mark the best results in terms of performance and compressibility with negligible drop in top-1 accuracy.

Model	Prec.[a]	Method[b]	Acc. (%)	Acc. drop	$\frac{\|W=0\|}{\|W\|}$ (%)[c]	Size (kB)	CR[d]
CIFAR-10							
VGG16	W4A16	**ECQ^x**	92.27	+1.55	41.39	4,446.39	13.48
	W4A16	**ECQ^x**	90.86	+0.14	91.95	933.99	64.17
	W4A16	**ECQ^x**	90.62	−0.10	94.67	584.16	102.59
	W4A16	ECQ	92.09	+1.37	29.88	4,658.01	12.87
	W4A16	ECQ	91.03	+0.31	88.03	1,246.27	48.09
	W4A16	ECQ	89.49	−1.23	93.97	585.40	102.37
	W2A16	**ECQ^x**	90.42	−0.30	83.23	1.394,52	42.98
	W2A16	ECQ	90.19	−0.53	81.58	1,486.76	40.31
Google Speech Commands							
MLP_GSC	W4A16	**ECQ^x**	88.95	+0.71	65.14	128.03	20.05
	W4A16	**ECQ^x**	88.34	+0.10	78.77	92.46	27.77
	W4A16	**ECQ^x**	87.89	−0.34	80.45	87.52	29.33
	W4A16	ECQ	88.71	+0.47	59.95	139.96	18.34
	W4A16	ECQ	88.32	+0.08	70.74	98.32	26.11
	W4A16	ECQ	86.84	−1.40	80.39	69.67	36.85
	W2A16	**ECQ^x**	87.46	−0.78	83.97	68.77	37.33
	W2A16	ECQ	87.72	−0.52	77.55	78.54	32.69
Pascal VOC							
ResNet18	W4A16	**ECQ^x**	73.13	−0.27	32.82	3,797.97	11.79
	W4A16	**ECQ^x**	72.78	−0.62	68.67	2,246.71	19.93
	W4A16	**ECQ^x**	72.48	−0.92	74.65	1,946.22	23.01
	W4A16	ECQ	72.95	−0.45	24.63	3,882.62	11.53
	W4A16	ECQ	72.56	−0.84	61.12	2,480.59	18.05
	W4A16	ECQ	71.74	−1.66	74.88	1,841.82	24.32

[a]WxAy indicates a quantization of weights and activations to x and y bit.
[b]ECQ refers to ECQ^x w/o LRP constraint.
[c]Sparsity, measured as the percentage of zero-valued parameters in the DNN.
[d]Compression ratio (full-precision size/compressed size) when applying the DeepCABAC codec of the ISO/IEC MPEG NNR standard [24].

6   Conclusion

In this chapter we presented a new entropy-constrained neural network quantization method (ECQ^x), utilizing weight relevance information from Layer-wise Relevance Propagation (LRP). Thus, our novel method combines concepts of explainable AI (XAI) and information theory. In particular, instead of only assigning weight values based on their distances to respective quantization clusters, the assignment function additionally considers weight relevances based on LRP. In detail, each weight's contribution to inference in interaction with the transformed data, as well as cluster information content is calculated and applied. For this approach, we first utilized the observation that a weight's magnitude does not necessarily correlate with its importance or relevance for a model's inference capability. Next, we verified this observation in a relevance vs. weight (magnitude) correlation analysis and subsequently introduce our ECQ^x method. As a result, smaller weight parameters that are usually omitted in a classical quantization process are preserved, if their relevance score indicates a stronger contribution to the overall neural network accuracy or performance.

The experimental results show that this novel ECQ^x method generates low bit width (2–5 bit) and sparse neural networks while maintaining or even improving model performance. Therefore, in particular the 2 and 4 bit variants are highly suitable for neural network hardware adaptation tasks. Due to the reduced parameter precision and high number of zero-elements, the rendered networks are also highly compressible in terms of file size, e.g., up to 103× compared to the full-precision unquantized DNN model, without degrading the model performance. Our ECQ^x approach was evaluated on different types of models and datasets (including Google Speech Commands, CIFAR-10 and Pascal VOC). The comparative results vs. state-of-the-art entropy-constrained-only quantization (ECQ) show a performance increase in terms of higher sparsity, as well as a higher compression. Finally, also hyperparameter optimization and bit width variation results were presented, from which the optimal parameter selection for ECQ^x was derived.

Acknowledgements. This work was supported by the German Ministry for Education and Research as BIFOLD (ref. 01IS18025A and ref. 01IS18037A), the European Union's Horizon 2020 programme (grant no. 965221 and 957059), and the Investitionsbank Berlin under contract No. 10174498 (Pro FIT programme).

References

1. Ancona, M., Ceolini, E., Öztireli, C., Gross, M.: Gradient-based attribution methods. In: Samek, W., Montavon, G., Vedaldi, A., Hansen, L.K., Müller, K.-R. (eds.) Explainable AI: Interpreting, Explaining and Visualizing Deep Learning. LNCS (LNAI), vol. 11700, pp. 169–191. Springer, Cham (2019). https://doi.org/10.1007/978-3-030-28954-6_9
2. Anders, C.J., Neumann, D., Samek, W., Müller, K.R., Lapuschkin, S.: Software for dataset-wide XAI: from local explanations to global insights with Zennit, CoRelAy, and ViRelAy. CoRR abs/2106.13200 (2021)
3. Bach, S., Binder, A., Montavon, G., Klauschen, F., Müller, K.R., Samek, W.: On pixel-wise explanations for non-linear classifier decisions by layer-wise relevance propagation. PLoS ONE **10**(7), e0130140 (2015)
4. Balduzzi, D., Frean, M., Leary, L., Lewis, J., Ma, K.W.D., McWilliams, B.: The shattered gradients problem: if ResNets are the answer, then what is the question? In: International Conference on Machine Learning, pp. 342–350. PMLR (2017)
5. Bengio, Y., Léonard, N., Courville, A.C.: Estimating or propagating gradients through stochastic neurons for conditional computation. CoRR abs/1308.3432 (2013)

6. Bhalgat, Y., Lee, J., Nagel, M., Blankevoort, T., Kwak, N.: LSQ+: improving low-bit quantization through learnable offsets and better initialization. In: Proceedings of the IEEE/CVF Conference on Computer Vision and Pattern Recognition (CVPR) Workshops, June 2020

7. Choi, Y., El-Khamy, M., Lee, J.: Towards the limit of network quantization. CoRR abs/1612.01543 (2016)

8. Courbariaux, M., Bengio, Y., David, J.P.: BinaryConnect: training deep neural networks with binary weights during propagations. In: Advances in Neural Information Processing Systems, pp. 3123–3131 (2015)

9. Dai, X., Yin, H., Jha, N.K.: Nest: a neural network synthesis tool based on a grow-and-prune paradigm. IEEE Trans. Comput. **68**(10), 1487–1497 (2019)

10. Deng, B.L., Li, G., Han, S., Shi, L., Xie, Y.: Model compression and hardware acceleration for neural networks: a comprehensive survey. Proc. IEEE **108**(4), 485–532 (2020)

11. Denil, M., Shakibi, B., Dinh, L., Ranzato, M., de Freitas, N.: Predicting parameters in deep learning. In: Advances in Neural Information Processing Systems, pp. 2148–2156 (2013)

12. Elsken, T., Metzen, J.H., Hutter, F.: Neural architecture search: a survey. J. Mach. Learn. Res. **20**(1), 1997–2017 (2019)

13. Everingham, M., Van Gool, L., Williams, C.K.I., Winn, J., Zisserman, A.: The PASCAL Visual Object Classes Challenge 2012 (VOC2012) Results. http://www.pascal-network.org/challenges/VOC/voc2012/workshop/index.html

14. Gholami, A., Kim, S., Dong, Z., Yao, Z., Mahoney, M.W., Keutzer, K.: A survey of quantization methods for efficient neural network inference. CoRR abs/2103.13630 (2021)

15. Guillemot, M., Heusele, C., Korichi, R., Schnebert, S., Chen, L.: Breaking batch normalization for better explainability of deep neural networks through layer-wise relevance propagation. CoRR abs/2002.11018 (2020)

16. Han, S., Mao, H., Dally, W.J.: Deep compression: compressing deep neural network with pruning, trained quantization and Huffman coding. In: 4th International Conference on Learning Representations (ICLR) (2016)

17. Han, S., Pool, J., Tran, J., Dally, W.: Learning both weights and connections for efficient neural network. In: Advances in Neural Information Processing Systems, vol. 28. Curran Associates, Inc. (2015)

18. He, K., Zhang, X., Ren, S., Sun, J.: Deep residual learning for image recognition. In: Proceedings of the IEEE Conference on Computer Vision and Pattern Recognition, pp. 770–778 (2016)

19. He, Y., Zhang, X., Sun, J.: Channel pruning for accelerating very deep neural networks. In: Proceedings of the IEEE International Conference on Computer Vision, pp. 1389–1397 (2017)

20. Hinton, G.E., Vinyals, O., Dean, J.: Distilling the knowledge in a neural network. arXiv abs/1503.02531 (2015)

21. Hoefler, T., Alistarh, D., Ben-Nun, T., Dryden, N., Peste, A.: Sparsity in deep learning: pruning and growth for efficient inference and training in neural networks (2021)

22. Horowitz, M.: 1.1 computing's energy problem (and what we can do about it). In: 2014 IEEE International Solid-State Circuits Conference Digest of Technical Papers (ISSCC), pp. 10–14 (2014)

23. Kingma, D.P., Ba, J.: Adam: a method for stochastic optimization (2014). arxiv:1412.6980 Comment: Published as a Conference Paper at the 3rd International Conference for Learning Representations, San Diego (2015)

24. Kirchhoffer, H., et al.: Overview of the neural network compression and representation (NNR) standard. IEEE Trans. Circuits Syst. Video Technol. 1–14 (2021). https://doi.org/10.1109/TCSVT.2021.3095970
25. Kohlbrenner, M., Bauer, A., Nakajima, S., Binder, A., Samek, W., Lapuschkin, S.: Towards best practice in explaining neural network decisions with LRP. In: 2020 International Joint Conference on Neural Networks (IJCNN), pp. 1–7. IEEE (2020)
26. Krizhevsky, A.: Learning Multiple Layers of Features from Tiny Images, April 2009
27. LeCun, Y., Denker, J.S., Solla, S.A.: Optimal brain damage. In: Advances in Neural Information Processing Systems, pp. 598–605 (1990)
28. Marban, A., Becking, D., Wiedemann, S., Samek, W.: Learning sparse & ternary neural networks with entropy-constrained trained ternarization (EC2T). In: The IEEE/CVF Conference on Computer Vision and Pattern Recognition (CVPR) Workshops, pp. 3105–3113, June 2020
29. Montavon, G., Binder, A., Lapuschkin, S., Samek, W., Müller, K.-R.: Layer-wise relevance propagation: an overview. In: Samek, W., Montavon, G., Vedaldi, A., Hansen, L.K., Müller, K.-R. (eds.) Explainable AI: Interpreting, Explaining and Visualizing Deep Learning. LNCS (LNAI), vol. 11700, pp. 193–209. Springer, Cham (2019). https://doi.org/10.1007/978-3-030-28954-6_10
30. Montavon, G., Samek, W., Müller, K.R.: Methods for interpreting and understanding deep neural networks. Digit. Signal Process. **73**, 1–15 (2018)
31. NVIDIA Turing GPU Architecture - Graphics Reinvented. Technical report, WP-09183-001_v01, NVIDIA Corporation (2018)
32. Park, E., Ahn, J., Yoo, S.: Weighted-entropy-based quantization for deep neural networks. In: 2017 IEEE Conference on Computer Vision and Pattern Recognition (CVPR), pp. 7197–7205 (2017)
33. Paszke, A., et al.: Automatic differentiation in pytorch (2017)
34. Sabih, M., Hannig, F., Teich, J.: Utilizing explainable AI for quantization and pruning of deep neural networks. CoRR abs/2008.09072 (2020)
35. Samek, W., Montavon, G., Lapuschkin, S., Anders, C.J., Müller, K.R.: Explaining deep neural networks and beyond: a review of methods and applications. Proc. IEEE **109**(3), 247–278 (2021)
36. Sandler, M., Howard, A., Zhu, M., Zhmoginov, A., Chen, L.C.: MobileNetV2: inverted residuals and linear bottlenecks. In: Proceedings of the IEEE Conference on Computer Vision and Pattern Recognition, pp. 4510–4520 (2018)
37. Schütt, K.T., Arbabzadah, F., Chmiela, S., Müller, K.R., Tkatchenko, A.: Quantum-chemical insights from deep tensor neural networks. Nat. Commun. **8**(1), 1–8 (2017)
38. Senior, A.W., et al.: Improved protein structure prediction using potentials from deep learning. Nature **577**(7792), 706–710 (2020)
39. Shannon, C.E.: A mathematical theory of communication. Bell Syst. Tech. J. **27**(3), 379–423 (1948)
40. Shrikumar, A., Greenside, P., Shcherbina, A., Kundaje, A.: Not just a black box: learning important features through propagating activation differences. CoRR abs/1605.01713 (2016)
41. Simonyan, K., Zisserman, A.: Very deep convolutional networks for large-scale image recognition. arXiv preprint arXiv:1409.1556 (2014)
42. Srivastava, N., Hinton, G., Krizhevsky, A., Sutskever, I., Salakhutdinov, R.: Dropout: a simple way to prevent neural networks from overfitting. J. Mach. Learn. Res. **15**(1), 1929–1958 (2014)

43. Sze, V., Chen, Y., Yang, T., Emer, J.S.: Efficient processing of deep neural networks: a tutorial and survey. Proc. IEEE **105**(12), 2295–2329 (2017)
44. Tan, M., et al.: MnasNet: platform-aware neural architecture search for mobile. In: Proceedings of the IEEE/CVF Conference on Computer Vision and Pattern Recognition, pp. 2820–2828 (2019)
45. Warden, P., Situnayake, D.: TinyML: Machine Learning with TensorFlow Lite on Arduino and Ultra-Low-Power Microcontrollers. O'Reilly Media (2020)
46. Warden, P.: Speech commands: a dataset for limited-vocabulary speech recognition. CoRR abs/1804.03209 (2018)
47. Wiedemann, S., et al.: DeepCABAC: a universal compression algorithm for deep neural networks. IEEE J. Sel. Top. Signal Process. **14**(4), 700–714 (2020)
48. Wiedemann, S., Marban, A., Müller, K.R., Samek, W.: Entropy-constrained training of deep neural networks. In: 2019 International Joint Conference on Neural Networks (IJCNN), pp. 1–8 (2019)
49. Wiedemann, S., Müller, K.R., Samek, W.: Compact and computationally efficient representation of deep neural networks. IEEE Trans. Neural Netw. Learn. Syst. **31**(3), 772–785 (2020)
50. Wiedemann, S., et al.: FantastIC4: a hardware-software co-design approach for efficiently running 4bit-compact multilayer perceptrons. IEEE Open J. Circuits Syst. **2**, 407–419 (2021)
51. Yeom, S.K., et al.: Pruning by explaining: a novel criterion for deep neural network pruning. Pattern Recogn. **115**, 107899 (2021)
52. Zhou, S., Ni, Z., Zhou, X., Wen, H., Wu, Y., Zou, Y.: DoReFa-Net: training low bitwidth convolutional neural networks with low bitwidth gradients. CoRR abs/1606.06160 (2016)
53. Zhu, C., Han, S., Mao, H., Dally, W.J.: Trained ternary quantization. In: International Conference on Learning Representations (ICLR) (2017)

A Whale's Tail - Finding the Right Whale in an Uncertain World

Diego Marcos[1]([✉]) [ID], Jana Kierdorf[2] [ID], Ted Cheeseman[3], Devis Tuia[4] [ID],
and Ribana Roscher[2,5] [ID]

[1] Wageningen University, Wageningen, The Netherlands
diego.marcos@wur.nl
[2] Institute of Geodesy and Geoinformation, University of Bonn, Bonn, Germany
{jkierdorf,ribana.roscher}@uni-bonn.de
[3] Happywhale, St Albans, UK
ted@happywhale.com
[4] Ecole Polytechnique Federale de Lausanne, Lausanne, Switzerland
devis.tuia@epfl.ch
[5] Data Science in Earth Observation, Technical University of Munich,
Ottobrunn, Germany

Abstract. Explainable machine learning and uncertainty quantification have emerged as promising approaches to check the suitability and understand the decision process of a data-driven model, to learn new insights from data, but also to get more information about the quality of a specific observation. In particular, heatmapping techniques that indicate the sensitivity of image regions are routinely used in image analysis and interpretation. In this paper, we consider a landmark-based approach to generate heatmaps that help derive sensitivity and uncertainty information for an application in marine science to support the monitoring of whales. Single whale identification is important to monitor the migration of whales, to avoid double counting of individuals and to reach more accurate population estimates. Here, we specifically explore the use of fluke landmarks learned as attention maps for local feature extraction and without other supervision than the whale IDs. These individual fluke landmarks are then used jointly to predict the whale ID. With this model, we use several techniques to estimate the sensitivity and uncertainty as a function of the consensus level and stability of localisation among the landmarks. For our experiments, we use images of humpback whale flukes provided by the Kaggle Challenge "Humpback Whale Identification" and compare our results to those of a whale expert.

Keywords: Attention maps · Sensitivity · Uncertainty · Whale identification

The work is partly funded by the German Federal Ministry of Education and Research (BMBF) in the framework of the international future AI lab "AI4EO – Artificial Intelligence for Earth Observation: Reasoning, Uncertainties, Ethics and Beyond" (Grant number: 01DD20001).

A. Holzinger et al. (Eds.): xxAI 2020, LNAI 13200, pp. 297–313, 2022.
https://doi.org/10.1007/978-3-031-04083-2_15

1   Introduction

For many scientific disciplines, reliability and trust in a machine learning result are of great importance, in addition to the prediction itself. Two key values that can contribute significantly to this are the interpretability and the estimation of uncertainty:

- An interpretation aims at the presentation of properties of a machine learning model (e.g., a decision process of a neural network) in a way that it is understandable to a human [21]. One possibility to obtain an interpretation is sensitivity analysis which provides information about how the models' output is affected by small or specifically chosen changes in the input [18].
- Uncertainty is the quantity of all possible changes in the output that result from uncertainties already included in the data (aleatoric/data uncertainty) or a lack of knowledge of the machine learning model (epistemic/model uncertainty) [6].

Both uncertainty quantification and sensitivity analysis have become a broad field of research in recent years, especially for developing methods to check the suitability and to better understand the decision-making process of a data-driven model [6, 21, 24]. However, so far, the two areas have usually been considered separately, although a joint consideration has clear benefits, since the analysis of sensitivity can often be considered as a part or first step towards uncertainty quantification.

In this chapter, we will consider a use case from marine science to demonstrate the usefulness of a joint use of sensitivity and uncertainty quantification in landmark-based identification. In particular, we look at the identification of whales by means of images of their fluke. Whale populations worldwide are threatened by commercial whaling, global warming, and the struggle for food in competition with the fishing industry [33]. A protection of whales is essentially supported by the reconstruction of the spatio-temporal migration of whales, which in turn is based on the (re)identification of whales. Individual whales can be identified by the shape of their whale flukes and their unique pigmentation [13]. Three features in particular play a crucial role for whale experts in distinguishing between individual whales (see Fig. 1):

- *Pigmentation-based features.* These features correspond to coloured patches on the fluke, forming unique patterns. They are very clearly visible to the human eye. They can change significantly within the first few years of whale life and in extremely cold water (for example, Antarctica, but also Greenland and the North Atlantic). They may be partially obscured by heavy diatom growth, characterized by a yellow-orange appearance of the fluke.
- *Fluke shape.* This feature is reliable and robust. The outer 20% of the tail may become more distorted and change over time, but the inner 80% and V-notch are reliable and stable. Although it is difficult to detect by the human eye, it has proven to be very useful for machine learning-based approaches [14, 15, 25].

Fig. 1. Important characteristics of a whale fluke.

- *Scars.* The surface of the fluke usually shows contrasting scars. However, the contrast can vary greatly and the scars may change over time. Certain scars grow with the whale, such as killer whale rake marks that form parallel lines or barnacle marks that form circles. In addition, lighting conditions can significantly affect the detectability of scars.

For whale monitoring, whale researchers often use geo-tagged photos with time and location information to reconstruct activities. Since manual analysis is too costly and thus a huge amount of data remained unused, current approaches focus on machine learning [14,15,25].

Despite the accuracy observed in recent competitions [29], limited effort has been devoted to actually quantify sensitivity in the prediction and identify sources of uncertainty. We argue that uncertainty identification remains a central topic requiring attention and propose a methodology based on landmarks and their spatial sensitivity and uncertainty to answer a number of scientific questions useful for experts in animal conservation. Specifically, we tackle the following questions:

- Which parts of the fluke are more consistently useful to identify whales? A whale fluke changes with time and therefore, characteristic features of a fluke may no longer be present and therefore not visualized in the interpretation tool results.
- Can landmarks together with uncertainty and sensitivity indicate the suitability of images for identification? Suitability is influenced, for example, by image quality, position, and size of the object, but also by the presence of relevant features.

These goals are formulated from the perspective of whale research, but are also intended to raise relevant questions from the perspective of machine learning, such as the usefulness of interpretation tools to improve models. In general, the task of re-identifying objects or living beings from images and is a common topic [2,16,26], and the approach and insights presented in this paper can also be applied to similar tasks from other fields.

2 Related Work

Self-explainable Deep Learning Models. Although the vast majority of methods to improve the interpretability and explainability of deep learning models are designed to work *post-hoc* [19, 28, 32], *i.e.* the important parts of the input are highlighted while the model itself remains unmodified, a few approaches aim at modifying the model so that its inherent interpretability is enhanced, also referred to as self-explainable models [23]. This has the advantage that the interpretation is actually part of the inference process, rather than being computed *a posteriori* by an auxiliary interpretation method, resolving potential trustworthiness issues of *post-hoc* methods [22]. The visual interpretation can be obtained, for example, by incorporating a global average pooling after the last convolutional layer of the model [39] or by levering a spatial attention mechanism [36]. Our self-explainable method is inspired by [36] and [38], and learns a fixed set of landmarks, along with their associated attention maps, in a weakly supervised setting by only using class labels. To gain further insight, the landmarks can be used for sensitivity analysis and uncertainty quantification.

Uncertainty Quantification. The field of uncertainty quantification has gained new popularity in recent years, especially for determining the uncertainty of complex models such as neural networks. In most applications, the predictive uncertainty is of interest, i.e. the uncertainty that affects the estimation from various sources of uncertainty, originating from the data itself (aleatoric uncertainty) and arising from the model (model uncertainty). These sources are often not negligible, especially in real-wold applications, and must be determined for a comprehensive statement about the reliability and accuracy of the result. Several works have been carried out such as [5, 30], which explore Monte Carlo dropout or quantify uncertainty analysing the softmax output of neural networks. [7, 12, 34] give comprehensive overviews of the field, where [6] specifically focuses on the applicability in real-world scenarios.

Sensitivity Analysis. This kind of analysis is usually considered in the context of explainable machine learning. Here, a set input variables, such as pixel values in an image region or a unit in some of the model's intermediate representations [3, 31], are perturbed, and the effect of such changes on the result is considered. This approach helps to understand the decision process and causes of uncertainties, and to gain insights into salient features that can be spatial, temporal or spectral. According to [21], sensitivity analysis approaches belong to interpretation tools, as they transform complex aspects such as model behavior into concepts understandable by a human [19, 24]. Many approaches use heatmaps that visualize the sensitivity of the output to perturbations of the input, the attention map of the classifier model, or the importance of the features [11]. These tools are extremely helpful and have been used recently to infer new scientific knowledge and discoveries and to improve the model [21, 27, 31]. Probably the best known principle is study of the effects of masking selected regions of the input, which is systematically applied in occlusion sensitivity maps [20]. For more details, including specific types of interpretation and further implementation, we refer to recent studies [1, 8, 9].

Sensitivity vs. Uncertainty. There are significant differences between the analysis of uncertainties and sensitivity, and previous applications mostly consider only one of the two. Sensitivity analysis focuses more on the input and the effect of modifications on the predictions, while uncertainty quantification focuses on the propagation of uncertainties in the model. Nevertheless, there are also strong correlations, as shown in [18]. Sensitivity analysis, for example, explores the causes and importance of specific uncertainties in the input data for the decision, while uncertainty analysis describes the whole set of possible outcomes. Both consider variations in the input and their influence on the output to derive statements for decision-making. Our work is based on the preliminary work of [14], in which occlusion sensitivity maps are created by systematically covering individual areas in images of whale flukes in order to identify the characteristic features of flukes for whale identification. Here, we propose to learn a set of compact attention maps such that each specializes in the detection of a fluke landmark. These learned landmarks are use to extend [14] by a combined analysis of the sensitivity of the classification to each landmark and their uncertainty.

3 Humpback Whale Data

3.1 Image Data

In this work, we use a set of humpback whale images from the Kaggle Challenge "Humpback Whale Identification". More specifically, we process their tails, called flukes (see Fig. 1). The data set consists of more than 67.000 images, in which 10.008 different whale individuals, i.e., 10.008 different classes, are represented. We pruned the dataset and used only the 1.646 classes that contained three or more images in the training set of the challenge. For our experiments, we restrict ourselves to use images in the training set because the test set does not provide reference information, as it is generally the case for Kaggle challenges. We split the images into a training set $\mathcal{X}_{\text{train}} = \{x_1, \ldots, x_N\}$ (9.408 images) and a test set $\mathcal{X}_{\text{test}} = \{x_1, \ldots, x_T\}$ (1.646 images, or one per class, i.e. a specific whale individual). The number of images per set is given by N and T, respectively. The set $\mathcal{X}_c = \{x_1, \ldots, x_R\}$ describes a subset that includes R images for one specific class c.

3.2 Expert Annotations

A domain expert participated to the study and provided human annotation of remarkable features helping in the discrimination of the whale individuals. For each annotation the expert was provided with a pair of images and asked to mark a set of features helping in discriminating whether the images were of the same individual or not. Three features are generally used by the expert (personal communication), who therefore provided three features per image analysed. Some examples are shown in Fig. 5a.

4   Methods

4.1   Landmark-Based Identification Framework

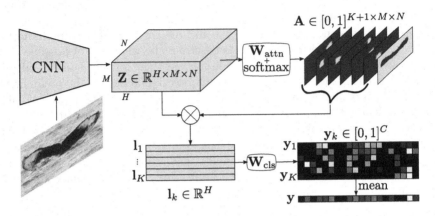

Fig. 2. Given the image of a fluke, we extract the feature tensor $\mathbf{Z}$ using a CNN. A set of compact attention maps $\mathbf{A}$, excluding a background map, is then used to extract localized features from $\mathbf{Z}$. These features are then averaged and used for classification into C classes, each corresponding to an individual whale.

We propose to learn a set of discriminant landmarks for whale identification such that the model uses evidence from each one separately in order to solve the task. The rationale behind this approach is twofold:

1. Each landmark will gather evidence from a different region of the image, effectively resulting in an ensemble of diverse classifiers, each using a different subset of the data. This independence between the different classifiers provides an improved uncertainty estimation.
2. Since landmarks are trained to attend to a small region of the image, it becomes very easy to visualize where the evidence is coming from with no further computation, thus inherently providing an enhanced level of interpretability.

In order to learn to detect informative landmarks without further supervision than the whale ID, we use an approach inspired by [38]. Likewise, we aim at learning to detect a fixed set of keypoints in the image to establish at which locations landmarks are to be extracted. Unlike [38], we do not use an hourglass-type architecture, but a standard classification CNN with a reduced downsampling rate in order to allow for a better spatial resolution. Another major difference is that we do not use any reconstruction loss and therefore need no decoding elements.

Given an image $\mathbf{X} \in \mathbb{R}^{3 \times MD \times ND}$ and a CNN with a downsampling factor D, the H-channel tensor resulting from applying the CNN to $\mathbf{X}$ is:

$$\mathbf{Z} = \text{CNN}(\mathbf{X}; \theta) \in \mathbb{R}^{H \times M \times N}. \tag{1}$$

We obtain the $K + 1$ attention maps, representing the K keypoints and the background, by applying a linear layer to each location of $\mathbf{Z}$, which is equivalent to a 1×1 convolutional filter parametrized by the weight matrix $\mathbf{W}_{\text{attn}} \in \mathbb{R}^{H \times (K+1)}$, followed by a channel-wise softmax:

$$\mathbf{A} = \text{softmax}(\mathbf{Z} * \mathbf{W}_{\text{attn}}) \in \mathbb{R}^{(K+1) \times M \times N}. \tag{2}$$

Each attention map $\mathbf{A}_k$, except for the $(K + 1)^{\text{th}}$, which captures the background, is applied to the tensor $\mathbf{Z}$ in order to obtain the corresponding landmark vector:

$$l_k = \sum_{u=1}^{M} \sum_{v=1}^{N} \mathbf{A}_k(u, v)\mathbf{Z}(u, v) \in \mathbb{R}^H. \tag{3}$$

Each landmark l_k undergoes a linear operation in order to generate the C classification scores, where C is the total number of classes, associated to it:

$$\mathbf{y}_k = l_k \mathbf{W}_{\text{class}} \in \mathbb{R}^C. \tag{4}$$

We apply different losses to the classification scores $\mathbf{y}$, the landmark feature vectors l and the attention maps $\mathbf{A}$. For the classification scores, we use a cross-entropy loss, providing the only gradients for learning the weights of the linear operator $\mathbf{W}_{\text{class}} \in \mathbb{R}^{H \times C}$:

$$\mathcal{L}_{\text{class}}(\mathbf{y}, c) = -\log\left(\frac{\exp(y(c))}{\exp(\sum_i y(i))}\right) \tag{5}$$

In addition, we make sure that landmark vectors are similar across images of the same individual. We use a triplet loss for each landmark k, which is computed on the landmark vector l_k^a, used as anchor in the triplet loss, a positive vector from the corresponding landmark stemming from an image of the same class, l_k^p, and a negative one from a different class l_k^n:

$$\mathcal{L}_{\text{triplet}}(l_k^a, l_k^p, l_k^n) = \max(\|l_k^a - l_k^p\|_2 - \|l_k^a - l_k^n\|_2 + 1, 0) \tag{6}$$

Regarding the losses applied to the landmark attention maps, which have the role of ensuring learning a good set of keypoints for landmark extraction, we apply two losses:

$$\mathcal{L}_{\text{conc}}(\mathbf{A}) = \frac{\sum_{k=1}^{K} \sigma_u^2(\mathbf{A_k}) + \sigma_v^2(\mathbf{A_k})}{K}, \tag{7}$$

which aims at encouraging each attention map to be concentrated around its center of mass by minimizing the variances of each attention map, $\sigma_u^2(\mathbf{A_k})$ and $\sigma_v^2(\mathbf{A_k})$, across both spatial dimensions and

$$\mathcal{L}_{max}(\mathbf{A}) = \frac{\sum_{k=1}^{K} 1 - max(\mathbf{A_k})}{K}, \tag{8}$$

which ensures that all landmarks are present in each image.

These four losses are combined as a weighted sum to obtain the final loss:

$$\mathcal{L} = \lambda_{class}\mathcal{L}_{class} + \lambda_{triplet}\mathcal{L}_{triplet} + \lambda_{conc}\mathcal{L}_{conc} + \lambda_{max}\mathcal{L}_{max}, \tag{9}$$

where λ_{class}, $\lambda_{triplet}$, λ_{conc} are scalar hyperparameters.

4.2 Uncertainty and Sensitivity Analysis

Patch-Based Occlusion Sensitivity Maps. Determining occlusion sensitivity maps is a strategy developed by [37] to evaluate the sensitivity of a trained model to partial occlusions in an input image. The maps visualize which regions contribute positively and which contribute negatively to the result. The approach is to systematically mask different regions for a given input image, choosing a rectangular patch in our case. Two parameters, namely patch size p and step size, are chosen by the user, and the choice affects the result in terms of precision and smoothness. In the area around position $\mathbf{u}$ occluded by the patch, the pixel-wise results of the classifier for each class are compared with the results obtained after part of the image was occluded. For the expected class c, the score s is predicted for the corresponding position u of the patch. The difference δs_{cu} is given by.

$$\delta s_{cu} = s_c - \tilde{s}_{cu} \tag{10}$$

where the original predicted score for each class is denoted by s_c and the predicted score based on occlusion is given by $\tilde{s}_{cu}$. Performing this for the entire image yields a heat map of occlusion sensitivity.

Landmark-Based Sensitivity Analysis. Similarly to the patch-based occlusion sensitivity maps presented previously, landmark-based sensitivity analysis eliminates individual landmarks, by setting all the elements in the corresponding feature vector l_k to zero, in order to analyze their effect on the output, allowing to understand the impact that each landmark has on the final score. In addition to this, we also measure the impact that removing a landmark has on the accuracy across the validation set. In both cases, the same landmark k is removed for all images in the test, thus preventing it from contributing to the final score. This allows us to probe the importance of each landmark across the whole test set.

Landmark-Based Uncertainty Analysis. Due to occlusions, unreliable fluke features or wrongly placed landmarks, different groups of landmarks in the same image may provide evidence for conflicting outputs. Similarly, each individual landmark detector may receive conflicting signals from the previous layer about where to place the landmark on the image. This disagreement can be used to In order to measure this disagreement, we perform two experiments applying different types of Monte Carlo dropout (i.e. test time dropout) to the landmarks.

Class Uncertainty Through Whole Landmark Dropout. We randomly choose half of the landmarks and use them to obtain a class prediction y_r. We perform this operation R times to obtain a collection of class predictions $\mathbf{R} = \{y_1, \ldots, y_R\}$. The agreement score a is then computed as the proportion of random draws that output the most frequently predicted class:

$$a = \frac{1}{R} \sum_{r=1}^{R} [y_r = \text{mode}(\mathbf{R})]. \qquad (11)$$

Landmark Spatial Uncertainty Through Feature Dropout. In this case we apply standard dropout to the feature tensor $\mathbf{Z}$, thus perturbing the landmark attention maps $\mathbf{A}$. Landmarks that have not been reliably detected will be more sensitive to these perturbations, resulting in higher spatial uncertainty.

5   Experiments and Results

Our experiments address landmark detection focusing on the uncertainty and sensitivity of landmarks, and compare to previous results from patch-based occlusion sensitivity maps from [14] by means of whale identification. Furthermore, the landmarks and occlusion sensitivity maps are compared to the domain knowledge of an expert.

Our method allows to easily reach conclusions at both the dataset level and the image level. For one particular image, due to the spatial compactness of the landmark attention maps, we can visualize the contribution of each landmark to the final classification score. In addition, the fact that each landmark tends to focus on the same fluke features across images allows us to analyze the importance of each landmark at the dataset level.

5.1   Experimental Setup

We use a modified classification CNN, a ResNet-18 [10], with reduced downsampling, by a factor of four, in order to preserve better spatial details. For the final loss we used the same weight for each of the sub-losses $\lambda_{\text{triplet}} = \lambda_{\text{conc}} = \lambda_{\text{max}} = \lambda_{\text{class}} = 1$. We use Adam as an optimizer, with the ResNet-18 model starting with a learning rate of 10^{-4}, while $\mathbf{W}_{\text{attn}}$ and $\mathbf{W}_{\text{class}}$ are optimized starting with a learning rate of 10^{-2}. After every epoch, the learning rates are divided by 2 if the validation accuracy decreases. No image pre-processing is used. The top-1 accuracy reaches 86% on the held-out validation set. For comparison, we trained the same base model without the attention mechanism, obtaining an accuracy of 82%, showing that the landmark-based attention mechanism does not penalize the model's performance.

For comparison, we use our previously computed occlusion sensitivity maps presented in [14], which were based on the data and scores of the classification framework of the second winner solution[1] of the Kaggle Challenge. For pre-processing, the framework applies two steps to the raw image. First, the chosen framework automatically performs image cropping in order to reduce the image content to the fluke of the whale. The cropped images are resized to an uniform size of 256 px × 512 px. In the second step, the framework performs standard-normalization on the input images. The architecture is based on ResNet-101 [10] utilizing triplet loss [35], ArcFace loss [4], and focal loss [17]. With this model, we reach a top-5 accuracy of 94.2%.

5.2   Uncertainty and Sensitivity Analysis of the Landmarks

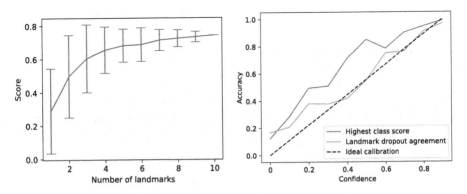

Fig. 3. Left: Average score and standard deviation by randomly selecting an increasing number of landmarks. **Right:** Expected accuracy as a function of two different confidence scores: the highest class score after softmax, and the agreement between 100 landmark dropout runs.

Figure 3 (left) shows the uncertainty of the predicted score, *i.e.* how much the result score varies when a certain number of landmarks is used. It can be seen that the uncertainty becomes smaller the more landmarks are used. The reason for this is that usually several features are used for identification - by the domain expert as well as by the neural network - and with increasing number of landmarks the possibility to cover several features increases. Figure 3 (right) displays the expected accuracy for varying levels of confidence estimates. We compare two estimates: the maximum softmax loss, in blue, and the agreement between 100 runs of MC landmark dropout with a dropout rate of 0.5, in orange. We can see that the latter follows more closely the behaviour of an ideally calibrated estimate (dashed line).

---

[1] 2nd place: https://github.com/SeuTao/Humpback-Whale-Identification-Challenge-2019_2nd_palce_solution.

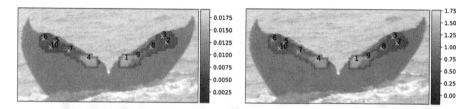

Fig. 4. Top: Average sensitivity heatmap rendered on the landmark locations of one image, representing the average reduction in the score of the correct class after removing each landmark. **Bottom**: Average loss in accuracy, in percent points, after removing each landmark. Photo CC BY-NC 4.0 John Calambokidis.

5.3 Heatmapping Results and Comparison with Whale Expert Knowledge

Figure 4 shows the mean landmark sensitivity (top), as well as the loss of accuracy after removing landmarks (bottom), calculated over the complete data set. When compared to the landmarks near the fluke tips, it can be seen that the landmarks near the notch change the score the most, and flip the classification towards the correct class the most often. This is consistent with the fact that the interior of a fluke changes rather little over time, while the fluke tips can change significantly over time. Also, the pose and activity of the whale when the images are captured might explain this behavior. It is worth noting that all the attention is concentrated along the trailing edge of the fluke. This may be due to the fact that it is the area of the fluke that is most reliably visible in the images, since the leading edge tends to be under water in a number of photos.

In the following, we examine the landmark-based and patch-based tools in terms of the features considered as important by the whale expert on individual images. We show the results on two pairs of images such that each pair belongs to the same individual. Figure 5a highlights the main areas the expert focused on in order to conclude whether they do belong to the same individual or not after inspecting both images side-to-side. Note the tendency of the expert of annotating just a small number of compact regions.

The heatmaps obtained using patch-based occlusion are shown in Fig. 5b. Although the fluke itself is recognised as being important to the classification, no particular area is highlighted, except for one case where the whole trailing edge appears to be important. In addition, some regions outside of the fluke seem to have a negative sensitivity, pointing at the possibility of an artifact in the dataset that is being used by the model. This was observed in previous publications [14], where authors concluded that patch-based occlusion was using the shape of the entire fluke, rather than specific, localised patterns.

The results of the landmark-based approach, in Fig. 5c, show more expert-like heatmaps, with the evidence for and against a match always located on the fluke and generally around the trailing edge and close to the notch. In each case, only a few small regions are responsible for the evidence in favor of assigning each pair to the same individual. However, although both the expert and the

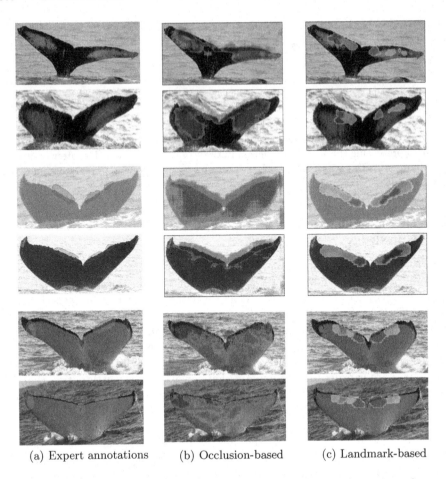

(a) Expert annotations   (b) Occlusion-based   (c) Landmark-based

Fig. 5. Heatmaps of attribution. Dark blue/red areas highlight the regions that are estimated to provide evidence for/against the match. The top two pairs are matching pairs (same individual) while the bottom one is not a match. (Color figure online)

landmark-based method have a tendency of pointing at the same general areas around the trailing edge with compact highlights, we do not observe a consistent overlap with the expert annotated images. This may be due to constraints in both the expert and the landmark-based highlights. Unlike the expert, the landmark-based approach tends to focus, by design, in the areas of the fluke that are most reliably visible. The expert, on the other hand, explores all visible fluke features and highlights them in a non-exhaustive manner. On the top image pair, a region that is also annotated by the expert on the left fluke provides most of the positive evidence, but a feature close to the leading edge is ignored. This is probably due to the model learning that the leading edge is less reliable, since it is under water in a large number of photos. On the middle pair, the area to the left of notch is assigned a negative sensitivity while being annotated as important by the

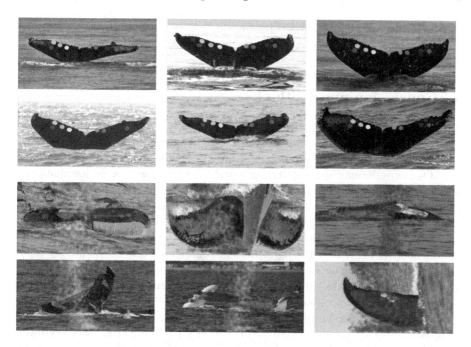

Fig. 6. Spatial uncertainty of each landmark on different whales determined by means of 500 dropout runs on the feature tensor $\mathbf{Z}$. Each disk represents the location of a landmark in one run and each of the ten landmarks is colored consistently across images. **Top:** The test images with the lowest uncertainty. **Bottom:** The test images with the highest uncertainty. (Color figure online)

expert. On the bottom pair we see that only the landmarks closest to the notch are used by the model to decide that the images do indeed belong to different individuals, while the expert has also annotated a region close to the fluke tip, which the landmark-based model systematically ignores, likely due to the fact, as with the leading edge, that the tips are less reliably visible in the images.

5.4   Spatial Uncertainty of Individual Landmarks

The visualizations in Fig. 6 display the six images in the test set with the lowest and with the highest uncertainty, each on a different individual. The colored disks represent the positions of each landmark across 500 random application of dropout, with a dropout probability of 0.5, to the feature tensor $\mathbf{Z}$. The colors are consistent (e.g. landmark 5, as seen in Fig. 4 is always represented in dark blue). The top rows tend to contain images with clearly visible flukes in a canonical pose. As we can see, the detected keypoints do behave as landmarks, each specializing in a particular part of the fluke, even if no particular element of the loss was designed to explicitly promote this behaviour. The bottom rows contain images with either substantial occlusions or uncommon poses. This shows how the spatial uncertainty uncovered by MC dropout can be used to detect unreliably located landmarks, which in turn can be used to find images with problematic poses and occlusions that are likely to be unsuitable for identification.

6 Conclusion and Outlook

In this work, we explore the use of landmark detection learning using only class labels (i.e. whale identities) and apply it to gain insights into which fluke parts are relevant to the model's decision in the context of cetacean individual identification. Our experiments show that, compared to patch-based occlusion mapping, our approach highlights regions in the images that are systematically located along the central part of the trailing edge of the fluke, which is the part most reliably visible in the images. At the same time, the landmarks highlight compact regions that are much more expert-like than the baseline OSM heatmaps. In addition, we show that the agreement of random subsets of the landmarks is a better estimate of the expected error rate than the softmax score. However, there seems to be little agreement between the specific regions chosen by the expert and the landmark-based highlights.

The use of landmarks makes it easy to match them across images, since each landmark develops a tendency to specialize on a particular region of the fluke. This allowed us to study their average importance for the whole validation set, leading us to conclude that the areas of the trailing edge right next to the notch tend to be the most relied upon. This is probably due to the to the higher temporal stability of the region around the notch, which is less exposed and thus less likely to develop scars, and to the fact that the trailing edge is the part of the fluke most often visible in the photos. Is also worth noting that the proposed method is inherently interpretable, thus not only guaranteeing that the generated heatmaps are relevant to the model's decision, but also doing so at a negligible computational cost, requiring to perform inference once and not using any gradient information. In addition, the accuracy obtained is noticeably higher than a model with the same base architecture but no attention mechanism.

In spite of these advantages, we also observed an inherent limitation of the method when compared to the expert annotations. Our landmark-based model requires to find all landmarks on each image, resulting in a tendency to only focus on the areas of the fluke that are most reliably visible and discarding those that are often occluded, such as the tips and the leading edge. Designing a model that is free to detect a varying number of landmarks is a potential path towards even more expert-like explanations.

References

1. Adadi, A., Berrada, M.: Peeking inside the black-box: a survey on explainable Artificial Intelligence (XAI). IEEE Access **6**, 52138–52160 (2018)
2. Andrew, W., Greatwood, C., Burghardt, T.: Aerial animal biometrics: individual friesian cattle recovery and visual identification via an autonomous UAV with onboard deep inference. In: IROS (2019)

3. Bau, D., Zhou, B., Khosla, A., Oliva, A., Torralba, A.: Network dissection: quantifying interpretability of deep visual representations. In: CVPR (2017)
4. Deng, J., Guo, J., Xue, N., Zafeiriou, S.: Arcface: additive angular margin loss for deep face recognition. In: CVPR, pp. 4690–4699 (2019)
5. Gal, Y., Ghahramani, Z.: Dropout as a Bayesian approximation: representing model uncertainty in deep learning. In: ICML, pp. 1050–1059 (2016)
6. Gawlikowski, J., et al.: A survey of uncertainty in deep neural networks. arXiv preprint arXiv:2107.03342 (2021)
7. Ghanem, R., Higdon, D., Owhadi, H. (eds.): Handbook of Uncertainty Quantification. Springer, Cham (2017). https://doi.org/10.1007/978-3-319-12385-1
8. Gilpin, L.H., Bau, D., Yuan, B.Z., Bajwa, A., Specter, M., Kagal, L.: Explaining explanations: an overview of interpretability of machine learning, May 2018, arXiv preprints arXiv:1806.00069
9. Guidotti, R., Monreale, A., Ruggieri, S., Turini, F., Giannotti, F., Pedreschi, D.: A survey of methods for explaining black box models. ACM Comput. Surv. **51**(5), 1–42 (2018)
10. He, K., Zhang, X., Ren, S., Sun, J.: Deep residual learning for image recognition. In: CVPR, pp. 770–778 (2016)
11. Hohman, F.M., Kahng, M., Pienta, R., Chau, D.H.: Visual analytics in deep learning: an interrogative survey for the next frontiers. IEEE Trans. Visual Comput. Graph. **25**(1), 1–20 (2018)
12. Hüllermeier, E., Waegeman, W.: Aleatoric and epistemic uncertainty in machine learning: an introduction to concepts and methods. Mach. Learn. **110**(3), 457–506 (2021). https://doi.org/10.1007/s10994-021-05946-3
13. Katona, S., Whitehead, H.: Identifying humpback whales using their natural markings. Polar Rec. **20**(128), 439–444 (1981)
14. Kierdorf, J., Garcke, J., Behley, J., Cheeseman, T., Roscher, R.: What identifies a whale by its fluke? on the benefit of interpretable machine learning for whale identification. In: ISPRS Annals, vol. 2, pp. 1005–1012 (2020)
15. Kniest, E., Burns, D., Harrison, P.: Fluke matcher: a computer-aided matching system for humpback whale (Megaptera novaeangliae) flukes. Mar. Mamm. Sci. **3**(26), 744–756 (2010)
16. Li, S., Li, J., Tang, H., Qian, R., Lin, W.: ATRW: a benchmark for Amur tiger re-identification in the wild. In: ACM International Conference on Multimedia, pp. 2590–2598 (2020)
17. Lin, T.Y., Goyal, P., Girshick, R., He, K., Dollár, P.: Focal loss for dense object detection. In: ICCV, pp. 2980–2988 (2017)
18. Loucks, D., Van Beek, E., Stedinger, J., Dijkman, J., Villars, M.: Model sensitivity and uncertainty analysis. Water Resources Systems Planning and Management, pp. 255–290 (2005)
19. Montavon, G., Samek, W., Müller, K.R.: Methods for interpreting and understanding deep neural networks. Digit. Sig. Process. **73**, 1–15 (2018)
20. Rajaraman, S., et al.: Understanding the learned behavior of customized convolutional neural networks toward malaria parasite detection in thin blood smear images. J. Med. Imaging **5**(3), 034501 (2018)
21. Roscher, R., Bohn, B., Duarte, M.F., Garcke, J.: Explainable machine learning for scientific insights and discoveries. IEEE Access **8**, 42200–42216 (2020)
22. Rudin, C.: Stop explaining black box machine learning models for high stakes decisions and use interpretable models instead. Nat. Mach. Intell. **1**(5), 206–215 (2019)

23. Samek, W., Montavon, G., Lapuschkin, S., Anders, C.J., Müller, K.R.: Explaining deep neural networks and beyond: a review of methods and applications. Proc. IEEE **109**(3), 247–278 (2021)
24. Samek, W., Müller, Klaus-R.: Towards explainable artificial intelligence. In: Samek, W., Montavon, G., Vedaldi, A., Hansen, L.K., Müller, K.-R. (eds.) Explainable AI: Interpreting, Explaining and Visualizing Deep Learning. LNCS (LNAI), vol. 11700, pp. 5–22. Springer, Cham (2019). https://doi.org/10.1007/978-3-030-28954-6_1
25. Schneider, S., Taylor, G.W., Linquist, S., Kremer, S.C.: Past, present and future approaches using computer vision for animal re-identification from camera trap data. Methods Ecol. Evol. **10**(4), 461–470 (2019)
26. Schofield, D., et al.: Chimpanzee face recognition from videos in the wild using deep learning. Sci. Adv. 5(9), **eaaw0736** (2019)
27. Schramowski, P., et al.: Right for the wrong scientific reasons: revising deep networks by interacting with their explanations. arXiv preprint arXiv:2001.05371 (2020)
28. Selvaraju, R.R., Cogswell, M., Das, A., Vedantam, R., Parikh, D., Batra, D.: Grad-cam: visual explanations from deep networks via gradient-based localization. In: ICCV, pp. 618–626 (2017)
29. Simoes, H., Meidanis, J.: Humpback whale identification challenge: a comparative analysis of the top solutions (2020)
30. Ståhl, N., Falkman, G., Karlsson, A., Mathiason, G.: Evaluation of uncertainty quantification in deep learning. In: Lesot, M.-J., et al. (eds.) IPMU 2020. CCIS, vol. 1237, pp. 556–568. Springer, Cham (2020). https://doi.org/10.1007/978-3-030-50146-4_41
31. Stomberg, T., Weber, I., Schmitt, M., Roscher, R.: Jungle-net: using explainable machine learning to gain new insights into the appearance of wilderness in satellite imagery. In: ISPRS Annals, vol. 3, pp. 317–324 (2021)
32. Sundararajan, M., Taly, A., Yan, Q.: Axiomatic attribution for deep networks. In: ICML, pp. 3319–3328. PMLR (2017)
33. Surma, S., Pitcher, T.J.: Predicting the effects of whale population recovery on northeast pacific food webs and fisheries: an ecosystem modelling approach. Fish. Oceanogr. **24**(3), 291–305 (2015)
34. Wang, H., Yeung, D.Y.: A survey on Bayesian deep learning. ACM Comput. Surv. (CSUR) **53**(5), 1–37 (2020)
35. Weinberger, K.Q., Saul, L.K.: Distance metric learning for large margin nearest neighbor classification. J. Mach. Learn. Res. **10**(2), 207–244 (2009)
36. Xu, K., et al.: Show, attend and tell: neural image caption generation with visual attention. In: ICML, pp. 2048–2057. PMLR (2015)
37. Zeiler, M.D., Fergus, R.: Visualizing and understanding convolutional networks. In: Fleet, D., Pajdla, T., Schiele, B., Tuytelaars, T. (eds.) ECCV 2014. LNCS, vol. 8689, pp. 818–833. Springer, Cham (2014). https://doi.org/10.1007/978-3-319-10590-1_53
38. Zhang, Y., Guo, Y., Jin, Y., Luo, Y., He, Z., Lee, H.: Unsupervised discovery of object landmarks as structural representations. In: CVPR, pp. 2694–2703 (2018)
39. Zhou, B., Khosla, A., Lapedriza, A., Oliva, A., Torralba, A.: Learning deep features for discriminative localization. In: CVPR, pp. 2921–2929 (2016)

Explainable Artificial Intelligence in Meteorology and Climate Science: Model Fine-Tuning, Calibrating Trust and Learning New Science

Antonios Mamalakis[1](✉)[iD], Imme Ebert-Uphoff[2,3][iD], and Elizabeth A. Barnes[1][iD]

[1] Department of Atmospheric Science, Colorado State University,
Fort Collins, CO, USA
amamalak@colostate.edu

[2] Department of Electrical and Computer Engineering, Colorado State University,
Fort Collins, CO, USA

[3] Cooperative Institute for Research in the Atmosphere, Colorado State University,
Fort Collins, CO, USA

Abstract. In recent years, artificial intelligence and specifically artificial neural networks (NNs) have shown great success in solving complex, nonlinear problems in earth sciences. Despite their success, the strategies upon which NNs make decisions are hard to decipher, which prevents scientists from interpreting and building trust in the NN predictions; a highly desired and necessary condition for the further use and exploitation of NNs' potential. Thus, a variety of methods have been recently introduced with the aim of attributing the NN predictions to specific features in the input space and explaining their strategy. The so-called eXplainable Artificial Intelligence (XAI) is already seeing great application in a plethora of fields, offering promising results and insights about the decision strategies of NNs. Here, we provide an overview of the most recent work from our group, applying XAI to meteorology and climate science. Specifically, we present results from satellite applications that include weather phenomena identification and image to image translation, applications to climate prediction at subseasonal to decadal timescales, and detection of forced climatic changes and anthropogenic footprint. We also summarize a recently introduced synthetic benchmark dataset that can be used to improve our understanding of different XAI methods and introduce objectivity into the assessment of their fidelity. With this overview, we aim to illustrate how gaining accurate insights about the NN decision strategy can help climate scientists and meteorologists improve practices in fine-tuning model architectures, calibrating trust in climate and weather prediction and attribution, and learning new science.

Keywords: eXplainable Artificial Intelligence (XAI) · Neural Networks (NNs) · Geoscience · Climate science · Meteorology · Trust · Black box models

A. Holzinger et al. (Eds.): xxAI 2020, LNAI 13200, pp. 315–339, 2022.
https://doi.org/10.1007/978-3-031-04083-2_16

1 Introduction

In the last decade, artificial neural networks (NNs) [38] have been increasingly used for solving a plethora of problems in the earth sciences [5,7,21,27,36,41, 60,62,68,72], including marine science [41], solid earth science [7], and climate science and meteorology [5,21,62]. The popularity of NNs stems partially from their high performance in capturing/predicting nonlinear system behavior [38], the increasing availability of observational and simulated data [1,20,56,61], and the increase in computational power that allows for processing large amounts of data simultaneously. Despite their high predictive skill, NNs are not interpretable (usually referred to as "black box" models), which means that the strategy they use to make predictions is not inherently known (as, in contrast, is the case for e.g., linear models). This may introduce doubt with regard to the reliability of NN predictions and it does not allow scientists to apply NNs to problems where model interpretability is necessary.

To address the interpretability issue, many different methods have recently been developed [3,4,32,53,69,70,73,75,77,84] in the emerging field of eXplainable Artificial Intelligence (XAI) [9,12,78]. These methods aim at a *post hoc* attribution of the NN prediction to specific features in the input domain (usually referred to as attribution/relevance heatmaps), thus identifying relationships between the input and the output that may be interpreted physically by the scientists. XAI methods have already offered promising results and fruitful insights into how NNs predict in many applications and in various fields, making "black box" models more transparent [50]. In the geosciences, physical understanding about how a model predicts is highly desired, so, XAI methods are expected to be a real game-changer for the further application of NNs in this field [79].

In this chapter, we provide an overview of the most recent studies from our group that implement XAI in the fields of climate science and meteorology. We focus here on outlining our work, the details of which we are more knowledgeable of, but we highlight that relevant work has been also established by other groups (see e.g., [18,34,52,74]). The first part of this overview presents results from direct application of XAI to solve various prediction problems that are of particular interest to the community. We start with XAI applications in remote sensing, specifically for image-to-image translation of satellite imagery to inform weather forecasting. Second, we focus on applications of climate prediction at a range of timescales from subseasonal to decadal, and last, we show how XAI can be used to detect forced climatic changes and anthropogenic footprint in observations and simulations. The second part of this overview explores ways that can help scientists gain insights about systematic strengths and weaknesses of different XAI methods and generally improve their assessment. So far in the literature, there has been no objective framework to assess how accurately an XAI method explains the strategy of a NN, since the ground truth of what the explanation should look like is typically unknown. Here, we discuss a recently introduced synthetic benchmark dataset that can introduce objectivity in assessing XAI methods' fidelity for weather/climate applications, which will lead to better understanding and implementation.

The overall aim of this chapter is to illustrate how XAI methods can be used to help scientists fine tune NN models that perform poorly, build trust in models that are successful and investigate new physical insights and connections between the input and the output (see Fig. 1).

i) XAI to guide the design of the NN architecture. One of the main challenges when using a NN is how to decide on the proper NN architecture for the problem at hand. We argue that XAI methods can be an effective tool for analysts to get insight into a flawed NN strategy and be able to revise it in order to improve prediction performance.

ii) XAI to help calibrate trust in the NN predictions. Even in cases when a NN (i.e., or any black model in general) exhibits a high predictive performance, it is not guaranteed that the underlying strategy that is used for prediction is correct. This has famously been depicted in the example of "clever Hans", a horse that was correctly solving mathematical sums and problems based on the reaction of the audience [35]. By using XAI methods, scientists can verify when a prediction is successful for the right reasons (i.e., they can test against "clever Hans" prediction models [35]), thus helping build model trust.

iii) XAI to help learn new science. XAI methods allow scientists to gain physical insights about the connections between the input variables and the predicted output, and generally about the problem at hand. In cases where the highlighted connections are not fully anticipated/understood by already established science, further research and investigation may be warranted, which can accelerate learning new science. We highlight though that XAI methods will most often motivate new analysis to learn and establish new science, but cannot prove the existence of a physical phenomenon, link or mechanism, since correlation does not imply causation.

The content of the chapter is mainly based on previously published work from our group [6,16,25,29,43,47,80], and is re-organized here to be easily followed by the non-expert reader. In Sect. 2, we present results from various XAI applications in climate science and meteorology. In Sect. 3, we outline a new framework to generate attribution benchmark datasets to objectively evaluate XAI methods' fidelity, and in Sect. 4, we state our conclusions.

2   XAI Applications

2.1   XAI in Remote Sensing and Weather Forecasting

As a first application of XAI, we focus on the field of remote sensing and short-term weather forecasting. When it comes to forecasting high impact weather hazards, imagery from geostationary satellites has been excessively used as a tool for situation awareness by human forecasters, since it supports the need for high spatial resolution and temporally rapid refreshing [40]. However, information from geostationary satellite imagery has less frequently been used in data

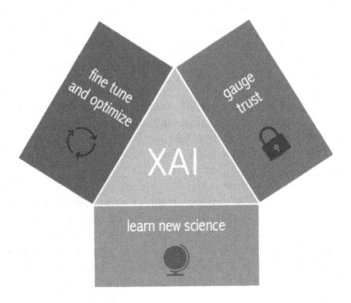

Fig. 1. XAI offers the opportunity for scientists to gain insights about the decision strategy of NNs, and help fine tune and optimize models, gauge trust and investigate new physical insights to establish new science.

assimilation or integrated into weather-forecasting numerical models, despite the advantages that these data could offer in improving numerical forecasts.

In recent work [25], scientists have used XAI to estimate precipitation over the contiguous United States from satellite imagery. These precipitating scenes that are typically produced by radars and come in the form of radar reflectivity can then be integrated into numerical models to spin up convection. Thus, the motivation of this research was to exploit the NNs' high potential in capturing spatial information together with the large quantity, high quality and low latency of satellite imagery, in order to inform numerical modeling and forecasting. This could be greatly advantageous for mitigation of weather hazards.

For their analysis, Hilburn et al. (2021) [25] developed a convolutional NN with a U-Net architecture (dubbed GREMLIN in the original paper). The inputs to the network were four-channel satellite images, each one containing brightness temperature and lightning information, over various regions around the US. As output, the network was trained to predict a single-channel image (i.e., an image-to-image translation application) that represents precipitation over the same region as the input, in the form of radar reflectivity and measured in dBZ. The network was trained against radar observations, and its overall prediction performance across testing samples was quite successful. Specifically, predictions from the GREMLIN model exhibited an overall coefficient of determination on the order of $R^2 = 0.74$ against the radar observations and a root mean squared difference on the order of 5.53 dBZ.

Apart from statistically evaluating the performance of GREMLIN predictions in reproducing reflectivity fields, it was also very important to assess the strategy upon which the model predicted. For this purpose, Hilburn et al. made use of a well-known XAI method, the Layer-wise Relevance Propagation (LRP [4]). Given an input sample and an output pixel, LRP reveals which features in the input contributed the most in deriving the value of the output. This is accomplished by sequentially propagating backwards the relevance from the output pixel to the neurons of the previous layers and eventually to the input features. So far, numerous different rules have been proposed in the literature as to how this propagation of relevance can be performed, and in this XAI application the alpha-beta rule was used [4], with alpha = 1 and beta = 0. The alpha-beta rule distinguishes between strictly positive and strictly negative pre-activations, which helps avoid the possibility of infinitely growing relevancies in the propagation phase, and it provides more stable results.

In Fig. 2, we show LRP results for GREMLIN for a specific sample, and a specific output pixel (namely, the central location of the shown sample), chosen for its close proximity to strong lightning activity. The first row of the figure shows the input channels and the corresponding desired output (i.e., the radar observation). The second row shows the LRP maps, highlighting which features in the input channels the neural network paid attention to in order to estimate the value of the chosen central output pixel for this sample.

The LRP results for the channel with lightning information show that the network focused only on regions where lightning was present in that channel. The LRP results for the other channels show that even in those channels the NN's attention was drawn to focus on regions where lightning was present. Hilburn et al. then performed a new experiment by modifying the input sample to have all lightning removed, that is, all the lightning values were set to zero. In this case, LRP highlighted that the network's focus shifted entirely in the first three input channels, as expected. More specifically, the focus shifted to two types of locations, namely, (i) cloud boundaries, or (ii) areas where the input channels had high brightness (cold temperatures), as can be seen by comparing the three leftmost panels of the first, and third row. In fact, near the center of the third-row panels, it can be seen that the LRP patterns represent the union of the cloud boundaries and the locations of strongest brightness in the first row. LRP vanishes further away from the center location, as it is expected considering the nature of the effective receptive field that corresponds to the output pixel.

The LRP results as presented above provide very valuable insight about how the network derives its estimates. Specifically, the results indicate the following strategy used by GREMLIN: whenever lightning is present near the output pixel, the NN primarily focuses on the values of input pixels where lightning is present, not only in the channel that contains the lightning information, but in all four input channels. It seems that the network has learned that locations containing lightning are good indicators of high reflectivity, even in the other input channels. When no lightning is present, the NN focuses primarily on cloud boundaries (locations where the gradient is strong) or locations of very cold cloud tops. The

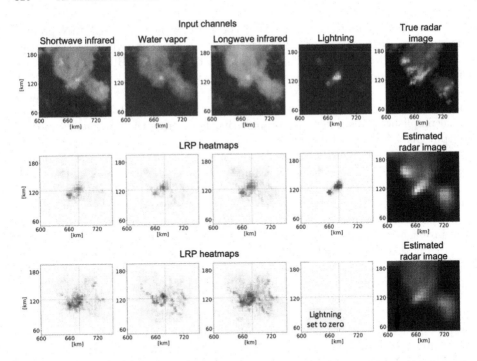

Fig. 2. LRP results for the GREMLIN model. (top) The four input channels and the corresponding observed radar image (ground truth). (middle) LRP results for the original four input channels and the chosen output pixel, and the prediction from GREMLIN. (bottom) The equivalent of the middle row, but after all lighting values were set to zero. Note that all images are zoomed into a region centered at the pixel of interest. Adapted from Hilburn et al., 2021 [25].

network seems to have learned that these locations have the highest predictive power for estimating reflectivity.

In this application of XAI in remote sensing, the obtained insights from LRP have given scientists the confidence that the network derived predictions based on a physically reasonable strategy and thus helped build more trust about its predictions. Moreover, if scientists wish to improve the model further by testing different model architectures, knowing how much physically consistent the different decision strategies of the models are offers a criterion to distinguish between models, which goes beyond prediction performance.

2.2 XAI in Climate Prediction

Similar to weather forecasting, climate prediction at subseasonal, seasonal and decadal timescales is among the most important challenges in climate science, with great societal risks and implications for the economy, water security, and ecosystem management for many regions around the world [8]. Typically, climate prediction draws upon sea surface temperature (SST) information (especially on seasonal timescales and beyond), which are considered as the principal forcing variable of the atmospheric circulation that ultimately drives regional climate [19, 23, 30, 44, 57]. SST information is used for prediction either through deterministic models (i.e., SST-forced climate model simulations) or statistical models which aim to exploit physically- and historically- established teleconnections of regional climate with large-scale modes of climate variability (e.g., the El Niño-Southern Oscillation, ENSO; [11, 17, 44, 45, 48, 49, 54, 59, 67]). Limits to predictive skill of dynamical models arise from incomplete knowledge of initial conditions, uncertainties and biases in model physics, and limits on computational resources that place constraints on the grid resolution used in operational systems. Similarly, empirical statistical models exhibit limited predictive skill, arising primarily from the complex and non-stationary nature of the relationship between large scale modes and regional climate.

To address the latter, in more recent years, data-driven machine learning methods that leverage information from the entire globe (i.e., beyond predefined climate indices) have been suggested in the literature and they have shown improvements in predictive skill [13, 76]. A number of studies have specifically shown the potential of neural networks in predicting climate across a range of scales, capitalizing on their ability to capture nonlinear dependencies (see e.g., [21]), while more recent studies have used XAI methods[1] to explain these networks and their strategies to increase trust and learn new science [47, 79, 80].

In the first study outlined herein, Mayer and Barnes (2021) [47] used XAI in an innovative way to show that NNs can identify when favorable conditions that lead to enhanced predictive skill of regional climate are present in the atmosphere (the so called "forecasts of opportunity") or not. More specifically, the authors based their analysis on the known climate teleconnections between the Madden-Julian Oscillation in the tropics (MJO; an eastward moving disturbance of convection in the tropical atmosphere) and the North Atlantic atmospheric pressure [10, 24]. When the MJO is active, it leads to a consistent and coherent modulation of the midlatitude climate on subseasonal timescales, and thus, corresponds to enhanced predictive potential for the midlatitudes. The question that Mayer and Barnes put forward was whether or not NNs can capture this inherent property of the climate system of exhibiting periods of enhanced predictability (i.e. forecasts of opportunity).

The authors used daily data of outgoing longwave radiation (OLR; a measure of convective activity) over the tropics of the globe and trained a fully connected NN to predict the sign of the 500 hPa geopotential height anomalies (a measure of atmospheric pressure) over the North Atlantic, 22 days later. Their results showed that when the network was assigning higher confidence to a prediction

---

[1] We note here that a newly introduced line of research in XAI that is potentially relevant for climate prediction applications is in the concept of causability. Although XAI is typically used to address transparency of AI, causability refers to the quality of an explanation and to the extend to which an explanation may allow the scientist to reach a specified level of causal understanding about the underlying dynamics of the climate system [26].

(i.e., the likelihood of either the positive or the negative geopotential height class was much higher than the opposite class), it was much more likely for that prediction to end up being correct. On the contrary, when the network was assigning low confidence to a prediction (i.e., the likelihoods of the positive or negative geopotential height classes were very similar), the predictive performance of the network was much poorer, almost identical to a random guess. This meant that the NN was able to correctly capture the presence of forecasts of opportunity in the climate system.

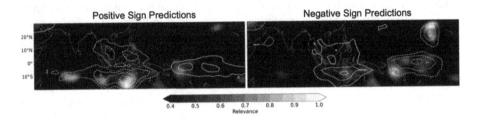

Fig. 3. Maps of LRP composites corresponding to the 10% most confident and correct predictions of positive and negative geopotential height anomalies. Contours indicate the corresponding composite fields of the outgoing longwave radiation with solid lines representing positive values and dashed lines negative values. Adapted from Mayer and Barnes et al., 2021 [47].

Mayer and Barnes continued in exploring which features over the tropics made the network highly confident during forecasts of opportunity, by using the LRP method. Figure 3 shows the LRP heatmaps for positive and negative, correctly predicted, anomalies of geopotential height over the North Atlantic. Note that only the top 10% of the most confident correct predictions were used for the LRP analysis (these predictions ought to represent cases of forecast of opportunity). As it is shown, LRP identified several sources of predictability over the southern Indian Ocean, the Maritime Continent and the western Pacific Ocean for positive predictions, and over the Maritime Continent, the western and central Pacific and over the western side of Hawaii for negative predictions. Judging by the OLR contours, the highlighted patterns correspond to dipoles of convection over the Indian Ocean and into the Maritime Continent in the first case and over the Maritime Continent and into the western Pacific in the second case. These patterns are consistent with the MJO structure and correspond to specific phases of the phenomenon, which in turn have been shown to be connected with the climate of the North Atlantic [10, 24]. Thus, the implementation of LRP in this problem confirms that the network correctly captured the MJO-modulated forecasts of opportunity on subseasonal scales, and it further builds trust for the network's predictive performance.

In a second climate prediction application, this time on decadal scales, Toms et al. (2021) [80] used simulated data from fully-coupled climate models and explored sources of decadal predictability in the climate system. Specifically,

Toms et al. used global SST information as the predictor, with the aim of predicting continental surface temperature around the globe; for each grid point over land, a separate dense network was used. In this way, by combining the large number of samples provided by the climate models (unrealistically large sample size compared to what is available in the observational record) and the ability of NNs to capture nonlinear dynamics, the authors were able to assess the predictability of the climate system in a nonlinear setting. Note that assessing predictability using observational records has been typically based on linear models of limited complexity to avoid overfitting, given the short sample sizes that are usually available [13,76]. Since the climate system is far from linear, the investigation by Toms et al. may be argued to provide a better estimate of predictability than previous work. The results showed that there are several regions where surface temperature is practically unpredictable, whereas there are also regions of high predictability, namely, "hotspots" of predictability, i.e., regions where the predictive skill is inherently high. The presence of hotspots of predictability is conceptually the same with the presence of forecasts of opportunity on subseasonal scales that was discussed in the previous application.

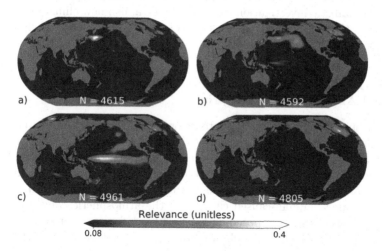

Fig. 4. Composite of LRP maps for the sea surface temperature (SST) field for accurate predictions of positive surface temperature anomalies at four locations across North America. The continental locations associated with the composites are denoted by the red dots in each panel. The LRP map for each sample is normalized between a value of 0 and 1 before compositing to ensure each prediction carries the same weight in the composite. The number of samples used in each composite is shown within each sub-figure. Adapted from Toms et al., 2021 [80].

Toms et al. explored the sources of predictability of surface temperature over North America by using the LRP method. Figure 4 shows the composite LRP maps that correspond to correctly predicted positive temperature anomalies over four different regions in the North America. One can observe that different

SST patterns are highlighted as sources of predictability for each of the four regions. Perhaps surprisingly, temperature anomalies over Central America are shown to be most associated with SST anomalies off the east coast of Japan (Fig. 4a), likely related to the Kuroshio Extension [58]. SST anomalies over the North-Central Pacific Ocean are associated with continental temperature anomalies along the west coast (Fig. 4b), while those within the tropical Pacific Ocean contribute to predictability across central North America (Fig. 4c). Lastly, the North Atlantic SSTs contribute predictability to all four regions, although their impacts are more prominent across the northeastern side of the continent (Fig. 4d). The highlighted patterns of predictability as assessed by LRP resemble known modes of SST variability, such as the El Niño-Southern Oscillation (e.g.,[55,81]), the Pacific Decadal Oscillation [46,54], and the Atlantic Multidecadal Oscillation [17]. These modes are known to affect hydroclimate over North America [11,17,48,49,54], thus, this application constitutes one more case where XAI methods can help scientists build model trust. More importantly, in this setting, physical insights can be extracted about sources of temperature predictability over the entire globe, by sequentially applying LRP to each of the trained networks. As Toms et al. highlight, such an analysis could motivate further mechanistic investigation to physically establish new climate teleconnections. Thus, this application also illustrates how XAI methods can help advance climate science.

2.3   XAI to Extract Forced Climate Change Signals and Anthropogenic Footprint

As a final application of XAI to meteorology and climate science, we consider studies that try to identify human-caused climatic changes (i.e. climate change signals) and anthropogenic footprint in observations or simulations. Detecting climate change signals has been recognized in the climate community as a signal-to-noise problem, where the warming "signal" arising from the slow (long timescales), human-caused changes in the atmospheric concentrations of greenhouse gases is superimposed on the background "noise" of natural climate variability [66]. By solely using observations, one cannot identify which climatic changes are happening due to anthropogenic forcing, since there is no way to strictly infer the possible contribution of natural variability to these observed changes. Hence, the state-of-the-art approach to quantify or to account for natural variability within the climate community is the utilization of large ensembles of climate model simulations (e.g., [14,28]). Specifically, researchers simulate multiple trajectories of the climate system, which start from slightly different initial states but share a common forcing (natural forcing or not). Under this setting, natural variability is represented by the range of the simulated future climates given a specific forcing, and the signal of the forced changes in the climate can be estimated by averaging across all simulations [51].

Utilizing these state-of-the-art climate change simulations, Barnes et al. (2020) [6] used XAI in an innovative way to detect forced climatic changes in temperature and precipitation. Specifically, the authors trained a fully connected NN to predict the year that corresponded to a given (as an input) map of annual-mean temperature (or precipitation) that had been simulated by a climate model. For the NN to be able to predict the year of each map correctly, it needs to learn to look and distinguish specific features of forced climatic change amidst the background natural variability and model differences. In other words, only robust (present in all models) and pronounced (not overwhelmed by natural variability) climate change signals arising from anthropogenic forcing would make the NN to distinguish between a year in the early decades versus late decades of the simulation. Climate change signals that are weak compared to the background natural variability or exhibit high uncertainty across different climate models will not be helpful to the NN.

In the way Barnes et al. have formed the prediction task, the prediction itself is of limited or no utility (i.e., there is no utility in predicting the year that a model-produced temperature map corresponds to; it is already known). Rather, the goal of the analysis is to explore which features help the NN distinguish each year and gain physical insight about robust signals of human-caused climate change. This means that the goal of the analysis lies on the explanation of the network and not the prediction. Barnes et al. trained the NN over the entire simulation period 1920–2099, using 80% of the climate model simulations and then tested on the remaining 20%. Climate simulations were carried out by 29 different models, since the authors were interested in extracting climate change signals that are robust across multiple climate models. Results showed that the NN was able to predict quite successfully the correct years that different temperature and precipitation maps corresponded to. Yet, the performance was lower for years before the 1960s and much higher for years well into the 21st century. This is due to the fact that the climate change signal becomes more pronounced with time, which makes it easier to distinguish amidst the background noise and the model uncertainty.

Next, Barnes et al. used LRP to gain insight into the forced climatic changes in the simulations that had helped the NN to correctly predict each year. Figure 5 shows the LRP results for the years 1975, 2035 and 2095. It can be seen that different areas are highlighted during different years, which indicates that the relative importance of different climate change signals varies through time. For example, LRP highlights the North Atlantic temperature to be a strong indicator of climate change during the late 20th and early 21st century, but not during the late 21st century. On the contrary, the Southern Ocean gains importance only throughout the 21st century. Similarly, the temperature over eastern China is highlighted only in the late 20th century, which likely reflects the aerosol forcing which acts to decrease temperature. Thus, the NN learned that strong cooling over China relatively to the overall warming of the world is an indicator for the corresponding temperature map to belong to the late 20th century.

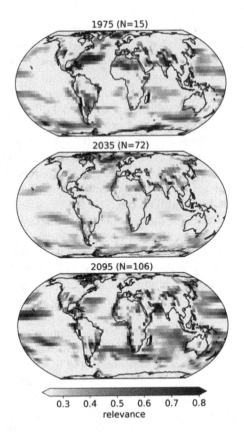

Fig. 5. LRP heatmaps for temperature input maps composited for a range of years when the prediction was deemed accurate. The years are shown above each panel along with the number of maps used in the composites. Darker shading denotes regions that are more relevant for the NN's accurate prediction. Adapted from Barnes et al., 2020 [6].

The above results (see original study by Barnes et al. for more information) highlight the importance and utility of explaining the NN decisions in this prediction task and the physical insights that XAI methods can offer. As we mentioned, in this analysis the explanation of the network was the goal, while the predictions themselves were not important. Generally, this application demonstrates that XAI methods constitute a powerful approach for extracting climate patterns of forced change amidst any background noise, and advancing climate change understanding.

A second application where XAI was used to extract the anthropogenic footprint was published by Keys et al. (2021) [29]. In that study, the authors aimed at constructing a NN to predict the global human footprint index (HFI) solely from satellite imagery. The HFI is a dimensionless metric that captures the extent to which humans have influenced the terrestrial surface of the Earth over a specific region (see e.g., [82,83]). Typically, the HFI is obtained by harmonizing eight different sub-indices, each one representing different aspects of human influence, like built infrastructure, population density, land use, land cover etc. So far, the process for establishing the HFI involves significant data analysis and modelling that does not allow for fast updates and continuous monitoring of the index, which means that large-scale, human-caused changes to the land surface may occur well before we are able to track them. Thus, estimating the HFI solely from satellite imagery that supports spatial resolution and temporally rapid refreshing can help improve monitoring of the human pressure on the Earth surface.

Keys et al. trained a convolutional NN to use single images of the land surface (Landsat; [22]) over a region to predict the corresponding Williams HFI [83]. The authors trained different networks corresponding to different areas around the world in the year 2000, and then used these trained networks to evaluate Landsat images from the year 2019. Results showed that the NNs were able to reproduce the HFI with high fidelity. Moreover, by comparing the estimated HFI in 2000 with the one in 2019, the authors were able to gain insight into the changes in the human pressure to the earth surface during the last 20 years. Patterns of change were consistent with a steady expansion of the human pressure into areas of previously low HFI or increase of density of pressure in regions with previously high HFI values.

Consequently, Keys et al. applied the LRP method for cases where the HFI increased significantly between the years 2000 and 2019. In this way, the authors aimed to gain confidence that the NN was focusing on the correct features in the satellite images to predict increases of the human footprint. As an example, in Fig. 6, we present the LRP results for a region over Texas, where wind farms were installed between the years 2000 and 2019; compare the satellite images in the left and middle panels of the figure. As shown in the LRP results, the NN correctly paid attention to the installed wind farm features in order to predict an increase of the HFI in the year 2019. By examining many other cases of increase in HFI, the authors reported that in most instances, the NN was found to place the highest attention to features that were clearly due to human activity, which provided them with confidence that the network performed with high accuracy for the right reasons.

3   Development of Attribution Benchmarks for Geosciences

As was illustrated in the previous sections, XAI methods have already shown their potential and been used in various climate and weather applications to provide valuable insights about NN decision strategies. However, many of these methods have been shown in the computer science literature to not honor desirable properties (e.g., "completeness" or "implementation invariance"; see [77]), and in general, to face nontrivial limitations for specific problem setups [2,15,31,63]. Moreover, given that many different methods have been proposed in the field of XAI (see e.g., [3,4,32,53,69,70,73,75,77,84] among others) with

Wind farm installation (Texas, USA)

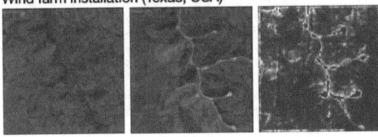

Fig. 6. Satellite imagery from the Global Forest Change dataset over Texas, USA, in (left) 2000 and (middle) 2019. (right) the most relevant features to the NN for its year-2019 prediction of the HFI, as estimated using LRP. Adapted from Keys et al., 2021 [29].

each one explaining the network in a different way, it is key to better understand differences between methods, both their relative strengths and weaknesses, so that researchers are aware which methods are more suitable to use depending on the model architecture and the objective of the explanation. Thus, thorough investigation and objective assessment of XAI methods is of vital importance.

So far, the assessment of different XAI methods has been mainly based on applying these methods to benchmark problems, where the scientist is expected to know what the attribution heatmaps should look like, hence, being able to judge the performance of the XAI method in question. Examples of benchmark problems in climate science include the classification of El Niño or La Niña years or seasonal prediction of regional hydroclimate [21,79]. In computer science, commonly used benchmark datasets for image classification problems are, among others, the MNIST or ImageNet datasets [39,64]. Although the use of such benchmark datasets help the scientist gain some general insight about the XAI method's efficiency, this is always based on the scientist's subjective visual inspection of the result and their prior knowledge and understanding of the problem at hand, which has high risk of cherry-picking specific samples/methods and reinforcing individual biases [37]. In classification tasks, for example, just because it might make sense to a human that an XAI method highlights the ears or the nose of a cat for an image successfully classified as "cat", this does not necessarily mean that this is the strategy the model in question is actually using, since there is no objective truth about the relative importance of these two or other features to the prediction. The actual importance of different features to the network's prediction is always case- or dataset-dependent, and the human perception of an explanation alone is not a solid criterion for assessing its trustworthiness.

With the aim of a more falsifiable XAI research [37], Mamalakis et al. (2021) [43] put forward the concept of *attribution benchmark datasets*. These are synthetic datasets (consisting of synthetic inputs and outputs) that have been designed and generated in a way so that the importance of each input feature to the prediction is objectively derivable and known *a priori*. This *a priori* known attribution can be used as ground truth for evaluating different XAI methods and identifying systematic strengths and weaknesses. The authors referred to such synthetic datasets as attribution benchmark datasets, to distinguish from benchmarks where no ground truth of the attribution/explanation is available. The framework was proposed for regression problems (but can be extended into classification problems too), where the input is a 2D field (i.e., a single-channel image); commonly found in geoscientific applications (e.g., [13, 21, 76, 79]). Below we briefly summarize the proposed framework and the attribution benchmark dataset that Mamalakis et al. used, and we present comparisons between different XAI methods that provide insights about their performance.

3.1 Synthetic Framework

Mamalakis et al. considered a climate prediction setting (i.e., prediction of regional climate from global 2D fields of SST; see e.g., [13, 76]), and generated N realizations of an input random vector $\mathbf{X} \in \Re^d$ from a multivariate Normal Distribution (see step 1 in Fig. 7); these are N synthetic inputs representing vectorized 2D SST fields. Next, the authors used a nonlinear function $F : \Re^d \to \Re$, which represented the physical system, to map each realization x_n into a scalar y_n, and generated the output random variable Y (see step 2 in Fig. 7); these synthetic outputs represented the series of the predictand climatic variable. Subsequently, the authors trained a fully-connected NN to approximate function F and compare the model attributions estimated by different XAI methods with the ground truth of the attribution. The general idea of this framework is summarized in Fig. 7, and although the dataset was inspired from a climate prediction setting, the concept of attribution benchmarks is generic and applicable to a large number of problem settings in the geosciences and beyond.

Regarding the form of function F that is used to generate the variable Y from $\mathbf{X}$, Mamalakis et al. claimed that it can be of an arbitrary choice, as long as it has such a form so that the importance/contribution of each of the input variables to the response Y is objectively derivable. The simplest form for F so that the above property is honored is when F is an *additively separable function*, i.e. there exist local functions C_i, with $i = 1, 2, ..., d$, so that:

$$F(\mathbf{X}) = F(X_1, X_2, ..., X_d) = C_1(X_1) + C_2(X_2) + ... + C_d(X_d) \qquad (1)$$

where, X_i is the random variable at grid point i, and the local functions C_i are nonlinear; if the local functions C_i are linear, Eq. 1 falls back to a trivial linear problem, which is not particularly interesting to benchmark a NN or an XAI method against. Mamalakis et al., defined the local functions to be piece-wise linear functions, with number of break points $K = 5$. The break points and the slopes between the break points were chosen randomly for each grid point (see the original paper for more information). Importantly, with F being an additively separable function as in Eq. 1, the relevance/contribution of each of the variables X_i to the response y_n for any sample n, is by definition equal to the value of the corresponding local function, i.e., $R_{i,n}^{true} = C_i(x_{i,n})$; that is when considering a zero baseline. This satisfies the basic desired property for F that any response can be objectively attributed to the input.

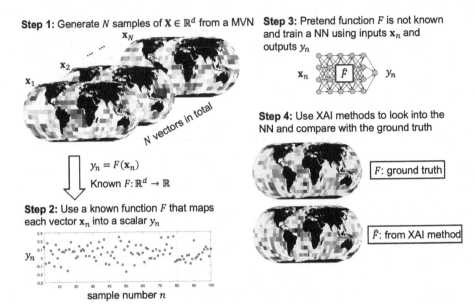

Step 1: Generate N samples of $\mathbf{X} \in \mathbb{R}^d$ from a MVN

$\mathbf{x}_N$

$\mathbf{x}_2$

$\mathbf{x}_1$

N vectors in total

$y_n = F(\mathbf{x}_n)$

Known $F: \mathbb{R}^d \to \mathbb{R}$

Step 2: Use a known function F that maps each vector $\mathbf{x}_n$ into a scalar y_n

y_n

sample number n

Step 3: Pretend function F is not known and train a NN using inputs $\mathbf{x}_n$ and outputs y_n

$\mathbf{x}_n$ $\hat{F}$ y_n

Step 4: Use XAI methods to look into the NN and compare with the ground truth

F: ground truth

$\hat{F}$: from XAI method

Fig. 7. Schematic overview of the framework to generate synthetic attribution benchmarks. In step 1, N independent realizations of a random vector $\mathbf{X} \in \mathbb{R}^d$ are generated from a multivariate Normal Distribution. In step 2, a response $Y \in \mathbb{R}^d$ to the synthetic input $\mathbf{X}$ is generated using a known nonlinear function F. In step 3, a fully-connected NN is trained using the synthetic data $\mathbf{X}$ and Y to approximate the function F. The NN learns a function $\hat{F}$. Lastly, in step 4, the attribution heatmaps estimated from different XAI methods are compared to the ground truth (that represents the function F), which has been objectively derived for any sample $n = 1, 2, ..., N$. Similar to Mamalakis et al., 2021 [43].

Mamalakis et al. generated $N = 10^6$ samples of input and output and trained a fully connected NN to learn the function F (see step 3 in Fig. 7), using the first 900,000 samples for training and the last 100,000 samples for testing. Apart from assessing the prediction performance, the testing samples were also used to assess the performance of different post hoc, local XAI methods. The sample size was on purpose chosen to be large compared to typical samples in climate prediction applications. In this way, the authors aimed to ensure that they could achieve an almost perfect training and establish a fair assessment of XAI methods; they wanted to ensure that any discrepancy between the ground truth of the attribution and the results of XAI methods came from systematic pitfalls in the XAI method and to a lesser degree from poor training of the NN. Indeed, the authors achieved a very high prediction performance, with the coefficient of determination of the NN prediction in the testing data being slightly higher than $R^2 = 99\%$, which suggests that the NN could capture 99% of the variance in Y.

3.2   Assessment of XAI Methods

For their assessment, Mamalakis et al. considered different post hoc, local XAI methods that have been commonly used in the literature. Specifically, the methods that were assessed included Gradient [71], Smooth Gradient [73], Input*Gradient [70], Intergradient Gradients [77], Deep Taylor [53] and LRP [4]. In Fig. 8, we present the ground truth and the estimated relevance heatmaps from the XAI methods (each heatmap is standardized by the corresponding maximum absolute relevance within the map). This sample corresponds to a response $y_n = 0.0283$, while the NN predicted 0.0301. Based on the ground truth, features that contributed positively to the response y_n occur mainly over the northern, eastern tropical and southern Pacific Ocean, the northern Atlantic Ocean, and the Indian Ocean. Features with negative contribution occur over the tropical Atlantic Ocean and the southern Indian Ocean.

The results from the method Gradient are not consistent at all with the ground truth. In the eastern tropical and southern Pacific Ocean, the method returns negative values instead of positive, and over the tropical Atlantic, positive values (instead of negative) are highlighted. The pattern (Spearman's) correlation is very small on the order of 0.13, consistent with the above observations. As theoretically expected, this result indicates that the *sensitivity* of the output to the input is not the same as the *attribution* of the output to the input [3]. The method Smooth Gradient performs poorly and similarly to the method Gradient, with a correlation coefficient on the order of 0.16.

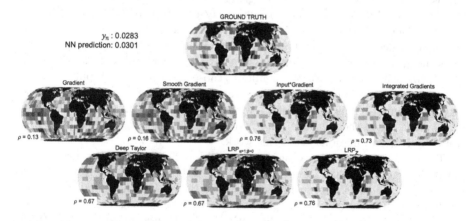

Fig. 8. Performance of different XAI methods. The XAI performance is assessed by comparing the estimated heatmaps to the ground truth. All heatmaps are standardized with the corresponding maximum (absolute) value. Red (blue) color corresponds to positive (negative) contribution to the response/prediction, with darker shading representing higher (absolute) values. The Spearman's rank correlation coefficient between each heatmap and the ground truth is also provided. Only for the methods Deep Taylor and LRP$_{\alpha=1,\beta=0}$, the correlation with the absolute ground truth is given. Similar to Mamalakis et al., 2021 [43]. (Color figure online)

Methods Input*Gradient and Integrated Gradients perform very similarly, both capturing the ground truth very closely. Indeed, both methods capture the positive patterns over eastern Pacific, northern Atlantic and the Indian Oceans, and to an extend the negative patterns over the tropical Atlantic and southern Indian Oceans. The Spearman's correlation with the ground truth for both methods is on the order of 0.75, indicating the very high agreement.

Regarding the LRP method, first, results confirm the arguments in [53,65], that the Deep Taylor leads to similar results with the $LRP_{\alpha=1,\beta=0}$, when a NN with ReLU activations is used. Second, both methods return only positive contributions. This was explained by Mamalakis et al. and is due to the fact that the propagation rule of $LRP_{\alpha=1,\beta=0}$ is performed based on the product of the relevance in the higher layer with a strictly positive number. Hence, the sign of the NN prediction is propagated back to all neurons and to all features of the input. Because the NN prediction is positive in Fig. 8, then it is expected that $LRP_{\alpha=1,\beta=0}$ (and Deep Taylor) returns only positive contributions (see also remarks by [33]). What is not so intuitive is the fact that the $LRP_{\alpha=1,\beta=0}$ seems to highlight all important features, independent of the sign of their contribution (compare with ground truth). Given that, by construction, $LRP_{\alpha=1,\beta=0}$ considers only positive preactivations [23], one might assume that it will only highlight the features that positively contribute to the prediction. However, the results in Fig. 8 show that the method highlights the tropical Atlantic Ocean with a positive contribution. This is problematic, since the ground truth clearly indicates that this region is contributing negatively to the response y_n in this example. The issue of $LRP_{\alpha=1,\beta=0}$ about highlighting all features independent of whether they are contributing positively or negatively to the prediction has been very recently discussed in other applications of XAI as well [33].

Lastly, when using the LRP_z rule, the attribution heatmap very closely captures the ground truth, and it exhibits a very high Spearman's correlation on the order of 0.76. The results are very similar to those of the methods Input*Gradient and Integrated Gradients, making these three methods the best performing ones for this example. This is consistent with the discussion in [2], which showed the equivalence of the methods Input*Gradient and LRP_z in cases of NNs with ReLU activation functions, as in this work.

To verify that the above insights are valid for the entire testing dataset and not only for the specific example in Fig. 8, we also generated the histograms of the Spearman's correlation coefficients between the XAI methods and the ground truth for all 100,000 testing samples (similarly to Mamalakis et al.). As shown in Fig. 9, methods Gradient and Smooth Gradient perform very poorly (both exhibit almost zero average correlation with the ground truth), while methods Input*Gradient and Integrated Gradients perform equally well, exhibiting an average correlation with the ground truth around 0.7. The LRP_z rule is seen to be the best performing among the LRP rules, with very similar performance to the Input*Gradient and Integrated Gradients methods (as theoretically expected for this model setting; see [2]). The corresponding average correlation coefficient is also on the order of 0.7. Regarding the $LRP_{\alpha=1,\beta=0}$ rule, we present two

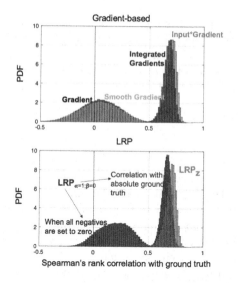

Fig. 9. Summary of the performance of different XAI methods. Histograms of the Spearman's correlation coefficients between different XAI heatmaps and the ground truth for 100,000 testing samples. Similar to Mamalakis et al., 2021 [43].

curves. The first curve (black curve in Fig. 9) corresponds to correlation with the ground truth after we have set all the negative contributions in the ground truth to zero. The second curve (blue curve) corresponds to correlation with the absolute value of the ground truth. For both curves we multiply the correlation value with -1 when the NN prediction was negative, to account for the fact that the prediction's sign is propagated back to the attributions. Results show that when correlating with the absolute ground truth (blue curve), the correlations are systematically higher than when correlating with the nonnegative ground truth (black curve). This verifies that the issue of $LRP_{\alpha=1,\beta=0}$ highlighting both positive and negative attributions occurs for all testing samples.

In general, these results demonstrate the benefits of attribution benchmarks for the identification of possible systematic pitfalls of XAI. The above assessment suggests that methods Gradient and Smooth Gradient may be suitable for estimating the sensitivity of the output to the input, but this is not necessarily equivalent to the attribution. When using the $LRP_{\alpha=1,\beta=0}$ rule, one should be cautious, keeping always in mind that, i) it might propagate the sign of the prediction back to all the relevancies of the input layer and ii) it is likely to mix positive and negative contributions. For the setup used here (i.e. to address the specific prediction task using a shallow, fully connected network), the methods Input*Gradient, Integrated Gradients, and the LRP_z rule all very closely captured the true function F and are the best performing XAI methods considered. However, this result does not mean that the latter methods are systematically better performing for all types of applications. For example, in a different prediction setting (i.e. for a different function F) and when using a deep convolutional neural network, the above methods have been found to provide relatively incomprehensible explanations due to gradient shattering [42]. Thus, no optimal method exists in general and each method's suitability depends on the type of the application and the adopted model architecture, which highlights the need to objectively assess XAI methods for a range of applications and develop best-practice guidelines.

4 Conclusions

The potential of NNs to successfully tackle complex problems in earth sciences has become quite evident in recent years. An important requirement for further application and exploitation of NNs in geoscience is their interpretability, and newly developed XAI methods show very promising results for this task. In this chapter we provided an overview of the most recent work from our group, applying XAI to meteorology and climate science. This overview clearly illustrates that XAI methods can provide valuable insights on the NN strategies, and that they are used in these fields under many different settings and prediction tasks, being beneficial for different scientific goals. For many applications that have been published in the literature, the ultimate goal is a highly-performing prediction model, and XAI methods are used by the scientists to calibrate their trust to the model, by ensuring that the decision strategy of the network is physically consistent (see e.g., [18, 21, 25, 29, 35, 47, 80]). In this way scientists can ensure that a high prediction performance is due to the right reasons, and that the network has learnt the true dynamics of the problem. Moreover, in many prediction applications, the explanation is used to help guide the design of the network that will be used to tackle the prediction problem (see e.g., [16]). As we showed, there are also applications where the prediction is not the goal of the analysis, but rather, the scientists are interested solely in the explanation. In this category of studies, XAI methods are used to gain physical insights about the dynamics of the problem or the sources of predictability. The highlighted relationships between the input and the output may warrant further investigation and advance our understanding, hence, establishing new science (see e.g., [6, 74, 79, 80]).

Independent of the goal of the analysis, an important aspect in XAI research is to better understand and assess the many different XAI methods that exist, in order to more successfully implement them. This need for objectivity in the XAI assessment arises from the fact that XAI methods are typically assessed without the use of any ground truth to test against and the conclusions can often be subjective. Thus, here we also summarized a newly introduced framework to generate synthetic attribution benchmarks to objectively test XAI methods [43]. In the proposed framework, the ground truth of the attribution of the output to the input is derivable for any sample and known a $priori$. This allows the scientist to objectively assess if the explanation is accurate or not. The framework is based on the use of additively separable functions, where the response $Y \in \Re$ to the input $\mathbf{X} \in \Re^d$ is the sum of local responses. The local responses may have any functional form, and independent of how complex that might be, the true attribution is always derivable. We believe that a common use and engagement of such attribution benchmarks by the geoscientific community can lead to a more cautious and accurate application of XAI methods to physical problems, towards increasing model trust and facilitating scientific discovery.

References

1. Agapiou, A.: Remote sensing heritage in a petabyte-scale: satellite data and heritage earth engine applications. Int. J. Digit. Earth **10**(1), 85–102 (2017)
2. Ancona, M., Ceolini, E., Öztireli, C., Gross, M.: Towards better understanding of gradient-based attribution methods for deep neural networks. arXiv preprint arXiv:1711.06104 (2017)
3. Ancona, M., Ceolini, E., Öztireli, C., Gross, M.: Gradient-based attribution methods. In: Samek, W., Montavon, G., Vedaldi, A., Hansen, L.K., Müller, K.-R. (eds.) Explainable AI: Interpreting, Explaining and Visualizing Deep Learning. LNCS (LNAI), vol. 11700, pp. 169–191. Springer, Cham (2019). https://doi.org/10.1007/978-3-030-28954-6_9
4. Bach, S., Binder, A., Montavon, G., Klauschen, F., Müller, K.R., Samek, W.: On pixel-wise explanations for non-linear classifier decisions by layer-wise relevance propagation. PLoS ONE **10**(7), e0130140 (2015)
5. Barnes, E.A., Hurrell, J.W., Ebert-Uphoff, I., Anderson, C., Anderson, D.: Viewing forced climate patterns through an AI lens. Geophys. Res. Lett. **46**(22), 13389–13398 (2019)
6. Barnes, E.A., Toms, B., Hurrell, J.W., Ebert-Uphoff, I., Anderson, C., Anderson, D.: Indicator patterns of forced change learned by an artificial neural network. J. Adv. Model. Earth Syst. **12**(9), e2020MS002195 (2020)
7. Bergen, K.J., Johnson, P.A., Maarten, V., Beroza, G.C.: Machine learning for data-driven discovery in solid earth geoscience. Science **363**(6433), eaau0323 (2019)
8. Ocean Studies Board, The National Academies of Sciences, Engineering, and Medicine et al.: Next Generation Earth System Prediction: Strategies for Subseasonal to Seasonal Forecasts. National Academies Press (2016)
9. Buhrmester, V., Münch, D., Arens, M.: Analysis of explainers of black box deep neural networks for computer vision: a survey. arXiv preprint arXiv:1911.12116 (2019)
10. Cassou, C.: Intraseasonal interaction between the Madden-Julian oscillation and the North Atlantic oscillation. Nature **455**(7212), 523–527 (2008)
11. Dai, A.: The influence of the inter-decadal pacific oscillation on us precipitation during 1923–2010. Clim. Dyn. **41**(3–4), 633–646 (2013)
12. Das, A., Rad, P.: Opportunities and challenges in explainable artificial intelligence (XAI): a survey. arXiv preprint arXiv:2006.11371 (2020)
13. DelSole, T., Banerjee, A.: Statistical seasonal prediction based on regularized regression. J. Clim. **30**(4), 1345–1361 (2017)
14. Deser, C., et al.: Insights from earth system model initial-condition large ensembles and future prospects. Nat. Clim. Chang. **10**(4), 277–286 (2020)
15. Dombrowski, A.K., Anders, C.J., Müller, K.R., Kessel, P.: Towards robust explanations for deep neural networks. Pattern Recogn. **121**, 108194 (2021)
16. Ebert-Uphoff, I., Hilburn, K.: Evaluation, tuning, and interpretation of neural networks for working with images in meteorological applications. Bull. Am. Meteor. Soc. **101**(12), E2149–E2170 (2020)

17. Enfield, D.B., Mestas-Nuñez, A.M., Trimble, P.J.: The Atlantic multidecadal oscillation and its relation to rainfall and river flows in the continental us. Geophys. Res. Lett. **28**(10), 2077–2080 (2001)

18. Gagne, D.J., II., Haupt, S.E., Nychka, D.W., Thompson, G.: Interpretable deep learning for spatial analysis of severe hailstorms. Mon. Weather Rev. **147**(8), 2827–2845 (2019)

19. Goddard, L., Mason, S.J., Zebiak, S.E., Ropelewski, C.F., Basher, R., Cane, M.A.: Current approaches to seasonal to interannual climate predictions. Int. J. Climatol. J. R. Meteorol. Soc. **21**(9), 1111–1152 (2001)

20. Guo, H.: Big earth data: a new frontier in earth and information sciences. Big Earth Data **1**(1–2), 4–20 (2017)

21. Ham, Y.G., Kim, J.H., Luo, J.J.: Deep learning for multi-year ENSO forecasts. Nature **573**(7775), 568–572 (2019)

22. Hansen, M.C., et al.: High-resolution global maps of 21st-century forest cover change. Science **342**(6160), 850–853 (2013)

23. Hao, Z., Singh, V.P., Xia, Y.: Seasonal drought prediction: advances, challenges, and future prospects. Rev. Geophys. **56**(1), 108–141 (2018)

24. Henderson, S.A., Maloney, E.D., Barnes, E.A.: The influence of the Madden-Julian oscillation on northern hemisphere winter blocking. J. Clim. **29**(12), 4597–4616 (2016)

25. Hilburn, K.A., Ebert-Uphoff, I., Miller, S.D.: Development and interpretation of a neural-network-based synthetic radar reflectivity estimator using GOES-R satellite observations. J. Appl. Meteorol. Climatol. **60**(1), 3–21 (2021)

26. Holzinger, A., Malle, B., Saranti, A., Pfeifer, B.: Towards multi-modal causability with graph neural networks enabling information fusion for explainable AI. Inf. Fusion **71**, 28–37 (2021). https://doi.org/10.1016/j.inffus.2021.01.008

27. Karpatne, A., Ebert-Uphoff, I., Ravela, S., Babaie, H.A., Kumar, V.: Machine learning for the geosciences: challenges and opportunities. IEEE Trans. Knowl. Data Eng. **31**(8), 1544–1554 (2018)

28. Kay, J.E., et al.: The community earth system model (CESM) large ensemble project: a community resource for studying climate change in the presence of internal climate variability. Bull. Am. Meteor. Soc. **96**(8), 1333–1349 (2015)

29. Keys, P.W., Barnes, E.A., Carter, N.H.: A machine-learning approach to human footprint index estimation with applications to sustainable development. Environ. Res. Lett. **16**(4), 044061 (2021)

30. Khan, M.Z.K., Sharma, A., Mehrotra, R.: Global seasonal precipitation forecasts using improved sea surface temperature predictions. J. Geophys. Res. Atmos. **122**(9), 4773–4785 (2017)

31. Kindermans, P.-J., et al.: The (un)reliability of saliency methods. In: Samek, W., Montavon, G., Vedaldi, A., Hansen, L.K., Müller, K.-R. (eds.) Explainable AI: Interpreting, Explaining and Visualizing Deep Learning. LNCS (LNAI), vol. 11700, pp. 267–280. Springer, Cham (2019). https://doi.org/10.1007/978-3-030-28954-6_14

32. Kindermans, P.J., et al.: Learning how to explain neural networks: PatternNet and PatternAttribution. arXiv preprint arXiv:1705.05598 (2017)

33. Kohlbrenner, M., Bauer, A., Nakajima, S., Binder, A., Samek, W., Lapuschkin, S.: Towards best practice in explaining neural network decisions with LRP. In: 2020 International Joint Conference on Neural Networks (IJCNN), pp. 1–7. IEEE (2020)

34. Lagerquist, R., McGovern, A., Homeyer, C.R., Gagne, D.J., II., Smith, T.: Deep learning on three-dimensional multiscale data for next-hour tornado prediction. Mon. Weather Rev. **148**(7), 2837–2861 (2020)

35. Lapuschkin, S., Wäldchen, S., Binder, A., Montavon, G., Samek, W., Müller, K.R.: Unmasking clever HANs predictors and assessing what machines really learn. Nat. Commun. **10**(1), 1–8 (2019)

36. Lary, D.J., Alavi, A.H., Gandomi, A.H., Walker, A.L.: Machine learning in geosciences and remote sensing. Geosci. Front. **7**(1), 3–10 (2016)

37. Leavitt, M.L., Morcos, A.: Towards falsifiable interpretability research. arXiv preprint arXiv:2010.12016 (2020)

38. LeCun, Y., Bengio, Y., Hinton, G.: Deep learning. Nature **521**(7553), 436–444 (2015)

39. LeCun, Y., Bottou, L., Bengio, Y., Haffner, P.: Gradient-based learning applied to document recognition. Proc. IEEE **86**(11), 2278–2324 (1998)

40. Line, W.E., Schmit, T.J., Lindsey, D.T., Goodman, S.J.: Use of geostationary super rapid scan satellite imagery by the storm prediction center. Weather Forecast. **31**(2), 483–494 (2016)

41. Malde, K., Handegard, N.O., Eikvil, L., Salberg, A.B.: Machine intelligence and the data-driven future of marine science. ICES J. Mar. Sci. **77**(4), 1274–1285 (2020)

42. Mamalakis, A., Barnes, E.A., Ebert-Uphoff, I.: Investigating the fidelity of explainable artificial intelligence methods for applications of convolutional neural networks in geoscience. arXiv preprint arXiv:2202.03407 (2022)

43. Mamalakis, A., Ebert-Uphoff, I., Barnes, E.A.: Neural network attribution methods for problems in geoscience: a novel synthetic benchmark dataset. arXiv preprint arXiv:2103.10005 (2021)

44. Mamalakis, A., Yu, J.Y., Randerson, J.T., AghaKouchak, A., Foufoula-Georgiou, E.: A new interhemispheric teleconnection increases predictability of winter precipitation in southwestern us. Nat. Commun. **9**(1), 1–10 (2018)

45. Mamalakis, A., Yu, J.Y., Randerson, J.T., AghaKouchak, A., Foufoula-Georgiou, E.: Reply to: a critical examination of a newly proposed interhemispheric teleconnection to southwestern us winter precipitation. Nat. Commun. **10**(1), 1–5 (2019)

46. Mantua, N.J., Hare, S.R.: The pacific decadal oscillation. J. Oceanogr. **58**(1), 35–44 (2002)

47. Mayer, K.J., Barnes, E.A.: Subseasonal forecasts of opportunity identified by an explainable neural network. Geophys. Res. Lett. **48**(10), e2020GL092092 (2021)

48. McCABE, G.J., Dettinger, M.D.: Decadal variations in the strength of ENSO teleconnections with precipitation in the western united states. Int. J. Climatol. J. R. Meteorol. Soc. **19**(13), 1399–1410 (1999)

49. McCabe, G.J., Palecki, M.A., Betancourt, J.L.: Pacific and Atlantic ocean influences on multidecadal drought frequency in the united states. Proc. Natl. Acad. Sci. **101**(12), 4136–4141 (2004)

50. McGovern, A., et al.: Making the black box more transparent: understanding the physical implications of machine learning. Bull. Am. Meteor. Soc. **100**(11), 2175–2199 (2019)

51. McKinnon, K.A., Poppick, A., Dunn-Sigouin, E., Deser, C.: An "observational large ensemble" to compare observed and modeled temperature trend uncertainty due to internal variability. J. Clim. **30**(19), 7585–7598 (2017)

52. Molina, M.J., Gagne, D.J., Prein, A.F.: A benchmark to test generalization capabilities of deep learning methods to classify severe convective storms in a changing climate. Earth and Space Science Open Archive ESSOAr (2021)

53. Montavon, G., Lapuschkin, S., Binder, A., Samek, W., Müller, K.R.: Explaining nonlinear classification decisions with deep Taylor decomposition. Pattern Recogn. **65**, 211–222 (2017)

54. Newman, M., et al.: The pacific decadal oscillation, revisited. J. Clim. **29**(12), 4399–4427 (2016)

55. Newman, M., Compo, G.P., Alexander, M.A.: Enso-forced variability of the pacific decadal oscillation. J. Clim. **16**(23), 3853–3857 (2003)

56. Overpeck, J.T., Meehl, G.A., Bony, S., Easterling, D.R.: Climate data challenges in the 21st century. Science **331**(6018), 700–702 (2011)

57. Palmer, T.N., Anderson, D.L.T.: The prospects for seasonal forecasting-a review paper. Q. J. R. Meteorol. Soc. **120**(518), 755–793 (1994)

58. Qiu, B., Chen, S.: Variability of the Kuroshio extension jet, recirculation gyre, and mesoscale eddies on decadal time scales. J. Phys. Oceanogr. **35**(11), 2090–2103 (2005)

59. Redmond, K.T., Koch, R.W.: Surface climate and streamflow variability in the western united states and their relationship to large-scale circulation indices. Water Resour. Res. **27**(9), 2381–2399 (1991)

60. Reichstein, M., Camps-Valls, G., Stevens, B., Jung, M., Denzler, J., Carvalhais, N., et al.: Deep learning and process understanding for data-driven earth system science. Nature **566**(7743), 195–204 (2019)

61. Reinsel, D., Gantz, J., Rydning, J.: The digitization of the world from edge to core. Framingham: International Data Corporation, p. 16 (2018)

62. Rolnick, D., et al.: Tackling climate change with machine learning. arXiv preprint arXiv:1906.05433 (2019)

63. Rudin, C.: Stop explaining black box machine learning models for high stakes decisions and use interpretable models instead. Nat. Mach. Intell. **1**(5), 206–215 (2019)

64. Russakovsky, O., et al.: ImageNet large scale visual recognition challenge. Int. J. Comput. Vis. **115**(3), 211–252 (2015)

65. Samek, W., Montavon, G., Binder, A., Lapuschkin, S., Müller, K.R.: Interpreting the predictions of complex ml models by layer-wise relevance propagation. arXiv preprint arXiv:1611.08191 (2016)

66. Santer, B.D., et al.: Separating signal and noise in atmospheric temperature changes: the importance of timescale. J. Geophys. Res. Atmos. **116**(D22), D22105 (2011)

67. Schonher, T., Nicholson, S.: The relationship between California rainfall and ENSO events. J. Clim. **2**(11), 1258–1269 (1989)

68. Shen, C.: A transdisciplinary review of deep learning research and its relevance for water resources scientists. Water Resour. Res. **54**(11), 8558–8593 (2018)

69. Shrikumar, A., Greenside, P., Kundaje, A.: Learning important features through propagating activation differences. In: International Conference on Machine Learning, pp. 3145–3153. PMLR (2017)

70. Shrikumar, A., Greenside, P., Shcherbina, A., Kundaje, A.: Not just a black box: learning important features through propagating activation differences. arXiv preprint arXiv:1605.01713 (2016)

71. Simonyan, K., Vedaldi, A., Zisserman, A.: Deep inside convolutional networks: visualising image classification models and saliency maps. arXiv preprint arXiv:1312.6034 (2013)

72. Sit, M., Demiray, B.Z., Xiang, Z., Ewing, G.J., Sermet, Y., Demir, I.: A comprehensive review of deep learning applications in hydrology and water resources. Water Sci. Technol. **82**(12), 2635–2670 (2020)

73. Smilkov, D., Thorat, N., Kim, B., Viégas, F., Wattenberg, M.: SmoothGrad: removing noise by adding noise. arXiv preprint arXiv:1706.03825 (2017)
74. Sonnewald, M., Lguensat, R.: Revealing the impact of global heating on north atlantic circulation using transparent machine learning. Earth and Space Science Open Archive ESSOAr (2021)
75. Springenberg, J.T., Dosovitskiy, A., Brox, T., Riedmiller, M.: Striving for simplicity: the all convolutional net. arXiv preprint arXiv:1412.6806 (2014)
76. Stevens, A., et al.: Graph-guided regularized regression of pacific ocean climate variables to increase predictive skill of southwestern us winter precipitation. J. Clim. **34**(2), 737–754 (2021)
77. Sundararajan, M., Taly, A., Yan, Q.: Axiomatic attribution for deep networks. In: International Conference on Machine Learning, pp. 3319–3328. PMLR (2017)
78. Tjoa, E., Guan, C.: A survey on explainable artificial intelligence (XAI): toward medical XAI. IEEE Trans. Neural Netw. Learn. Syst. **32**, 4793–4813 (2020)
79. Toms, B.A., Barnes, E.A., Ebert-Uphoff, I.: Physically interpretable neural networks for the geosciences: applications to earth system variability. J. Adv. Model. Earth Syst. **12**(9), e2019MS002002 (2020)
80. Toms, B.A., Barnes, E.A., Hurrell, J.W.: Assessing decadal predictability in an earth-system model using explainable neural networks. Geophys. Res. Lett. **48**, e2021GL093842 (2021)
81. Trenberth, K.E.: The definition of El Nino. Bull. Am. Meteor. Soc. **78**(12), 2771–2778 (1997)
82. Venter, O., et al.: Sixteen years of change in the global terrestrial human footprint and implications for biodiversity conservation. Nat. Commun. **7**(1), 1–11 (2016)
83. Williams, B.A., et al.: Change in terrestrial human footprint drives continued loss of intact ecosystems. One Earth **3**(3), 371–382 (2020)
84. Zeiler, M.D., Fergus, R.: Visualizing and understanding convolutional networks. In: Fleet, D., Pajdla, T., Schiele, B., Tuytelaars, T. (eds.) ECCV 2014. LNCS, vol. 8689, pp. 818–833. Springer, Cham (2014). https://doi.org/10.1007/978-3-319-10590-1_53

An Interdisciplinary Approach to Explainable AI

Varieties of AI Explanations Under the Law. From the GDPR to the AIA, and Beyond

Philipp Hacker[1(✉)] and Jan-Hendrik Passoth[2]

[1] Chair for Law and Ethics of the Digital Society,
European New School of Digital Studies, European University Viadrina,
Große Scharnstraße 59, 15230 Frankfurt, Oder, Germany
hacker@europa-uni.de

[2] Chair of Sociology of Technology, European New School of Digital Studies,
European University Viadrina, Große Scharnstraße 59,
15230 Frankfurt, Oder, Germany
passoth@europa-uni.de
https://www.europeannewschool.eu/law-ethics.html
https://www.europeannewschool.eu/sociology-of-technology.html

Abstract. The quest to explain the output of artificial intelligence systems has clearly moved from a mere technical to a highly legally and politically relevant endeavor. In this paper, we provide an overview of legal obligations to explain AI and evaluate current policy proposals. In this, we distinguish between different functional varieties of AI explanations - such as multiple forms of enabling, technical and protective transparency - and show how different legal areas engage with and mandate such different types of explanations to varying degrees. Starting with the rights-enabling framework of the GDPR, we proceed to uncover technical and protective forms of explanations owed under contract, tort and banking law. Moreover, we discuss what the recent EU proposal for an Artificial Intelligence Act means for explainable AI, and review the proposal's strengths and limitations in this respect. Finally, from a policy perspective, we advocate for moving beyond mere explainability towards a more encompassing framework for trustworthy and responsible AI that includes actionable explanations, values-in-design and co-design methodologies, interactions with algorithmic fairness, and quality benchmarking.

Keywords: Artificial intelligence · Explainability · Regulation

1 Introduction

Sunlight is the best disinfectant, as the saying goes. Therefore, it does not come as a surprise that transparency constitutes a key societal desideratum vis-à-vis complex, modern IT systems in general [67] and artificial intelligence (AI)

© The Author(s) 2022
A. Holzinger et al. (Eds.): xxAI 2020, LNAI 13200, pp. 343–373, 2022.
https://doi.org/10.1007/978-3-031-04083-2_17

in particular [18,74]. As in the case of very similar demands concerning other forms of opaque or, at least from an outsider perspective, inscrutable decision making processes of bureaucratic systems, transparency is seen as a means of making decisions more understandable, more contestable, or at least more rational. More specifically, explainability of AI systems generally denotes the degree to which an observer may understand the causes of the system's output [15,64]. Various technical implementations of explainability have been suggested, from truth maintenance systems for causal reasoning in the case of symbolic reasoning systems that were developed mainly from the 1970s s to the 1990s s to layerwise relevance propagation methods for neural networks today. Importantly, observers, and with them the adequate explanations for a specific context, may vary [3, p. 85].

In recent years, the quest for transparent and explainable AI has not only spurred a vast array of research efforts in machine learning [3,82, and the chapters in this volume for an overview], but it has also emerged at the heart of many ethics and responsible design proposals [43,45,66,68] and has nurtured a vivid debate on the promises and limitations of advanced machine learning models for various high-stakes scenarios [12,37,88].

1.1 Functional Varieties of AI Explanations

Importantly, from a normative perspective, different arguments can be advanced to justify the need for transparency in AI systems [3]. For example, given its relation to human autonomy and dignity, one may advance a 'deontological' conception viewing transparency as an aim in itself [17,92,104]. Moreover, research suggests that explanations may satisfy the curiosity of counterparties, their desire for learning or control, or fulfill basic communicative standards of dialogue and exchange [59,62,64]. From a legal perspective, however, it is submitted that three major functional justifications for demands of AI explainability may be distinguished: enabling, technical, and protective varieties. All of them subscribe to an 'instrumentalist' approach conceiving of transparency as a means to achieve technically or normatively desirable ends.

First, explainability of AI is seen as a prerequisite for empowering those affected by its decisions or charged with reviewing them ('enabling transparency'). On the one hand, explanations are deemed crucial to afford due process to the affected individuals [23] and to enable them to effectively exercise their subjective rights vis-à-vis the (operators of the) AI system [89] ('rights-enabling transparency'). Similarly, other parties such as NGOs, collective redress organizations or supervisory authorities may use explanations to initiate legal reviews, e.g. by inspecting AI systems for unlawful behavior such as manipulation or discrimination [37, p. 55]('review-enabling transparency'). On the other hand, information about the functioning of AI systems may facilitate informed choice of the affected persons about whether and how to engage with the models or the offers they accompany and condition. Such 'decision-enabling transparency' seeks to support effective market choice, for example by switching contracting partners [14, p. 156].

Second, with respect to technical functionality, explainability may help fine-tune the performance (e.g., accuracy) of the system in real-world scenarios and evaluate its generalizability to unseen data [3,47,57,79]. In this vein, it also acts as a catalyst for informed decision making, though not of the affected persons, but rather of the technical operator or an expert auditor of the system. That approach may hence be termed 'technical transparency', its explanations being geared toward a technically sophisticated audience. Beyond model improvements, a key aim here is to generate operational and institutional trust in the AI system [37, p. 54], both in the organization operating the AI system and beyond in the case of third-party reviews and audits.

Third, technical improvements translate into legal relevance to the extent that they contribute to reducing normatively significant risks. Hence, technically superior performance may lead to improved safety (e.g., AI in robots; medical AI), reduced misallocation of resources (e.g., planning and logistics tools), or better control of systemic risks (e.g., financial risk modelling). This third variety could be dubbed 'protective transparency', as it seeks to harness explanations to guard against legally relevant risks.

These different types of legally relevant, functional varieties of AI explanations are not mutually exclusive. For example, technical explanations may, to the extent available, also be used by collective redress organizations or supervisory authorities in a review-enabling way. Nonetheless, the distinctions arguably provide helpful analytical starting points. As we shall see, legal provisions compelling transparency are responsive to these different strands of justification to varying degrees. It should not be overlooked, however, that an excess of sunlight can be detrimental as well, as skeptics note: explainability requirements may not only impose significant and sometimes perhaps prohibitive burdens on the use of some of the most powerful AI systems, but also offer affected persons the option to strategically "game the system" and accrue undeserved advantages [9]. This puts differentiated forms of accountability front and center: to whom - users, affected persons, professional audit experts, legitimized rights protection organizations, public authorities - should an AI system be transparent? Such limitations need to be considered by the regulatory framework as well.

1.2   Technical Varieties of AI Explanations

From a technical perspective, in turn, it seems uncontroversial that statements about AI and explainability, as well as the potential trade-off with accuracy, must be made in a context- and model-specific way [57,81][3, p. 100]. While some types of ML models, such as linear or logistic regressions or small decision trees [22,47,57], lend themselves rather naturally to global explanations about the feature weights for the entire model (often called ex ante interpretability), such globally valid statements are much harder to obtain for other model types, particularly random forests or deep neural networks [57,79,90]. In recent years, such complex model types have been the subject of intense technical research to provide for, at the minimum, local explanations of specific decisions ex post, often by way of sensitivity analysis [31,60,79]. One specific variety of local explanations seeks to provide counterfactuals, i.e., suggestions for minimal changes of the input data to achieve a more desired output [64,97]. Counterfactuals are a variety of contrastive explanations, which seek to convey reasons for the concrete output ('fact') in relation to another, possible output ('foil') and which have recently gained significant momentum [65,77]. Other methods have sought to combine large numbers of local explanations to approximate a global explanatory model of the AI system by way of overall feature relevance [16,55], while other scholars have sought to fiercely defend the benefits of designing models that are interpretable ex ante rather than explainable ex post [81].

1.3   Roadmap of the Paper

Arguably, much of this research has been driven, at least implicitly, by the assumption that explainable AI systems would be ethically desirable and perhaps even legally required [47]. Hence, this paper seeks to provide an overview of explainability obligations flowing from the law proper, while engaging with the functional and technical distinctions just introduced. The contemporary legal debate has its roots in an interpretive battle over specific norms of the GDPR [89,96], but has recently expanded beyond the precincts of data protection law to other legal fields, such as contract and tort law [42,84]. As this paper will show, another important yet often overlooked area which might engender incentives to provide explanations for AI models is banking law [54]. Finally, the question of transparency has recently been taken up very prominently by the regulatory proposals at the EU level, particularly in the Commission proposal for an Artificial Intelligence Act (AIA). It should be noted that controversies and consultations about how to meaningfully regulate AI systems are still ongoing processes and that the questions of what kind of explainability obligations follow already from existing regulations and which obligations should - in the future - become part of AI policy are still very much in flux. This begs the question of the extent to which these diverging provisions and calls for explainability properly take into account the usability of that information for the recipients, in other words: the actionability of explainable AI (XXAI), which is also at the core of this volume.

Against this background, the paper will use the running example of credit scoring to investigate whether positive law mandates, or at least sets incentives for, the provision of actionable explanations in the use of AI tools, particularly in settings involving private actors (Sect. 2); to what extent the proposals for AI regulation at the EU level will change these findings (Sect. 3); and how regulation and practice could go beyond such provisions to ensure actionable explanations and trustworthy AI (Sect. 4). In all of these sections, the findings will be linked to the different (instrumentalist) functions of transparency, which are taken up to varying degrees by the different provisions and proposals. Figure 1 below provides a quick overview of the relations between functions and several existing legal acts surveyed in this paper; Fig. 2 (in Sect. 3) connects these functions to the provisions of the planned AIA.

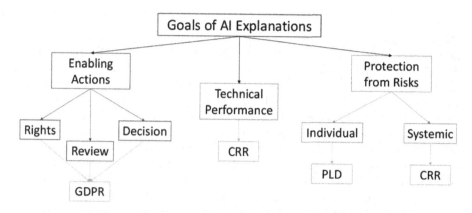

Fig. 1. Overview of the functions of different EU law instruments concerning AI explanations; abbreviations: GDPR: General Data Protection Regulation; CRR: Capital Requirements Regulation; PLD: Product Liability Directive

2 Explainable AI Under Current Law

The quest for explainable AI interacts with existing law in a number of ways. The scope of this paper will be EU law, and for the greatest part the law governing exchange between private parties more particularly (for public law, see, e.g. [14, 2.2]). Most importantly, and bridging the public-privates divide, the GDPR contains certain rules, however limited and vague, which might be understood as an obligation to provide explanations of the functioning of AI models (Sect. 2.1.). Beyond data protection law, however, contract and tort law (Sect. 2.2) and banking law (Sect. 2.3) also provide significant incentives for the use of explainable AI (XAI).

2.1 The GDPR: Rights-Enabling Transparency

In the GDPR, whether a subjective right to an explanation of AI decisions exists or not has been the object of a long-standing scholarly debate which, until this day, has not been finally settled [36,61,89,96]. To appreciate the different perspectives, let us consider the example of AI-based credit scoring. Increasingly, startups use alternative data sets and machine learning to compute credit scores, which in turn form the basis of lending decisions (see, e.g., [34,54]). If a particular person receives a specific credit score, the question arises if, under the GDPR, the candidate may claim access to the feature values used to make the prediction, to the weights of the specific features in his or her case (local explanation), or even to the weights of the features in the model more generally (global explanation). For example, the person might want to know what concrete age and income values were used to predict the score, to what extent age or income contributed to the prediction in the specific case, and how the model generally weights these features.

So far, there is no guidance by the Court of Justice of the European Union (CJEU) on precisely this question. However, exactly this case was decided by the German Federal Court for Private Law (BGH) in 2014 (BGH, Case VI ZR 156/13 = MMR 2014, 489). The ruling came down not under the GDPR, but its predecessor (the 1995 Data Protection Directive) and relevant German data protection law. In substance, however, the BGH noted that the individual information interest of the plaintiff needed to be balanced against the legitimate interests of the German credit scoring agency (Schufa) to keep its trade secrets, such as the precise score formula for credit scoring, hidden from the view of the public, lest competitors free ride on its know-how. In weighing these opposing interests, the BGH concluded that the plaintiff did have a right to access its personal data processed for obtaining the credit score (the feature values), but not to obtain information on the score formula itself, comparison groups, or abstract methods of calculation. Hence, the plaintiff was barred from receiving either a local or a global explanation of its credit score.

2.1.1 Safeguards for Automated Decision Making

How would such a case be decided under the GDPR, particularly if an AI-based scoring system was used? There are two main normative anchors in the GDPR that could be used to obtain an explanation of the score, and hence more generally of the output of an AI system. First, Article 22 GDPR regulates the use of automated decision making in individual cases. That provision, however, is subject to several significant limitations. Not only does its wording suggest that it applies only to purely automated decisions, taken independently of even negligible human interventions (a limitation that could potentially be overcome by a more expansive interpretation of the provision, see [96]); more importantly, the safeguards it installs in Article 22(3) GDPR for cases of automated decision making list 'the right to obtain human intervention on the part of the controller, to express his or her point of view and to contest the decision' - but not the right to an explanation. Rather, such a right is only mentioned in Recital 71 GDPR, which provides additional interpretive guidance for Article 22(3) GDPR. Since, however, only the Articles of the regulation, not the recitals, constitute binding law, many scholars are rightly skeptical whether the CJEU would deduce a right to an explanation (of whatever kind) directly from Article 22(3) GDPR [84,96].

2.1.2 Meaningful Information About the Logic Involved

A second, much more promising route is offered by different provisions obliging the data controller (i.e., the operator of the AI system) to provide the data subject not only with information on the personal data processed (the feature values), but also, at least in cases of automated decision making, with 'meaningful information about the logic involved' (Art. 13(2)(f), Art. 14(2)(g), Art. 15(1)(h) GDPR).

A Rights-Enabling Conception of Meaningful Information. Since the publication of the GDPR, scholars have intensely debated what these provisions mean for

AI systems (see, e.g. for overviews [20, 49]. For instance, in our running example, we may more concretely ask whether a duty to disclose local or global weights of specific features exists in the case of credit scoring. Some scholars stress the reference to the concept of 'logic', which to them suggests that only the general architecture of the system must be divulged, but not more specific information on features and weights [73, para. 31c][103]. A more convincing interpretation, in our view, would take the purpose of the mentioned provisions into account. Hence, from a teleological perspective, the right to meaningful information needs to be read in conjunction with the individual rights the GDPR confers in Art. 16 et seqq. [89]. Such a rights-enabling instrumentalist approach implies that information will only be meaningful, to the data subject, if it facilitates the exercise of these rights, for example the right to erasure, correction, restriction of processing or, perhaps most importantly, the contestation of the decision pursuant to Article 22(3) GDPR. An overarching view of the disclosure provisions forcing meaningful information and the safeguards in Article 22(3) GDPR therefore suggests that, already under current data protection law, the information provided must be actionable to fulfill its enabling function. Importantly, this directly relates to the quest of XXAI research seeking to provide explanations that enable recipients to meaningfully reflect upon and intervene in AI-powered decision-making systems.

Hence, in our view, more concrete explanations may have to be provided if information about the individual features and corresponding weights are necessary to formulate substantive challenges to the algorithmic scores under the GDPR's correction, erasure or contestation rights. Nevertheless, as Article 15(4) GDPR and more generally Article 16 of the Charter of Fundamental Rights of the EU (freedom to conduct the business) suggest, the information interests of the data subject must still be balanced against the secrecy interests of the controller, and their interest in protecting the integrity of scores against strategic gaming. In this reading, a duty to provide actionable yet proportionate information follows from Art. 13(2)(f), Art. 14(2)(g) and Art. 15(1)(h) GDPR, read in conjunction with the other individual rights of the data subject.

Application to Credit Scores. In the case of AI-based credit scores, such a regime may be applied as follows. In our view, meaningful information will generally imply a duty to provide local explanations of individual cases, i.e., the disclosure of at least the most important features that contributed to the specific credit score of the applicant. This seems to be in line with the (non-binding) interpretation of European privacy regulators (Article 29 Data Protection Working Party, 2018, at 25–26). Such information is highly useful for individuals when exercising the mentioned rights and particularly for contesting the decision: if, for example, it turns out that the most important features do not seem to be related in any plausible way to creditworthiness or happen to be closely correlated with attributes protected under non-discrimination law, the data subject will be in a much better position to contest the decision in a substantiated way. Furthermore, if only local information is provided, trade secrets are implicated to a much lesser extent than if the entire score formula was disclosed; and possibilities to 'game the system' are significantly reduced. Finally, such local explanations can increasingly be provided even for complex models, such as deep neural networks, without loss of accuracy [31,79].

On the other hand, meaningful information will generally not demand the disclosure of global explanations, i.e., of weights referring to the entire model. While this might be useful for individual complainants to detect, for example, whether their case represents an outlier (i.e., features were weighted differently in the individual case than generally in the model), the marginal benefit of a global explanation vis-à-vis a local explanation seems outweighed by the much more significant impact on trade secrets and incentives to innovation if weights for an entire model need to be disclosed. Importantly, such a duty to provide global explanations would also significantly hamper the use of more complex models, such as deep neural networks (cf. [14, p. 162]. While such technical limitations do not generally speak against certain interpretations of the law (see, e.g., BVerfG NJW 1979, 359, para. 109 - Kalkar), they seem relevant here because such models may, in a number of cases, perform better in the task of credit scoring than simpler but globally explainable models. If this premise holds, another provision of EU law becomes relevant. More accurate models allow to fulfill the requirements of *responsible lending* to a better extent (see Sect. 2.3 for details): if models more correctly predict creditworthiness, loans will be handed out more often only to persons who are indeed likely to repay the loan. Since this is a core requirement of the post-financial crisis framework of EU credit law, it should be taken into account in the interpretation of the GDPR in cases of credit scoring as well (see, for such overarching interpretations of different areas of EU law, CJEU, Case C-109/17, Bankia, para. 49; [38]).

Ultimately, for local and global explanations alike, a compromise between information interests and trade secrets might require the disclosure of weights not in a highly granular, but in a 'noisy' fashion (e.g., providing relevance intervals instead of specific percentage numbers) [6, para. 54]. Less mathematically trained persons often disregard or have trouble cognitively processing probability information in explanations [64] so that the effective information loss for recipients would likely be limited. Noisy weights, or simple ordinal feature ranking by importance, would arguably convey a measure enabling meaningful evaluation and critique while safeguarding more precise information relevant for the competitive advantage of the developer of the AI system, and hence for incentives to innovation. Such less granular information could be provided whenever the confidentiality of the information is not guaranteed; if the information is treated confidentially, for example in the framework of a specific procedure in a review or audit, more precise information might be provided without raising concerns about unfair competition. The last word on these matters will, of course, have the CJEU. It seems not unlikely, though, that the Court would be open to an interpretation guaranteeing actionable yet proportionate information. This would correspond to a welcome reading of the provisions of the GDPR with a view to due process and the exercise of subjective rights by data subjects (rights-enabling transparency).

2.2   Contract and Tort Law: Technical and Protective Transparency

In data protection law, as the preceding section has shown, much will depend on the exact interpretation of the vague provisions of the GDPR, and on the extent to which these provisions can be applied even if humans interact with AI systems in more integrated forms of decision making. These limitations should lead us to consider incentives for actionable AI explanations in other fields of the law, such as contract and tort law. This involves particularly product liability (Sect. 2.2.1), and general negligence standards under contract and tort law (Sect. 2.2.2). Clearly, under freedom of contract, parties may generally contract for specific explanations that the provider of an AI system may have to enable. In the absence of such explicit contractual clauses, however, the question arises to what extent contract and tort law still compel actionable explanations. As we shall see, in these areas, the enabling instrumentalist variety of transparency (due process, exercise of rights) is to a great extent replaced by a more technical and protective instrumentalist approach focusing on trade-offs with accuracy and safety.

2.2.1   Product Liability

In product liability law, the first persevering problem is the extent to which it applies to non-tangible goods such as software. Article 2 of the EU Product Liability Directive (PLD), passed in 1985, defines a product as any movable, as well as electricity. While an AI system embedded in a physical component, such as a robot, clearly qualifies as a product under Article 2, this is highly contested for a standalone system such as, potentially, a credit scoring application (see [84, 99]). In the end, at least for professionally manufactured software, one will have to concede that it exhibits defect risks similar to traditional products and entails similar difficulties for plaintiffs in proving them, which speaks strongly in favor of applying the PLD, at least by analogy, to such software independently of any embeddedness in a movable component [29, p. 43]. A proposal by the EU Commission on that question, and on liability for AI more generally, is expected for 2022.

Design Defects. As it currently stands, the PLD addresses producers by providing those harmed by defective products with a claim against them (Art. 1 PLD). There are different types of defects a product may exhibit, the most important in the context of AI being a design defect. With respect to the topic of this paper, one may therefore ask if the lack of an explanation might qualify as a design defect of an AI system. This chiefly depends on the interpretation of the concept of a design defect.

In EU law, two rivaling interpretations exist: the consumer expectations test and the risk-utility test. Article 6 PLD at first glance seems to enshrine the former variety by holding that a 'product is defective when it does not provide the safety which a person is entitled to expect'. The general problem with this formulation is that it is all but impossible to objectively quantify legitimate consumer expectations [99]. For example, would the operator of an AI system, the affected person, or the public in general be entitled to expect explanations, and if so, which ones?

Product safety law is often understood to provide minimum standards in this respect [100, para. 33]; however, exact obligations on explainability of AI are lacking so far in this area, too (but see Annex I, Point 1.7.4.2.(e) of the Machinery Directive 2006/42 and Sect. 3). Precisely because of these uncertainties, many scholars prefer the risk-utility test which has a long-standing tradition in US product liability law (see § 402A Restatement (Second) of Torts). Importantly, it is increasingly used in EU law as well [86][99, n. 48] and was endorsed by the BGH in its 2009 Airbag decision[1]. Under this interpretation, a design defect is present if the cost of a workable alternative design, in terms of development and potential reduced utility, is smaller than the gain in safety through this alternative design. Hence, the actually used product and the workable alternative product must be compared considering their respective utilities and their risks [94, p. p. 246].

With respect to XAI, it must hence be asked if an interpretable tool would have provided additional safety through the explanation, and if that marginal benefit is not outweighed by additional costs. Such an analysis, arguably, aligns with a technical and protective instrumentalist conception of transparency, as a means to achieve safety gains. Importantly, therefore, the analysis turns not only on the monetary costs of adding explanations to otherwise opaque AI systems, but it must also consider whether risks are really reduced by the provision of an explanation.

The application of the risk-utility test to explainability obligations has, to our knowledge, not been thoroughly discussed in the literature yet (for more general discussions, see [87, p. 1341, 1375][42]. Clearly, XAI may be *helpful*, in evidentiary terms, for producers in showing that there was no design defect involved in an accident [19, p. 624][105, p. 217]; but is XAI *compulsory* under the test? The distinguishing characteristic of applying a risk-utility test to explainable AI seems to be that the alternative (introducing explainability) does not necessarily reduce risk overall: while explanations plausibly lower the risk of misapplication of the AI system, they might come at the expense of accuracy. Therefore, in our view, the following two cases must be distinguished:

1. The explainable model exhibits the same accuracy as the original, non-explainable model (e.g., ex post local explanation of a DNN). In that case, only the expected gain in safety, from including explanations, must be weighed against potential costs of including explanations, such as longer run time, development costs, license fees etc. Importantly, as the BGH specified in its Airbag ruling, the alternative model need not only be factually ready for use, but its use must also be normatively reasonable and appropriate for the producer[2]. This implies that, arguably, trade secrets must be considered in the analysis, as well. Therefore, it seems sensible to assume that, as in data protection law, a locally (but not a globally) explainable model must be chosen,

---

[1] BGH, 16.6.2009, VI ZR 107/08, BGHZ 181, 253 para 18.
[2] BGH, 16.6.2009, VI ZR 107/08, BGHZ 181, 253 para 18.

unless the explainable add-on is unreasonably expensive. Notably, the more actionable explanations are in the sense of delivering clear cues for operators, or affected persons, to minimize safety risks, the stronger the argument that such explanations indeed must be provided to prevent a design defect.

2. Matters are considerably more complicated if including explanations lowers the accuracy of the model (e.g., switching to a less powerful model type): in this case, it must first be assessed whether explanations enhance safety overall, by weighing potential harm from lower accuracy against potential prevention of harm from an increase in transparency. If risk is increased, the alternative can be discarded. If, however, it can be reasonably expected that the explanations entail a risk reduction, this reduction must be weighed against any additional costs the inclusion of explainability features might entail, as in the former case (risk-utility test). Again, trade secrets and incentives for innovation must be accounted for, generally implying local rather than global explanations (if any).

Importantly, in both cases, product liability law broadens the scope of explanations vis-à-vis data protection law. While the GDPR focuses on the data subject as the recipient of explanations, product liability more broadly considers any explanations that may provide a safety benefit, targeting therefore particularly the operators of the AI systems who determine if, how and when a system is put to use. Hence, under product liability law producers have to consider to what extent explanations may help operators safely use the AI product.

Product Monitoring Obligations. Finally, under EU law, producers are not subject to product monitoring obligations once the product has been put onto the market. However, product liability law of some Member States does contain such monitoring obligations (e.g., Germany[3]). The producers, in this setting, have to keep an eye on the product to become aware of emerging safety risks, which is particularly important with respect to AI systems whose behavior might change after being put onto the market (e.g., via online learning). Arguably, explanations help fulfill this monitoring obligation. This, however, chiefly concerns explanations provided to the producer itself. If these are not shared with the wider public, trade secrets may be guarded; therefore, one might argue that even global explanations may be required. However, again, this would depend on the trade-off with the utility of the product as producers cannot be forced to put less utile products on the market unless the gain in safety, via local or global explanations, exceeds the potentially diminished utility.

Results. In sum, product liability law targets the producer as the responsible entity, but primarily focuses on explanations provided to the party controlling the safety risks of the AI system in the concrete application context, typically the operator. To the extent that national law contains product monitoring obligations, however, explanations to the producer may have to be provided as well.

---

[3] BGH, 17.3.1981, VI ZR 286/78 - Benomyl.

In all cases, the risk reduction facilitated by the explanations must be weighed against the potentially reduced utility of the AI system. In this, product liability law aligns itself with technical and protective transparency. It generates pressure to offer AI systems with actionable explanations by targeting the supply side of the market (producers).

2.2.2   General Negligence Standards

Beyond product liability, general contract and tort law define duties of care that operators of devices, such as AI systems, need to fulfill in concrete deployment scenarios. Hence, it reaches the demand side of the market. While contract law covers cases in which the operator has a valid (pre-)contractual agreement with the harmed person (e.g., a physician with a patient; the bank with a credit applicant), tort law steps in if such an agreement is missing (e.g., autonomous lawnmower and injured pedestrian). However, the duties of care that relate to the necessary activities for preventing harm to the bodily integrity and the assets of other persons are largely equivalent under contract and tort law (see, e.g., [5, para 115]. In our context, this raises the question: do such duties of care require AI to be explainable, even if any specific contractual obligations to this end are lacking?

From Error Reversal to Risk-Adequate Choice. Clearly, if the operator notices that the AI system is bound to make or has made an error, she has to overrule the AI decision to avoid liability [33,42,84]. Explanations geared toward the operator will often help her notice such errors and make pertaining corrections [80, p. 23][31]. For example, explanations could suggest that the system, in the concrete application, weighted features in an unreasonable manner and might fail to make a valid prediction [71,79]. What is unclear, however, is whether the duty of care more generally demands explanations as a necessary precondition for using AI systems.

While much will depend on the concrete case, at least generally, the duty of care under both contract and tort law comprises monitoring obligations for operators of potentially harmful devices. The idea is that those who operate and hence (at least partially) control the devices in a concrete case must make reasonable efforts to control the risks the devices pose to third parties (cf. [101, para. 459]). The scope of that obligation is similar to the one in product liability, but not directed toward the producer, but rather the operator of the system: they must do whatever is factually possible and normatively reasonable and appropriate to prevent harm by monitoring the system. Hence, to the extent possible the operator arguably has to choose, at the moment of procurement, an AI system that facilitates risk control. Again, this reinforces technical and protective transparency in the name of safety gains. If an AI system providing actionable explanations is available, such devices must therefore be chosen by the operator over non-explainable systems under the same conditions as in product liability law (i.e., if the explanation leads to an overall risk reduction justifying additional costs). For example, the operator need not choose an explainable system if the price difference to a non-explainable system constitutes an unreasonable burden. Note, however, that the operator, if distinct from the producer, cannot claim that trade secrets speak against an explainable version.

Alternative Design Obligations? Nonetheless, we would argue that the operator is not under an obligation to redesign the AI system, i.e., to actively install or use explanation techniques not provided by the producer, unless this is economically and technically feasible with efforts proportionate to the expected risk reduction. Rather, the safety obligations of the operator will typically influence the initial procurement of the AI system on the market. For example, if there are several AI-based credit scoring systems available the operator would have to choose the system with the best risk utility trade-off, taking into account explainability on both sides of the equation (potential reduction in utility and potential reduction of risk). Therefore, general contract and tort law sets incentives to use explainable AI systems similar to product liability, but with a focus on actions by, and explanations for, the operator of the AI system.

Results. The contractual and tort-law duty of care therefore does not, other than in product liability, primarily focus on a potential alternative design of the system, but on prudently choosing between different existing AI systems on the market. Interpreted in this way, general contract and tort law generate market pressure toward the offer of explainable systems by targeting the demand side of the market (operators). Like product liability, however, they cater to technical and protective transparency.

2.3   Banking Law: More Technical and Protective Transparency

Finally, banking law provides for detailed regulation governing the development and application of risk scoring models. It therefore represents an under-researched, but in fact highly relevant area of algorithmic regulation, particularly in the case of credit scoring (see, e.g., [54]). Conceptually, it is intriguing because the quality requirements inherent in banking law fuse technical transparency with yet another legal and economic aim: the control of systemic risk in the banking sector.

2.3.1   Quality Assurance for Credit Models

Significant regulatory experience exists in this realm because econometric and statistical models have long since been used to predict risk in the banking sector, such as creditworthiness of credit applicants [25]. In the wake of the financial crisis following the collapse of the subprime lending market, the EU legislator has enacted encompassing regulation addressing systemic risks stemming from the banking sector. Since inadequate risk models have been argued to have contributed significantly to the scope and the spread of the financial crisis [4, p. 243–245], this area has been at the forefront of the development of internal compliance and quality regimes - which are now considered for AI regulation as well.

In general terms, credit institutions regulated under banking law are required to establish robust risk monitoring and management systems (Art. 74 of Directive 2013/36). More specifically, a number of articles in the Capital Requirements Regulation 575/2013 (CRR) set out constraints for the quality assurance of banking scoring models. Perhaps most importantly, Article 185 CRR compels banks to validate the score quality ('accuracy and consistency') of models for internal rating and risk assessment, via a continuous monitoring of the functioning of these models. Art. 174 CRR, in addition, specifies that: statistical models and 'other mechanical methods' for risk assessments must have good predictive power (lit. a); input data must be vetted for accuracy, completeness, appropriateness and representativeness (lit. b, c); models must be regularly validated (lit. d) and combined with human oversight (lit. e) (see [58, para. 1]; cf. [26, para. 249]; [21, paras. 68, 256]; for similar requirement for medical products, see [84]).

These provisions foreshadow many of the requirements the AIA proposed by the EU Commission now seeks to install more broadly for the regulation of AI. However, to the extent that AI-based credit scoring is used by banks, these provisions - other than the AIA - already apply to the respective models. While the responsible lending obligation contained in Article 8 of the Consumer Credit Directive 2008/48 only spells out generic duties to conduct creditworthiness assessments before lending decisions, Articles 174 and 185 CRR have complemented this obligation with a specific quality assurance regime. Ultimately, more accurate risk prediction is supposed to not only spare lenders and borrowers the transaction costs of default events, but also and perhaps even more importantly to rein in systemic risk in the banking sector by mitigating exposure. This, in turn, aims at reducing the probability of severe financial crises.

2.3.2 Consequences for XAI

What does this entail for explainable AI in the banking sector? While accuracy (and model performance more generally) may be verified on the test data set in supervised learning settings without explanations relating to the relevant features for a prediction, explainability will, as mentioned, often be a crucial element for validating the generalizability of models beyond the test set (Art. 174(d) CRR), and for enabling human review (Art. 174(e) CRR). In its interpretive guidelines for supervision and model approval, the European Banking Authority (EBA) therefore stipulates that banks must 'understand the underlying models used', particularly in the case of technology-enabled credit assessment tools [26, para. 53c]. More specifically, it cautions that consideration should be given to developing interpretable models, if necessary for appropriate use of the model [26, para. 53d].

Hence, the explainability of AI systems becomes a real compliance tool in the realm of banking law, an idea we shall return to in the discussion of the AIA. In banking law, explainability is intimately connected to the control of systemic risk via informed decision making of the individual actors. One might even argue that both local and global explainability are required under this perspective: local explainability helps determine accuracy in individual real-world cases for which no ground truth is available, and global explanations contribute to the verification of the consistency of the scoring tool across various domains and scenarios. As these explanations are generated internally and only shared with supervisory authorities, trade secrets do not stand in the way.

The key limitation of these provisions is that they apply only to banks in the sense of banking law (operating under a banking license), but not to other institutions not directly subject to banking regulation, such as mere credit rating agencies [7]. Nevertheless, the compliance and quality assurance provisions of banking law seem to have served as a blue print for current AI regulation proposals such as the EU Artificial Intelligence Act (esp. Art. 9, 14, 15 and 17), to which we now turn.

3   Regulatory Proposals at the EU Level: The AIA

The AIA, proposed by the EU Commission in April 2021, is set to become a cornerstone of AI regulation not only in the EU, but potentially with repercussions on a global level. Most notably, it subscribes to a risk-based approach and therefore categorically differentiates between several risk categories for AI. Figure 2 offers a snapshot of the connections between the functions of transparency and various Articles of the AIA.

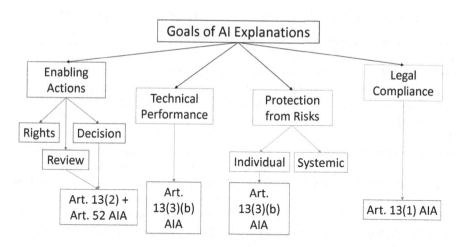

Fig. 2. Overview of the functions of different Articles of the AIA transparency provisions

3.1   AI with Limited Risk: Decision-Enabling Transparency (Art. 52 AIA)?

For specific AI applications with limited risk, Article 52 AIA spells out transparency provisions in an enabling but highly constrained spirit (see also [38,95]).

Thus, the providers of AI systems interacting with humans, of emotion recognition systems, biometric categorization systems and of certain AI systems meant to manipulate images, audio recordings or videos (e.g., deep fakes) need to disclose the fact that an AI system is operating and, in the last case, that content was manipulated. Transparency, in this sense, does not relate to the inner workings of the respective AI systems, but merely to their factual use and effects.

The aim of these rules arguably is also of an enabling nature, but primarily with respect to informed choice, or rather informed avoidance (decision-enabling transparency), not the exercise of rights. Whether these rules will have any meaningful informational and behavioral effect on affected persons, however, must at least be doubted. A host of studies document rational as well as boundedly rational ignorance of standard disclosures in digital environments [1,13,72]. But regardless of the individual benefit, the more or less complete information about the use of low-risk AI systems alone is indirectly helpful in providing overviews and insights to civil society initiatives or journalistic projects, for example. Moreover, in the specific case of highly controversial AI applications such as emotion recognition or remote biometric identification, compulsory disclosure might, via coverage by media and watchdogs, engender negative reputational effects for the providers, which may lead some of them to reconsider the use of such systems in the first place.

3.2 AI with High Risk: Encompassing Transparency (Art. 13 AIA)?

The regulatory environment envisioned by the AIA is strikingly different for high-risk AI applications. Such applications are supposed to be defined via a regularly updated Annex to the AIA and, according to the current proposal, comprise a wide variety of deployment scenarios, from remote biometric identification to employment and credit scoring contexts, and from the management of critical infrastructure to migration and law enforcement (see Annex III AIA). In this regard, the question of the process of updating the AIA Annex is still open in terms of participation and public consultation. The requirements for low-risk AI systems to at least document the use and effects of the selected technologies, however, leads us to expect case-related disputes about whether an AI application should be classified as high risk, in which stakeholder representatives, civil and human rights protection initiatives, and manufacturers and users of technologies will wrestle with each other. This public struggle can also be seen as a rights-enabling transparency measure.

3.2.1 Compliance-Oriented Transparency

For such high-risk applications, Article 13 AIA spells out a novel transparency regime that might be interpreted as seeking to fuse, to varying degrees, the several instrumentalist approaches identified in this paper, while notably foregrounding another goal of transparency: legal compliance.

Hence, Article 13(1) AIA mandates that high-risk AI systems be 'sufficiently transparent to enable users to interpret the system's output and use it appropriately'. In this, an 'appropriate type and degree of transparency' must be ensured. The provision therefore acknowledges the fundamentally different varieties of explanations that could be provided for AI systems, such as local, global or counterfactual explanations; or more or less granular information on feature weights. The exact scope and depth of the required transparency is further elaborated upon in Article 13(3) AIA and will need to be determined in a context-specific manner. Nothing in the wording of Article 13, however, suggests that global explanations, which may be problematic for complex AI systems, must be provided on a standard basis. However, explanations must be faithful to the model in the sense that they need to be an, at least approximately, correct reconstruction of the internal decision making parameters: explanation and explanandum need to match [57]. For example, local ex post explanations would have to verifiably and, within constraints, accurately measure feature relevance (or other aspects) *of the used model.*

Notably, with respect to the general goal of transparency, the additional explanatory language in Article 13(1) AIA introduces a specific and arguably novel variety of transparency instrumentalism geared toward effective and compliant application of AI systems in concrete settings. In fact, Article 13(1) AIA defines a particular and narrow objective for appropriate transparency under the AIA: facilitating the fulfillment of the obligations providers and users have under the very AIA (Chap. 3 = Art. 16–29). Most notably, any reference to rights of users or affected persons is lacking; rather, Article 29 AIA specifies that users may only deploy the AI system within the range of intended purposes specified by the provider and disclosed under Article 13(2) AIA. Hence, transparency under the AIA seems primarily directed toward compliance with the AIA itself, and not towards the exercise of rights affected persons might have. In this sense, the AIA establishes a novel, self-referential, compliance-oriented type of transparency instrumentalism.

3.2.2 Restricted Forms of Enabling and Protective Transparency

For specific applications, the recitals, however, go beyond this restrained compliance conception and hold that, for example in the context of law enforcement, transparency must facilitate the exercise of fundamental rights, such as the right to an effective remedy or a fair trial (Recital 38 AIA). This points to a more encompassing rights-enabling approach, receptive of demands for contestability, which stands in notable tension, however, with the narrower, compliance-oriented wording of Article 13(1) AIA. To a certain extent, however, the information provided under Article 13 AIA will facilitate audits by supervisory authorities, collective redress organizations or NGOs ('review-enabling transparency').

Furthermore, the list of specific items that need to be disclosed under Article 13(3) AIA connects to technical and protective instrumentalist conceptions of transparency (see also [41]). Hence, Article 15 AIA mandates appropriate levels of accuracy, as well as robustness and cybersecurity, for high-risk AI systems. According to Article 13(3)(b)(ii) AIA, the respective metrics and values need to be disclosed. In this, the AIA follows the reviewed provisions of banking law in installing a quality assurance regime for AI models whose main results need to be disclosed. As mentioned, this also facilitates legal review: if the disclosed performance metrics suggest a violation of the requirements of Article 15 AIA, the supervisory authority may exercise its investigative and corrective powers. The institutional layout of this oversight and supervisory regime however is still not fully defined: The sectoral differentiation of AI applications in the AIA's risk definitions on the one hand suggest an equally sectoral organization of supervisory authorities; the technical and procedural expertise needed for such oversight procedures on the other hand calls for a less distributed supervisory regime.

Similarly, Article 10 AIA installs a governance regime for AI training data, whose main parameters, to the extent relevant for the intended purpose, also need to be divulged (Art. 13(3)(b)(v) AIA). Any other functionally relevant limitations and predetermined changes must be additionally informed about (Art. 13(3)(b)(iii), (iv), (c) and (e)). Finally, disclosure also extends to human oversight mechanisms required under Article 14 AIA - like the governance of training data another transplant from the reviewed provisions on models in banking law. Such disclosures, arguably, cater to protective transparency as they seek to guard against use of the AI system beyond its intended purpose, its validated performance or in disrespect of other risk-minimizing measures.

Hence, transparency under Article 13 is intimately linked to the requirements of human oversight specified in Article 14 AIA. That provision establishes another important level of protective transparency: high-risk AI applications need to be equipped with interface tools enabling effective oversight by human persons to minimize risks to health, safety and fundamental rights. Again, as discussed in the contract/tort and banking law sections, local explanations particularly facilitate monitoring and the detection of inappropriate use or anomalies engendering such risks (cf. Art. 14(4)(a) AIA). While it remains a challenge to implement effective human oversight in AI systems making live decisions (e.g., in autonomous vehicles), the requirement reinforces the focus of the AIA on transparency vis-à-vis professional operators, not affected persons.

3.3 Limitations

The transparency provisions in the AIA in several ways represent steps in the right direction. For example, they apply, other than the GDPR rules reviewed, irrespective of whether decision making is automated or not and of whether personal data is processed or not. Furthermore, the inclusion of a quality assurance regime should be welcomed and even be (at least partially) expanded to non-high-risk applications, as disclosure of pertinent performance metrics may be of substantial signaling value for experts and the market. Importantly, the rules of the future AIA (and of the proposed Machinery Regulation) will likely at least generally constitute minimum thresholds for the avoidance of design defects in product liability law (see Sect. 2.2.1), enabling decentralized private enforcement next to the public enforcement foreseen in the AIA. Nonetheless, the transparency provisions of the AIA are subject to significant limitations.

First and foremost, self-referential compliance and protective transparency seems to detract from meaningful rights-enabling transparency for affected persons. Notably, the transparency provisions of Article 13 AIA are geared exclusively toward the users of the system, with the latter being defined in Article 3(4) AIA as anyone using the system with the exception of consumers. While this restriction has the beneficial effect of sparing consumers obligations and liability under the AIA (cf. [102]), for example under Article 29 AIA, it has the perhaps unintended and certainly significant effect of excluding non-professional users from the range of addressees of explanations and disclosure [27,91]. Therefore, the enabling variety of transparency, invoked in lofty words in Recital 38 AIA, is missing from the Articles of the AIA and will in practice be largely relegated to other, already existing legal acts - such as the transparency provisions of the GDPR reviewed above. In this sense, the AIA does not make any significant contribution to extending or sharpening the content of the requirement to provide 'meaningful information' to data subjects under the GDPR. In this context, information facilitating a review in terms of potential bias with respect to protected groups is missing, too.

Second, this focus on professional users and presumed experts continues in the long list of items to be disclosed under Article 13(3) AIA. While performance metrics, specifications about training data and other disclosures do provide relevant information to sophisticated users to determine whether the AI system might present a good fit to the desired application, such information will only rarely be understandable and actionable for users without at least a minimal training in ML development or practice. In this sense, transparency under the AIA might be described as transparency 'by experts for experts', likely leading to information overload for non-experts. The only exception in this sense is the very reduced, potentially decision-enabling transparency obligation under Article 52 AIA.

Third, despite the centrality of transparency for trustworthy AI in the communications of the EU Commission (see, e.g., European Commission, 2020), the AIA contains little incentive to actually disclose information about the inner workings of an AI system to the extent that they are relevant and actionable for affected persons. Most of the disclosure obligations refer either to the mere fact that an AI system of a specific type is used (Art. 52 AIA) or to descriptions of technical features and metrics (Art. 13(3) AIA). Returning briefly to the example of credit scoring, the only provision potentially impacting the question of whether local or even global explanations of the scores (feature weights) are compulsory is the first sentence of Article 13(1) AIA. According to it, users (i.e., professionals at the bank or credit scoring agency) must be able to interpret the system's output. The immediate reference, in the following sentence, to the obligations of users under Article 29 AIA, however, detracts from a reading that would engage Article 13 AIA to provide incentives for clear and actionable explanations beyond what is already contained in Articles 13–15 GDPR. The only interpretation potentially suggesting local, or even global, explanations is the connection to Article 29(4) AIA. Under this provision, users have to monitor the system to decide whether use according to the instructions may nonetheless lead to significant risks. One could argue that local explanations could be conducive to and perhaps even necessary for this undertaking to the extent that they enable professional users to determine if the main features used for the prediction were at least plausibly related to the target, or likely rather an artifact of the restrictions of training, e.g., of overfitting on training data (cf. [79]). Note, however, that for credit institutions regulated under banking law, the specific provisions of banking law take precedence over Article 29(4) and (5) AIA.

Fourth, while AI systems used by banks will undergo a conformity assessment as part of the supervisory review and evaluation process already in place for banking models (Art. 43(2)(2) AIA), the providers of the vast majority of high-risk AI systems will be able to self-certify the fulfilment of the criteria listed in the AIA, including the transparency provisions in Art. 13 (see Art. 43(2)(1) AIA). The preponderance of such self-assessment may result from an endeavor to exonerate regulatory agencies and to limit the regulatory burden for providers, but it clearly reduces enforcement pressure and invites sub-optimal compliance with the already vague and limited transparency provisions (cf. also [91,95]).

In sum, the AIA provides for a plethora of information relevant for sophisticated users, in line with technical transparency, but will disappoint those that had hoped for more guidance on and incentives for meaningful explanations enabling affected persons to review and contest the output of AI systems.

4   Beyond Explainability

As the legal overview has shown, different areas of law embody different conceptions of AI explainability. Perhaps most importantly, however, if explanations are viewed as a social act enabling a dialogical exchange and laying the basis for goal-oriented actions of the respective recipients, it will often not be sufficient to just provide them with laundry lists of features, weights or model architectures. There is a certain risk that the current drive toward explainable AI, particularly if increasingly legally mandated, generates information that does not justify the transaction costs it engenders. Hence, computer science and the law have to go beyond mere explainability toward interactions that enable meaningful agency of the respective recipients [103], individually, but even more so by strengthening the ability of stakeholder organizations or civil and human rights organizations. This includes a push for actionable explanations, but also for connections to algorithmic fairness, to quality benchmarking and to co-design strategies in an attempt to construct responsible, trustworthy AI [3,45].

4.1   Actionable Explanations

The first desideratum, therefore, is for explanations to convey actionable information, as was stressed throughout the article. Otherwise, for compliance reasons and particularly under the provisions of the AIA, explanations might be provided that few actors actually cognitively process and act upon. This implies a shift from a focus on the technical feasibility of explanations toward, with at least equal importance, the recipient-oriented design of the respective explanations.

4.1.1 Cognitive Optimization

Generally, to be actionable, explanations must be designed such that information overload is avoided, keeping recipients with different processing capabilities in mind. This is a lesson that can be learned from decades of experience with the disclosure paradigm in US and EU consumer law: most information is flatly ignored by consumers [8, 72]. To stand a chance of being cognitively processed, the design of explanations must thus be recipient-oriented. In this, a rich literature on enhancing the effectiveness of privacy policies and standard information in consumer and capital markets law can be exploited [10, 64]. Information, in this sense, must be cognitively optimized for the respective recipients, and the law, or at least the implementing guidelines, should include rules to this effect.

To work, explanations likely must be salient and simple [93] and include visualizations [48]. Empirical studies indeed show that addressees prefer simple explanations [78]. Furthermore, when more complex decisions need to be explained, information could be staggered by degree of complexity. Research on privacy policies, for example, suggests that multi-layered information may bridge the gap between diverging processing capacities of different actors [83]. Hence, simple and concise explanations could be given first, with more detailed, expert-oriented explanations provided on a secondary level upon demand. For investment information, this has already been implemented with the mandate on a Key Investor Document in EU Regulation 1286/2014 (PRIIPS Regulation) (see also [54, p.540]). Finally, empirical research again shows that actionable explanations tend to be contrastive, a concept increasingly explored in AI explanations as well [64, 65].

Hence, there are no one-size-fits-all explanations; rather, they need to be adapted to different contexts and addressees. What the now classic literature on privacy policies suggests is that providing information is only one element of a more general privacy awareness and privacy-by-design strategy [44] that takes different addressees, practical needs and usable tools into account: A browser-plugin notifying about ill-defined or non-standard privacy settings can be more helpful for individual consumers than a detailed and descriptive walk-through of specific privacy settings. A machine-readable and standardized format for reviewing and monitoring privacy settings, however, is helpful for more technical reviews by privacy advocacy organizations. The 'ability to respond' to different contexts and addressees therefore is a promising path towards 'response-able' [51] AI. One particular strategy might be to let affected persons choose foils (within reasonable constraints) and generate contrastive explanations bridging the gap between fact and foil.

4.1.2 Goal Orientation

Beyond these general observations for cognitive optimization, actionable explanations should be clearly linked to the respective goals of the explanations. If the objective is to enable an understanding of the decision by affected persons and to permit the exercise of rights or meaningful review (rights- or review-enabling transparency), shortlists of the most relevant features for the decision ought to be required [79][12, for limitations]. This facilitates, inter alia, checks for plausibility and discrimination. Importantly, such requirements have, in some areas, already been introduced into EU law by recent updates of consumer and business law. Under the new Art. 6a of the Consumer Rights Directive and the new Art. 7(4a) of the Unfair Commercial Practices Directive, online marketplaces will shortly need to disclose the main parameters for any ranking following a search query, and their relative importance. Art. 5 of the P2B Regulation 2019/1150 equally compels online intermediaries and search engines to disclose the main parameters of ranking and their relative importance. However, these provisions require global, not local explanations [37, p.52][14, p.161].

This not only generates technical difficulties for more complex AI systems, but the risk that consumers will flatly ignore such global explanations is arguably quite high. Rather, in our view, actionable information should focus on local explanations for individual decisions. Such information not only seems to be technically easier to provide, but it is arguably more relevant, particularly for the exercise of individual rights. From a review-enabling perspective, local information could be relevant as well for NGOs, collective redress organizations and supervisory authorities seeking to prosecute individual rights violations. In this sense, a collective dimension of individual transparency emerges (cf. also [46]). On the downside, however, local feature relevance information may produce a misleading illusion of simplicity; in non-linear models, even small input changes may alter principal reason lists entirely [12,57].

If, therefore, the goal is not to review or challenge the decision, but to facilitate market decisions and particularly to create spaces for behavioral change of affected persons (decision-enabling transparency), for example to improve their credit score, counterfactual or contrastive information might serve the purpose better [65,97]. In the example of credit scoring, this could set applicants toward the path of credit approval. Such information could be problematic, however, if the identified features merely correlate with creditworthiness, but are not causal for it. In this case, the risk of applicants trying to 'game the system' by artificially altering non-causal features are significant (e.g., putting felt tips under furniture as predictors of creditworthiness [85, p.71]). Moreover, in highly dimensional systems with many features, many counterfactuals are possible, making it difficult to choose the most relevant one for the affected person [97, p.851]. In addition, some counterfactually relevant features may be hard or impossible to change (e.g., age, residence) [50]. In these cases, local shortlists of the most relevant features [79] or minimal intervention advice [50] might be more helpful.

Overall, research for the type of explanation with the best fit for each context will have to continue; it will benefit from cross-fertilization with social science research on the effectiveness of information more generally and explanations more particularly [64] as well as with research in science & technology studies on organizational, institutional and cultural contextualization of decision support, explanations, and accountability. Ultimately, a context-dependent, goal-oriented mix of explanations (e.g., relevance shortlist combined with counterfactual explanation) might best serve the various purposes explanations have to fulfil in concrete settings. In this, a critical perspective drawing on the limitations of the disclosure paradigm in EU market law (see, e.g., [11,39]) should be helpful to prevent information overload and to limit disclosure obligations to what is meaningfully oriented to the respective goals of the explanations.

4.2 Connections to Algorithmic Fairness

Transparency, and explanations such as disclosure of the most relevant features of an AI output, may serve yet another goal: non-discrimination in algorithmic decision making. A vast literature deals with tools and metrics to implement non-discrimination principles at the level of AI models to facilitate legal compliance [52,76,106]. Explanations may reinforce such strategies by facilitating bias detection and prevention, both by affected persons and review institutions. For example, in the case of credit scoring, disclosure of the most important features (local explanations) could help affected persons determine to what extent the decision might have been driven by variables closely correlated with protected attributes [3]. Such cross-fertilization between bias detection and explanations could be termed 'fairness-enabling transparency' and should constitute a major research goal from a legal and technical perspective.

In a similar vein, Sandra Wachter and colleagues have convincingly advocated for the disclosure of summary statistics showing the distribution of scores between different protected groups [98]. As one of the authors of this contribution has argued, such disclosures might in fact already be owed under the current GDPR disclosure regime (Art. 13(2)(f), Art. 14(2)(g), Art. 15(1)(h) GDPR: information about the 'significance and envisaged consequences' of processing, see [40, p.1173–1174]). In addition, Art. 13(3)(b)(iv) AIA proposes the disclosure of a high-risk AI system's 'performance as regards the persons or groups of persons on which the system is intended to be used". While one could interpret this as a mandate for differential statistics concerning protected groups, such an understanding is unlikely to prevail, in the current version of the AIA, as a reference to protected attributes in the sense of antidiscrimination law is patently lacking. Fairness-enabling transparency, such as summary statistics showing distributions between protected groups, to the extent available, thus constitutes an area that should be included in the final version of the AIA.

4.3 Quality Benchmarking

Finally, technical and protective transparency closely relates to (the disclosure of) quality standards for AI systems. These metrics, in turn, also enable regulatory review and are particularly important, as seen, in banking law [54, p.561–563]. Two aspects seem to stand out at the intersection of explanations and quality benchmarking:

First, an absolute quality control, such as the one installed in Art. 174/185 CRR, could be enshrined for all AI applications, at least in medium- and high-stakes settings (transcending the ultimately binary logic of the AIA with respect to risk classification). In these settings, quality assurance might be considered as important as, or even more important than, mere explainability. Quality control would include, but not be limited to, explanations facilitating decisions about the generalizability of the model (e.g., local explanations). Importantly, the disclosure of performance metrics would also spur workable competition by enabling meaningful comparison between different AI systems. Notably, relevant quality assurance provisions in the AIA (Art. 10/15 AIA) are limited to high-risk applications. An update of the AIA might draw inspiration from banking law in working toward a quality assurance regime for algorithmic decision making in which the monitoring of field performance and the assessment of the generalizability of the model via explainability form an important regulatory constraint not only for high-risk but also for medium-risk applications, at least.

Second, understanding the risks and benefits of, and generating trust in, AI systems should be facilitated by testing the quality of AI models against the benchmark of traditional (non-AI-based) methods (relative quality control). For example, a US regulator, the Consumer Financial Protection Bureau, ordered a credit scoring startup working with alternative data to provide such an analysis. The results were promising: according to the analysis, AI-based credit scoring was able to deliver cheaper credit and improved access, both generally and with respect to many different consumer subgroups [30][35, p.42]. To the extent that the analysis is correct, it shows that AI, if implemented properly and monitored rigorously, may provide palpable benefits not only to companies using it, but to consumers and affected persons as well. Communicating such benefits by benchmarking reports seems a sensible way to enable more informed market decisions, to facilitate review and to generate trust - strengthening three important pillars of any explainability regime for AI systems.

4.4   Interventions and Co-design

Such ways of going beyond the already existing and currently proposed forms of transparency obligations by developing formats and methods to produce actionable explanations, by connecting transparency and explainability issues to questions of algorithmic fairness and new or advanced forms of quality benchmarking and control are, as favorable as they are, mainly ex post mechanisms aiming at helping affected persons, users, NGOs or supervisory authorities to evaluate and act upon the outcomes of AI systems in use. They can inform market decisions, help affected persons to claim rights or enable regular oversight and supervision, but they do not intervene in the design and implementation of complex AI systems. Linking to two distinct developments of inter- and transdisciplinary research can help to further develop forms of intervention and co-design:

First, methods and formats for 'values-in-design' [53,70] projects have been developed in other areas of software engineering, specifically in human computer interaction (HCI) and computer-supported collaborative work (cscw) setups that traditionally deal with heterogenous user groups as well as with a diverse set of organizational and contextual requirements due to the less domain-specific areas of application of these software systems (see [32] for an overview). Formats and methods include the use of software engineering artifacts to make normative requirements visible and traceable or the involvement of affected persons, stake-holders, or spokespersons in requirements engineering, evaluation and testing [32,75]. Technical transparency as discussed above can support the transfer and application of such formats and methods to the co-design of AI systems [2] with global explanations structuring the process and local explanations supporting concrete co-design practices.

Second, these methodological advances have been significantly generalized and advanced under the 2014–2020 Horizon 2020 funding scheme, moving from 'co-design to ELSI co-design' [56] and leading to further developing tools, methods and approaches designed for research on SwafS ('Science with and for Society') into a larger framework for RRI ('Responsible Research and Innovation') [28]. In AI research, specifically in projects aiming to improve accountability or transparency, a similar, but still quite disconnected movement towards 'Responsible AI' [24] has gained momentum, tackling very similar questions of stakeholder integration, formats for expert/non-expert collaboration, domain-knowledge evaluation or contestation and reversibility that have been discussed within the RRI framework with a focus on energy technologies, biotechnologies or genetic engineering. This is a rich resource to harvest for further steps towards XAI by adding addressee orientation, contestability criteria or even, reflexively, tools to co-design explanations through inter- and transdisciplinary research [63,69].

5   Conclusion

This paper has sought to show that the law, to varying degrees, mandates or incentivizes different varieties of AI explanations. These varieties can be distinguished based on their respective functions or goals. When affected persons are the addressees, explanations should be primarily rights-enabling or decision-enabling. Explanations for operators or producers, in turn, will typically facilitate technical improvements and functional review, fostering the mitigation of legally relevant risks. Finally, explanations may enable legal review if perceived by third parties, such as NGOs, collective address organizations or supervisory authorities.

The GDPR, arguably, subscribes to a rights-enabling transparency regime under which local explanations may, depending on the context, have to be provided to individual affected persons. Contract and tort law, by contrast, strive for technical and protective transparency under which the potential trade-off between performance and explainability takes center stage: any potentially reduced accuracy or utility stemming from enforcing explanations must be weighed against the potential safety gains such explanations enable. Explanations are required only to the extent that this balance is positive. Banking law, finally, endorses a quality assurance regime in which transparency contributes to the control of systemic risk in the banking sector. Here, even global explanations may be required. The proposal for the AIA, in turn, is primarily geared toward compliance-oriented transparency for professional operators of AI systems. From a rights-enabling perspective, this is a significant limitation.

These legal requirements, however, can be interpreted to increasingly call for actionable explanations. This implies moving beyond mere laundry lists of relevant features toward cognitively optimized and goal-oriented explanations. Multi-layered or contrastive explanations are important elements in such a strategy. Tools, methods and formats from various values-in-design approaches as well as those developed under the umbrella term of 'responsible research and innovation' can help co-designing such systems and explanations.

Finally, an update of the AIA should consider fairness-enabling transparency, which seeks to facilitate the detection of potential bias in AI systems, as well as broader provisions for quality benchmarking to facilitate informed decisions by affected persons, to enable critical review and the exercise of rights, and to generate trust in AI systems more generally.

References

1. Acquisti, A., Taylor, C., Wagman, L.: The economics of privacy. J. Econ. Liter. **54**(2), 442–92 (2016)
2. Aldewereld, H., Mioch, T.: Values in design methodologies for AI. In: Polyvyanyy, A., Rinderle-Ma, S. (eds.) CAiSE 2021. LNBIP, vol. 423, pp. 139–150. Springer, Cham (2021). https://doi.org/10.1007/978-3-030-79022-6_12
3. Arrieta, A.B., et al.: Explainable artificial intelligence (XAI): concepts, taxonomies, opportunities and challenges toward responsible AI. Inf. Fusion **58**, 82–115 (2020)
4. Avgouleas, E.: Governance of Global Financial Markets: The Law, The Economics, The Politics. Cambridge University Press, Cambridge (2012)
5. Bachmann, G.: Commentary on §241 BGB, in: Münchener Kommentar zum BGB. BECK, Munich, 8th ed. (2019)
6. Bäcker, M.: Commentary on Art. 13 GDPR, in: Kühling/Buchner, DS- GVO Commentary. BECK, Munich, 3rd ed. (2020)
7. BaFin: Rolle der Aufsicht bei der Verwendung von Kreditscores. BaFin J. 22–24 (2019)
8. Bakos, Y., Marotta-Wurgler, F., Trossen, D.D.: Does anyone read the fine print? Consumer attention to standard- form contracts. J. Leg. Stud. **43**, 1–35 (2014)
9. Bambauer, J., Zarsky, T.: The algorithm game. Notre Dame L. Rev. **94**, 1 (2018)
10. Bar-Gill, O.: Smart disclosure: promise and perils. Behav. Public Policy **5**, 238–251 (2021)
11. Bar-Gill, O., Ben-Shahar, O.: Regulatory techniques in consumer protection: a critique of European consumer contract law. Common Market Law Rev. **50**, 109–126 (2013)
12. Barocas, S., Selbst, A.D., Raghavan, M.: The hidden assumptions behind counterfactual explanations and principal reasons. In: Proceedings of the 2020 Conference on Fairness, Accountability, and Transparency, pp. 80–89 (2020)
13. Ben-Shahar, O., Chilton, A.S.: Simplification of privacy disclosures: an experimental test. J. Leg. Stud. **45**(S2), S41–S67 (2016)
14. Bibal, A., Lognoul, M., de Streel, A., Frénay, B.: Legal requirements on explainability in machine learning. Artif. Intell. Law **29**(2), 149–169 (2020). https://doi.org/10.1007/s10506-020-09270-4

15. Biran, O., Cotton, C.V.: Explanation and justification in machine learning: a survey. In: IJCAI-17 Workshop on Explainable AI (XAI), vol. 8, no. 1, pp. 8–13 (2017)
16. Breiman, L.: Random forests. Mach. Learn. **45**(1), 5–32 (2001)
17. Brownsword, R.: From Erewhon to AlphaGo: for the sake of human dignity, should we destroy the machines? Law Innov. Technol. **9**(1), 117–153 (2017)
18. Burrell, J.: How the machine 'thinks': understanding opacity in machine learning algorithms. Big Data Soc. **3**(1) (2016)
19. Cabral, T.S.: Liability and artificial intelligence in the EU: assessing the adequacy of the current product liability directive. Maastricht J. Eur. Compar. Law **27**(5), 615–635 (2020)
20. Casey, B., Farbangi, A., Vogl, R.: Rethinking explainable machines: the GDPR's 'right to explanation' debate and the rise of algorithmic audits in enterprise. Berkeley Technol. Law J. **34**, 143 (2019)
21. CEBS (Committee of the European Banking Supervisors): Guidelines on the implementation, validation and assessment of Advanced Measurement (AMA) and Internal Ratings Based (IRB) Approaches (2006)
22. Chen, J.M.: Interpreting linear beta coefficients alongside feature importances. Mach. Learn. (2021)
23. Citron, D.K., Pasquale, F.: The scored society: due process for automated predictions. Washington Law Rev. **89**(1) (2014)
24. Dignum, V.: Responsible Artificial Intelligence: How to Develop and Use AI in a Responsible Way. Artificial Intelligence: Foundations, Theory, and Algorithms, Springer, Cham (2019). https://doi.org/10.1007/978-3-030-30371-6
25. Dumitrescu, E.I., Hué, S., Hurlin, C.: Machine learning or econometrics for credit scoring: let's get the best of both worlds. Working Paper (2021)
26. EBA (European Banking Authority): Guidelines on loan origination and monitoring (2020)
27. Ebers, M., Hoch, V.R., Rosenkranz, F., Ruschemeier, H., Steinrötter, B.: The European commission's proposal for an artificial intelligence act-a critical assessment by members of the robotics and AI law society (rails). J **4**(4), 589–603 (2021)
28. European Commission: Responsible research and innovation Europe's ability to respond to societal challenges (2012)
29. Expert Group on Liability and New Technologies: New Technologies Formation, Liability for Artificial Intelligence and other emerging digital technologies. Technical report (2019)
30. Fickling, P.A., Watkins, P.: An update on credit access and the Bureau's first No - Action Letter (2019)
31. Fisher, A.J., Rudin, C., Dominici, F.: All models are wrong, but many are useful: learning a variable's importance by studying an entire class of prediction models simultaneously. J. Mach. Learn. Res. **20**(177), 1–81 (2019)
32. Friedman, B., Hendry, D.G., Borning, A.: A survey of value sensitive design methods. Found. Trends Human-Comput. Interact. **11**(2), 63–125 (2017)
33. Froomkin, A.M., Kerr, I.R., Pineau, J.: When AIs outperform doctors: confronting the challenges of a tort-induced over-reliance on machine learning. Ariz. Law Rev. **61**, 33 (2019)
34. Fuster, A., Goldsmith-Pinkham, P., Ramadorai, T., Walther, A.: Predictably unequal? The effects of machine learning on credit markets. Working Paper (2020)
35. Gillis, T.B.: The input fallacy. Minnesota Law Rev. (forthcoming) (2021)

36. Goodman, B., Flaxman, S.: EU regulations on algorithmic decision-making and a "right to explanation". WHI (2016)

37. Grochowski, M., Jabłonowska, A., Lagioia, F., Sartor, G.: Algorithmic transparency and explainability for EU consumer protection: unwrapping the regulatory premises. Crit. Anal. Law **8**(1), 43–63 (2021)

38. Hacker, P.: Manipulation by algorithms. exploring the triangle of unfair commercial practice, data protection, and privacy law. Eur. Law J. (forthcoming). https://doi.org/10.1111/eulj.12389

39. Hacker, P.: The behavioral divide: a critique of the differential implementation of behavioral law and economics in the US and the EU. Eur. Rev. Contract Law **11**(4), 299–345 (2015)

40. Hacker, P.: Teaching fairness to artificial intelligence: existing and novel strategies against algorithmic discrimination under EU law. Common Market Law Rev. **55**(4), 1143–1186 (2018)

41. Hacker, P.: Europäische und nationale Regulierung von Künstlicher Intelligenz. NJW (Neue Juristische Wochenschrift), pp. 2142–2147 (2020)

42. Hacker, P., Krestel, R., Grundmann, S., Naumann, F.: Explainable AI under contract and tort law: legal incentives and technical challenges. Artif. Intell. Law **28**(4), 415–439 (2020). https://doi.org/10.1007/s10506-020-09260-6

43. Hagendorff, T.: The ethics of AI ethics: an evaluation of guidelines. Mind. Mach. **30**, 99–120 (2020)

44. Hansen, M.: Data protection by design and by default à la European general data protection regulation. In: Lehmann, A., Whitehouse, D., Fischer-Hübner, S., Fritsch, L., Raab, C. (eds.) Privacy and Identity 2016. IAICT, vol. 498, pp. 27–38. Springer, Cham (2016). https://doi.org/10.1007/978-3-319-55783-0_3

45. High-Level Expert Group on Artificial Intelligence: Ethics guidelines for trustworthy AI (2019)

46. Hildebrandt, M.: Privacy as protection of the incomputable self: from agnostic to agonistic machine learning. Theoret. Inquiries Law **20**(1), 83–121 (2019)

47. Holzinger, A., Biemann, C., Pattichis, C.S., Kell, D.B.: What do we need to build explainable AI systems for the medical domain? arXiv preprint arXiv:1712.09923 (2017)

48. Jolls, C.: Debiasing through law and the first Amendment. Stanford Law Rev. **67**, 1411 (2015)

49. Kaminski, M.E.: The right to explanation, explained. Berkeley Technol. Law J. **34**, 189 (2019)

50. Karimi, A.H., Schölkopf, B., Valera, I.: Algorithmic recourse: from counterfactual explanations to interventions. In: Proceedings of the 2021 ACM Conference on Fairness, Accountability, and Transparency (2021)

51. Kenney, M.: Fables of response-ability: feminist science studies as didactic literature. Catalyst: Feminism Theory Technosci. **5**(1), 1–39 (2019)

52. Kleinberg, J., Ludwig, J., Mullainathan, S., Rambachan, A.: Algorithmic fairness. In: AEA Papers and Proceedings, vol. 108, pp. 22–27 (2018)

53. Knobel, C., Bowker, G.C.: Values in design. Commun. ACM **54**(7), 26–28 (2011)

54. Langenbucher, K.: Responsible AI-based credit scoring – a legal framework. Eur. Bus. Law Rev. **31**(4), 527–572 (2020)

55. Lapuschkin, S., Wäldchen, S., Binder, A., Montavon, G., Samek, W., Müller, K.R.: Unmasking clever HANS predictors and assessing what machines really learn. Nat. Commun. **10**(1), 1–8 (2019)

56. Liegl, M., Oliphant, R., Buscher, M.: Ethically aware IT design for emergency response: from co-design to ELSI co-design. In: Proceedings of the ISCRAM 2015 Conference (2015)

57. Lipton, Z.C.: The mythos of model interpretability: in machine learning, the concept of interpretability is both important and slippery. Queue **16**(3), 31–57 (2018)

58. Loch, F.: Art. 174, Boos/Fischer/Schulte- Mattler (eds.), VO (EU) 575/2013, 5th ed. (2016)

59. Lombrozo, T.: The structure and function of explanations. Trends Cogn. Sci. **10**(10), 464–470 (2006)

60. Lundberg, S.M., Lee, S.I.: A unified approach to interpreting model predictions. In: Advances in Neural Information Processing Systems, vol. 30, pp. 4765–4774 (2017)

61. Malgieri, G., Comandé, G.: Why a right to legibility of automated decision-making exists in the general data protection regulation. International Data Privacy Law (2017)

62. Malle, B.F.: How the Mind Explains Behavior: Folk Explanations, Meaning, and Social Interaction. MIT Press, Cambridge (2004)

63. Mendez Fernandez, D., Passoth, J.H.: Empirical software engineering. From discipline to interdiscipline. J. Syst. Softw. **148**, 170–179 (2019)

64. Miller, T.: Explanation in artificial intelligence: insights from the social sciences. Artif. Intell. **267**, 1–38 (2019)

65. Miller, T.: Contrastive explanation: a structural-model approach. arXiv preprint arXiv:1811.03163 (2020)

66. Mittelstadt, B.D., Allo, P., Taddeo, M., Wachter, S., Floridi, L.: The ethics of algorithms: mapping the debate. Big Data Soc. **32** (2016)

67. Moore, J.D., Swartout, W.: Explanation in expert systems: a survey, Information Sciences Institute Tech Report. Technical report ISI/RR-88-228 (1988)

68. Müller, H., Mayrhofer, M.T., Van Veen, E.B., Holzinger, A.: The ten commandments of ethical medical AI. Computer **54**(07), 119–123 (2021)

69. Müller, P., Passoth, J.-H.: Engineering collaborative social science toolkits. STS methods and concepts as devices for interdisciplinary diplomacy. In: Karafillidis, A., Weidner, R. (eds.) Developing Support Technologies. BB, vol. 23, pp. 137–145. Springer, Cham (2018). https://doi.org/10.1007/978-3-030-01836-8_13

70. Nissenbaum, H.: Values in the design of computer systems. Computers in Society (March), pp. 38–39 (1998)

71. N.N.: Editorial, towards trustable machine learning. Nat. Biomed. Eng. **2**, 709–710 (2018)

72. Obar, J.A., Oeldorf-Hirsch, A.: The biggest lie on the internet: ignoring the privacy policies and terms of service policies of social networking services. Inf. Commun. Soc. **23**(1), 128–147 (2020)

73. Paal, B., Hennemann, M.: Commentary on Art. 13, in Paal/Pauly (eds.), Datenschutz-Grundverordnung. Kommentar. BECK, Munich, 3rd ed. (2021)

74. Pasquale, F.: The Black Box Society. Harvard University Press, Cambridge (2015)

75. Passoth, J.H.: Die Demokratisierung des Digitalen. Konrad Adenauer Stiftung: Analysen & Argumente **424**, 1–13 (2021)

76. Pessach, D., Shmueli, E.: Algorithmic fairness. arXiv preprint arXiv:2001.09784 (2020)

77. Rathi, S.: Generating counterfactual and contrastive explanations using SHAP. arXiv preprint arXiv:1906.09293 (2019)

78. Read, S.J., Marcus-Newhall, A.: Explanatory coherence in social explanations: a parallel distributed processing account. J. Pers. Soc. Psychol. **65**(3), 429 (1993)

79. Ribeiro, M.T., Singh, S., Guestrin, C.: "Why should i trust you?" Explaining the predictions of any classifier. In: Proceedings of the 22nd ACM SIG KDD International Conference on Knowledge Discovery and Data Mining, pp. 1135–1144 (2016)

80. Ronan, H., Junklewitz, H., Sanchez, I.: Robustness and explainability of artificial intelligence. JRC Technical Report 13 (2020)

81. Rudin, C.: Stop explaining black box machine learning models for high stakes decisions and use interpretable models instead. Nat. Mach. Intell. 1(5), 206–215 (2019)

82. Samek, W., Montavon, G., Vedaldi, A., Hansen, L.K., Müller, K.-R. (eds.): Explainable AI: Interpreting, Explaining and Visualizing Deep Learning. LNCS (LNAI), vol. 11700. Springer, Cham (2019). https://doi.org/10.1007/978-3-030-28954-6

83. Schaub, F., Balebako, R., Durity, A.L.: A design space for effective privacy notices. In: Eleventh Symposium On Usable Privacy and Security ({SOUPS} 2015), pp. 1–17 (2015)

84. Schneeberger, D., Stöger, K., Holzinger, A.: The European legal framework for medical AI. In: Holzinger, A., Kieseberg, P., Tjoa, A.M., Weippl, E. (eds.) CD-MAKE 2020. LNCS, vol. 12279, pp. 209–226. Springer, Cham (2020). https://doi.org/10.1007/978-3-030-57321-8_12

85. Schröder, T.: Programming fairness. Max Planck Research, pp. 68–73 (2019)

86. Seehafer, A., Kohler, J.: Künstliche Intelligenz: Updates für das Produkthaftungsrecht? EuZW, pp. 213–218 (2020)

87. Selbst, A.D.: Negligence and AI's human user. BUL Rev. 100, 1315 (2020)

88. Selbst, A.D., Barocas, S.: The intuitive appeal of explainable machines. Fordham Law Rev. 87, 1085 (2018)

89. Selbst, A.D., Powles, J.: Meaningful information and the right to explanation. Int. Data Priv. Law 7(4), 233 (2017)

90. Simonyan, K., Vedaldi, A., Zisserman, A.: Deep inside convolutional networks: visualising image classification models and saliency maps. In: In Workshop at International Conference on Learning Representations (2014)

91. Smuha, N.A., et al.: How the EU can achieve legally trustworthy AI: a response to the European commission's proposal for an artificial intelligence act. Available at SSRN (2021)

92. Strandburg, K.J.: Adjudicating with Inscrutable Decision Tools. MIT Press (2021, forthcoming)

93. Sunstein, C.R.: Simpler: The Future of Government. Simon & Schuster, Manhattan (2013)

94. Toke, M.J.: Restatement (third) of torts and design defectiveness in American products liability law. Cornell J. Law Public Policy 5(2), 239 (1996)

95. Veale, M., Borgesius, F.Z.: Demystifying the draft EU artificial intelligence act-analysing the good, the bad, and the unclear elements of the proposed approach. Comput. Law Rev. Int. 22(4), 97–112 (2021)

96. Wachter, S., Mittelstadt, B., Floridi, L.: Why a right to explanation of automated decision-making does not exist in the general data protection regulation. Int. Data Priv. Law 7(2), 76–99 (2017)

97. Wachter, S., Mittelstadt, B., Russell, C.: Counterfactual explanations without opening the black box: automated decisions and the GDPR. Harvard J. Law Technol. 31, 841 (2018)

98. Wachter, S., Mittelstadt, B., Russell, C.: Why fairness cannot be automated: bridging the gap between EU non-discrimination law and AI. Comput. Law Secur. Rev. (2021, forthcoming)
99. Wagner, G.: Robot liability. In: Liability for Artificial Intelligence and the Internet of Things. Nomos Verlagsgesellschaft mbH & Co. KG (2019)
100. Wagner, G.: Commentary on §3 ProdHaftG, in: Münchener Kommentar zum BGB. BECK, Munich, 8th ed. (2020)
101. Wagner, G.: Commentary on §823 BGB, in: Münchener Kommentar zum BGB. BECK, Munich, 8th ed. (2020)
102. Wendehorst, C.: Strict liability for AI and other emerging technologies. J. Eur. Tort Law **11**(2), 150–180 (2020)
103. Wischmeyer, T.: Artificial intelligence and transparency: opening the black box. In: Wischmeyer, T., Rademacher, T. (eds.) Regulating Artificial Intelligence, pp. 75–101. Springer, Cham (2020). https://doi.org/10.1007/978-3-030-32361-5_4
104. Zarsky, T.Z.: Transparent Predictions. U. Ill. L. Rev, p. 1503 (2013)
105. Zech, H.: Künstliche Intelligenz und Haftungsfragen. ZfPW, pp. 198–219 (2019)
106. Zehlike, M., Hacker, P., Wiedemann, E.: Matching code and law: achieving algorithmic fairness with optimal transport. Data Min. Knowl. Disc. **34**(1), 163–200 (2019). https://doi.org/10.1007/s10618-019-00658-8

Towards Explainability for AI Fairness

Jianlong Zhou[1,2]([envelope]) [iD], Fang Chen[1], and Andreas Holzinger[2,3] [iD]

[1] Human-Centered AI Lab, University of Technology Sydney, Sydney, Australia
{Jianlong.Zhou,Fang.Chen}@uts.edu.au
[2] Human-Centered AI Lab, Medical University Graz, Graz, Austria
Andreas.Holzinger@medunigraz.at
[3] xAI Lab, Alberta Machine Intelligence Institute, Edmonton, Canada

Abstract. AI explainability is becoming indispensable to allow users to gain insights into the AI system's decision-making process. Meanwhile, fairness is another rising concern that algorithmic predictions may be misaligned to the designer's intent or social expectations such as discrimination to specific groups. In this work, we provide a state-of-the-art overview on the relations between explanation and AI fairness and especially the roles of explanation on human's fairness judgement. The investigations demonstrate that fair decision making requires extensive contextual understanding, and AI explanations help identify potential variables that are driving the unfair outcomes. It is found that different types of AI explanations affect human's fairness judgements differently. Some properties of features and social science theories need to be considered in making senses of fairness with explanations. Different challenges are identified to make responsible AI for trustworthy decision making from the perspective of explainability and fairness.

Keywords: Fairness · Explainable AI · Explainability · Machine learning

1 Introduction

Artificial Intelligence (AI) including Machine Learning (ML) algorithms are increasingly shaping people's daily lives by making decisions with ethical and legal impacts in various domains such as banking, insurance, medical care, criminal justice, predictive policing, and hiring [43,44]. While AI-informed decision making can lead to faster and better decision outcomes, however, AI algorithms such as deep learning often use complex learning approaches and even their designers are often unable to understand why AI arrived at a specific decision. Therefore, AI remains a black box that makes it hard for users to understand why a decision is made or how the data is processed for the decision making [8,44,45]. Because of the black box nature of AI models, the deployment of AI algorithms especially in high stake domains usually requires testing and verification for reasonability by domain experts not only for safety but also for legal reasons [35]. Users also want to understand reasons behind specific AI-informed

A. Holzinger et al. (Eds.): xxAI 2020, LNAI 13200, pp. 375–386, 2022.
https://doi.org/10.1007/978-3-031-04083-2_18

decisions. For example, high-stake domains require explanations of AI before any critical decisions, computer scientists use explanations to refine and further improve performance of AI algorithms, and AI explanations can also improve the user experience of a product or service by helping end-users trust that the AI is making good decisions [7]. As a result, the issue of AI explanation has experienced a significant surge in interest from the international research community to various application domains, ranging from agriculture to human health and is becoming indispensable in addressing ethical concerns and fostering trust and confidence in AI systems [20,42,43].

Furthermore, AI algorithms are often trained on a large amount of historical data, which may not only replicate, but also amplify existing biases or discrimination in historical data. Therefore, due to such biased input data or faulty algorithms, unfair AI-informed decision making systems have been proven to systematically reinforce discrimination such as racial/gender biases in AI-informed decision making. These drive a distrust in and fear the use of AI in public discussions [41].

In addition, the wide use of AI in almost every aspect of our life implies that with great powers comes great responsibility. Fairness shows that an AI system exhibits certain desirable ethical characteristics, such as being bias-free, diversity-aware, and non-discriminatory. While explanations to an AI system provide human-understandable interpretations of the inner working of the system and decisions. Both fairness and explanation are important components for building "Responsible AI". For example, the fair treatment and/or fair outcome are important ethical issues that need to be considered in the algorithmic hiring decision making. How the decisions made by an algorithmic process can be explained in a transparent and compliant way is also necessary for ethical use of AI in the hiring [36]. Therefore, both fairness and explanations are important ethical issues that can be used to promote user trust in AI-informed decision making (see Fig. 1).

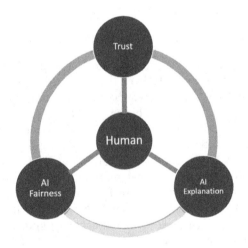

Fig. 1. Relations among AI fairness, AI explanation, and trust.

Previous research found that AI explanations are not only for human to understand the AI system, but also provide an interface for human in the loop, enabling them to identify and address fairness and other issues [12]. Furthermore, differences in AI outcomes amongst different groups in AI-informed decision making can be justified and explained via different attributes in some cases [27]. When these differences are justified and explained, the discrimination is not considered to be illegal [22]. Therefore, explanation and fairness have close relations in AI-informed decision making (as highlighted in orange colour in Fig. 1). Taken the talent recruiting as an example, disproportional recruitment rates for males and females may be explainable by the fact that more males may have higher education, and if males and females are treated equally, it will introduce reverse discrimination, which may be undesirable as well [22]. In another example on the annual income analysis [2], males have a higher annual income than females on average in the data. However, this does not mean that there is a discrimination to females in the annual income because females have fewer work hours than males per week on average. Therefore, the explanation to the difference of the annual income between males and females with the use of work hours per week helps the outcomes of annual income acceptable, legal and fair [22]. It shows that fairness and explanation are tightly related to each other. Therefore, it is significant to understand how AI explanations impact the fairness judgement or how the AI fairness enhances AI explanations. This paper aims to investigate state-of-the-art research in these areas and identifies key research challenges. The contributions of the paper include:

- The relations between explanability and AI fairness are identified as one of significant components for the responsible use of AI and trustworthy decision making.
- A systematic analysis on the explanabillitty and AI fairness to learn the current status of explanability for the human's fairness judgement;
- The challenges and future research directions on the explanability for AI fairness are identified.

2   Fairness

Fairness has become a key element in developing socio-technical AI systems when AI is used in various decision making tasks. In the context of decision-making, fairness is defined as the absence of any prejudice or favoritism towards an individual or a group based on their inherent or acquired characteristics [27, 33]. An unfair algorithm is one whose decisions are skewed toward a particular group. Fairness can be considered from at least four aspects [10]: 1) protected attributes such as race, gender, and their proxies, are not explicitly used to make decisions; 2) common measures of predictive performance (e.g., false positive and false negative rates) are equal across groups defined by the protected attributes; 3) outcomes are independent of protected attributes; and 4) treat similarly risky people similarly.

There are two potential sources of unfairness in machine learning outcomes: those arising from biases in data and those arising from algorithms. Mehrabi et al. [27] summarised 23 types of data biases that may result in fairness issues in machine learning: historical bias, representation bias, measurement bias, evaluation bias, aggregation bias, population bias, Simpson's paradox, longitudinal data fallacy, sampling bias, behavioural bias, content production bias, linking bias, temporal bias, popularity bias, algorithmic bias, user interaction bias, social bias, emergent bias, self-selection bias, omitted variable bias, cause-effect bias, observer bias, and funding bias. Different kinds of discrimination that may occur in algorithmic decision making are also categorised by Mehrabi et al. [27] such as direct discrimination, indirect discrimination, systemic discrimination, statistical discrimination, explainable discrimination, and unexplainable discrimination. Different metrics have been developed to measure AI fairness quantitatively and various approaches have been proposed to mitigate AI biases [6]. For example, statistical parity difference is defined as the difference of the rate of favorable outcomes received by the unprivileged group to the privileged group, and equal opportunity difference is defined as the difference of true positive rates between the unprivileged and the privileged groups. The true positive rate is the ratio of true positives to the total number of actual positives for a given group.

Since the disconnection between the fairness metrics and practical needs of society, politics, and law [21], Lee et al. [24] presented that the relevant contextual information should be considered in an understanding of a model's ethical impact, and fairness metrics should be framed within a broader view of ethical concerns to ensure their adoption for a contextually appropriate assessment of each algorithm.

As AI is often used by humans and/or for human-related decision making, people's perception of fairness is required to be taken into account when designing and implementing AI-informed decision making systems [38]. Following this, people's perception of fairness has been investigated along four dimensions: 1) algorithmic predictors, 2) human predictors, 3) comparative effects (human decision-making vs. algorithmic decision-making), and 4) consequences of AI-informed decision making [38].

3   AI Explanation

The AI explainability has been reviewed thoroughly in recent years [7,44], which are based on the explanation-generation approaches, the type of explanation, the scope of explanation, the type of model it can explain or combinations of these methods as well as others [1]. For example, explanation methods can be grouped into pre-model, in-model, and post-model methods by considering when explanations are applicable; there are also intrinsic and post-hoc explanation methods by considering whether explainability is achieved through constraints imposed on the AI model directly (intrinsic) or by applying explanation methods that analyse the model after training (post-hoc). Other types of explanations include model-specific and model-agnostic methods, as well as global and local explanation methods.

Miller [28] emphasised the importance of social science in AI explanations and found that 1) Explanations are contrastive and people do not ask why an event happened, but rather why this event happened instead of another event; 2) Explanations are selected in a biased manner. People are adept at selecting one or two causes from an infinite number of causes to be the explanation, which could be influenced by certain cognitive biases; 3) Probabilities probably don't matter. Explanations with statistical generalisations are unsatisfying and the causal explanation for the generalisation itself is usually effective; 4) Explanations are social. They are a transfer of knowledge to people and act as part of a conversation or interaction with people. Therefore, explanations are not just the presentation of associations and causes to predictions, they are contextual.

Wang et al. [39] highlighted three desirable properties that ideal AI explanations should satisfy: 1) improve people's understanding of the AI model, 2) help people recognize the model uncertainty, and 3) support people's calibrated trust in the model. Therefore, different approaches are investigated to evaluate whether and to what extent the offered explainability achieves the defined objective [44]. Objective and subjective metrics are proposed to evaluate the quality of explanations, such as clarity, broadness, simplicity, completeness, and soundness of explanations, as well as user trust. For example, Schmidt and Biessmann [34] presented a quantitative measure for the quality of explanation methods based on how faster and accurate decisions indicate intuitive understanding, i.e. the information transfer rate which is based on mutual information between human decisions and model predictions. [34] also argued that a trust metric must capture cases in which humans are too biased towards the decisions of an AI system and overly trust the system, and presented a quantitative measure for trust by considering the quality of AI models (see Eq. 1).

$$T = \frac{MI_{\hat{y}}}{MI_y} \qquad (1)$$

where T is the trust metric, $MI_{\hat{y}}$ is the mutual information between human decisions and model predictions and MI_y is the mutual information between human decisions and true labels.

Despite the extensive investigations of AI explanations, they still face different challenges [29]. For example, similar to AI models, uncertainty is inherently associated with explanations because they are computed from training data or models. However, many AI explanation methods such as feature importance-based approaches provide explanations without quantifying the uncertainty of the explanation. Furthermore, AI explanations, which should ideally reflect the true causal relations [17], mostly reflect statistical correlation structures between features instead.

4   Explanation for AI Fairness

As discussed previously, fairness and explanation are strongly dependent. Deciding an appropriate notion of fairness to impose on AI models or understanding whether a model is making fair decisions require extensive contextual understanding and domain knowledge. Shin and Park [37] investigated the role of Fairness, Accountability, and Transparency (FAT) in algorithmic affordance. It showed that FAT issues are multi-functionally related, and user attitudes about FAT are highly dependent on the context in which it takes place and the basis who is looking at. It also showed that topics regarding FAT are somehow related and overlapping, making them difficult to distinguish or separate. It demonstrated the heuristic role of FAT regarding their fundamental links to trust.

4.1 Explanation Guarantees Fairness

The explanation of the decision making is a way to gain insights and guarantee fairness to all groups impacted by AI-related decisions [13]. Lee et al. [24] argued that explanations may help identify potential variables that are driving the unfair outcomes. It is unfair if decisions were made without explanations or with unclear, untrusted, and unverifiable explanations [32]. For example, Begley et al. [5] introduced explainability methods for fairness based on the Shapley value framework for model explainability [25]. The proposed fairness explanations attribute a model's overall unfairness to individual input features, even the model does not operate on protected/sensitive attributes directly.

Warner and Sloan [40] argued that effective regulation to ensure fairness requires that AI systems be transparent. While explainability is one of approaches to acquire transparency. The explainability requires that an AI system provides a human-understandable explanation of why any given decision was reached in terms of the training data used, the kind of decision function, and the particular inputs for that decision. Different proxy variables of fairness are presented for the effective regulation of AI transparency in [40].

4.2 Influence of Explanation on Perception of Fairness

Baleis et al. [3] showed that transparency, trust and individual moral concepts demonstrably have an influence on the individual perception of fairness in AI applications. Dodge et al. [12] investigated the impact of four types of AI explanations on human's fairness judgments of AI systems. The four types of explanations are input influence-based explanation, demographic-based explanation, sensitivity-based explanation, and case-based explanation. It showed that case-based explanation is generally less fair. It was found that local explanations are more effective than global explanations for case-specific fairness issues. Sensitivity-based explanations are the most effective for the fairness issue of disparate impact.

4.3 Fairness and Properties of Features

Grgic-Hlaca et al. [14] proposed to understand why people perceive certain features as fair or unfair to be used in algorithms based on a case study of a criminal risk estimation tool for the use to help make judicial decisions. Eight properties of features are identified, which are reliability, relevance, volitionality, privacy, causes outcome, causes vicious cycle, causes disparity in outcomes, and caused by sensitive group membership. It was found that people's concerns on the unfairness of an input feature are not only discrimination, but also other consideration of latent properties such as the relevance of the feature to the decision making scenario and the reliability with which the feature can be assessed. In a further study, Grgic-Hlaca et al. [15] proposed measures for procedural fairness (the fairness of the decision making process) that consider the input features used in the decision process in the context of criminal recidivism. The analysis examined to what extent the perceived fairness of a characteristic is influenced by additional knowledge about increasing the accuracy of the prediction. It was found that input features that were classified as fairer were those that improved the accuracy of prediction and those features as more unfair that led to discrimination against certain feature holders of people.

4.4 Fairness and Counterfactuals

The use of counterfactuals has become one of popular approaches for AI explanation and making sense of algorithmic fairness [4,26,44], which can require an incoherent theory of what social categories are [23].

However, it was argued that the social categories may not admit counterfactual manipulation, and hence may not appropriately satisfy the demands for evaluating the truth or falsity of counterfactuals [23], which can lead to misleading results. Therefore, the approaches used for algorithmic explanations to make sense of fairness also need to consider social science theories to support AI fairness and explanations.

A good example of the use of counterfactuals [18] is algorithmic risk assessment [11]. Algorithmic risk assessments are increasingly being used to help experts make decisions, for example, in medicine, in agriculture or criminal justice. The primary purpose of such AI-based risk assessment tools is to provide decision-relevant information for actions such as medical treatments, irrigation measures or release conditions, with the aim of reducing the likelihood of the respective adverse event such as hospital readmission, crop drying, or criminal recidivism. The advantage of the principle of machine learning, namely learning from large amounts of historical data, is precisely counterproductive, even dangerous [19], here.

Because such algorithms reflect the risk from decision-making policies of the past – but not the current actual conditions. To cope with this problem, [11] presents a new method for estimating the proposed metrics that uses doubly robust estimation and shows that only under strict conditions can fairness be provided simultaneously according to the standard metric and the counterfactual metric. Consequently, fairness-enhancing methods that aim for parity in a standard fairness metric can cause greater imbalance in the counterfactual analogue.

5 Discussion

With the increasing use of AI in people's daily lives for various decision making tasks, the fairness of AI-informed decisions and explanation of AI for decision making are becoming significant concerns for the responsible use of AI and trustworthy decision making. This paper focused on the relations between explanation and AI fairness and especially the roles of explanation on AI fairness. The investigations demonstrated that fair decision making requires extensive contextual understanding. AI explanations help identify potential variables that are driving the unfair outcomes. Different types of AI explanations affect human's fairness judgements differently. Certain properties of features such as the relevance of the feature to the decision making scenario and the reliability with which the feature can be assessed affect human's fairness judgements. In addition, social science theories need to be considered in making sense of fairness with explanations. However, there are still challenges. For example,

- Despite the requirements of the extensive contextual understanding for the fair decision making, it is hard to decide what contextual understanding is the appropriate to boost fair decision making.
- There are various types of explanations. It is significant to decide what explanations that can promote the human's fairness judgement on decision making as expected. While the human's fairness judgement is highly related to users themselves, it is a challenge to justify what explanations are the best for human's fairness judgement.
- Since AI is applied in various sectors and scenarios, it is important to understand whether different application sectors or scenarios affect the effectiveness of explanations on the human's judgement on perception in decision making.

Investigating AI fairness explanations requires a multidisciplinary approach and must include research on machine learning [9], human-computer interaction [31] and social science [30] – regardless of the application domain - because the domain expert must always be involved and can bring valuable knowledge and contextual understanding [16].

All this provides us with clues for developing effective approaches to responsible AI and trustworthy decision-making in all future work processes.

6 Conclusion

The importance of fairness is undisputed. In this paper, we have explored the relationships between explainability, or rather explanation, and AI fairness, and in particular the role of explanation in AI fairness. We first identified the relationships between explanation and AI fairness as one of the most important components for the responsible use of AI and trustworthy decision-making. The systematic analysis of explainability and AI fairness revealed that fair decision-making requires a comprehensive contextual understanding, to which AI explanations can contribute. Based on our investigation, we were able to identify several other challenges regarding the relationships between explainability and AI fairness. We ultimately argue that the study of AI fairness explanations requires an important multidisciplinary approach, which is necessary for a responsible use of AI and for trustworthy decision-making - regardless of the application domain.

Acknowledgements. The authors declare that there are no conflict of interests and the work does not raise any ethical issues. Parts of this work have been funded by the Austrian Science Fund (FWF), Project: P-32554, explainable AI.

References

1. Arya, V., et al.: One explanation does not fit all: a toolkit and taxonomy of AI explainability techniques. arXiv:1909.03012 [cs, stat] (2019)
2. Asuncion, A., Newman, D.: UCI machine learning repository (2007). https://archive.ics.uci.edu/ml/index.php
3. Baleis, J., Keller, B., Starke, C., Marcinkowski, F.: Cognitive and emotional response to fairness in AI - a systematic review (2019). https://www.semanticscholar.org/paper/Implications-of-AI-(un-)fairness-in-higher-the-of-Marcinkowski-Kieslich/231929b1086badcbd149debb0abefc84cdb85665
4. Barocas, S., Selbst, A.D., Raghavan, M.: The hidden assumptions behind counterfactual explanations and principal reasons. In: Proceedings of the 2020 Conference on Fairness, Accountability, and Transparency, FAT* 2020, pp. 80–89 (2020)
5. Begley, T., Schwedes, T., Frye, C., Feige, I.: Explainability for fair machine learning. CoRR abs/2010.07389 (2020). https://arxiv.org/abs/2010.07389
6. Bellamy, R.K.E., et al.: AI fairness 360: an extensible toolkit for detecting, understanding, and mitigating unwanted algorithmic bias. CoRR abs/1810.01943 (2018). http://arxiv.org/abs/1810.01943
7. Carvalho, D.V., Pereira, E.M., Cardoso, J.S.: Machine learning interpretability: a survey on methods and metrics. Electronics **8**(8), 832 (2019)
8. Castelvecchi, D.: Can we open the black box of AI? Nat. News **538**(7623), 20 (2016)
9. Chouldechova, A., Roth, A.: The frontiers of fairness in machine learning. Commun. ACM **63**(5), 82–89 (2020). https://doi.org/10.1145/3376898
10. Corbett-Davies, S., Goel, S.: The measure and mismeasure of fairness: a critical review of fair machine learning. CoRR abs/1808.00023 (2018). http://arxiv.org/abs/1808.00023
11. Coston, A., Mishler, A., Kennedy, E.H., Chouldechova, A.: Counterfactual risk assessments, evaluation, and fairness. In: Proceedings of the 2020 Conference on Fairness, Accountability, and Transparency (FAT 2020), pp. 582–593 (2020). https://doi.org/10.1145/3351095.3372851
12. Dodge, J., Liao, Q.V., Zhang, Y., Bellamy, R.K.E., Dugan, C.: Explaining models: an empirical study of how explanations impact fairness judgment. In: Proceedings of the 24th International Conference on Intelligent User Interfaces, IUI 2019, pp. 275–285 (2019)
13. Ferreira, J.J., de Souza Monteiro, M.: Evidence-based explanation to promote fairness in AI systems. In: CHI2020 Fair and Responsible AI Workshop (2020)
14. Grgic-Hlaca, N., Redmiles, E.M., Gummadi, K.P., Weller, A.: Human perceptions of fairness in algorithmic decision making: a case study of criminal risk prediction. In: Proceedings of the 2018 World Wide Web Conference, WWW 2018, pp. 903–912 (2018)

15. Grgic-Hlaca, N., Zafar, M.B., Gummadi, K.P., Weller, A.: Beyond distributive fairness in algorithmic decision making: feature selection for procedurally fair learning. In: Proceedings of the Thirty-Second AAAI Conferenceon Artificial Intelligence (AAAI-18), pp. 51–60 (2018)

16. Holzinger, A.: Interactive machine learning for health informatics: when do we need the human-in-the-loop? Brain Inform. **3**(2), 119–131 (2016). https://doi.org/10.1007/s40708-016-0042-6

17. Holzinger, A., Carrington, A., Mueller, H.: Measuring the quality of explanations: the system causability scale (SCS). KI - Kuenstliche Intell. **34**(2), 193–198 (2020)

18. Holzinger, A., Malle, B., Saranti, A., Pfeifer, B.: Towards multi-modal causability with graph neural networks enabling information fusion for explainable AI. Inf. **71**(7), 28–37 (2021). https://doi.org/10.1016/j.inffus.2021.01.008

19. Holzinger, A., Weippl, E., Tjoa, A.M., Kieseberg, P.: Digital transformation for sustainable development goals (SDGs) - a security, safety and privacy perspective on AI. In: Holzinger, A., Kieseberg, P., Tjoa, A.M., Weippl, E. (eds.) CD-MAKE 2021. LNCS, vol. 12844, pp. 1–20. Springer, Cham (2021). https://doi.org/10.1007/978-3-030-84060-0_1

20. Holzinger, K., Mak, K., Kieseberg, P., Holzinger, A.: Can we trust machine learning results? artificial intelligence in safety-critical decision support. ERCIM News **112**(1), 42–43 (2018)

21. Hutchinson, B., Mitchell, M.: 50 years of test (un)fairness: Lessons for machine learning. In: Proceedings of the Conference on Fairness, Accountability, and Transparency, FAT* 2019, pp. 49–58 (2019)

22. Kamiran, F., Žliobaitė, I.: Explainable and non-explainable discrimination in classification. In: Custers, B., Calders, T., Schermer, B., Zarsky, T. (eds.) Discrimination and Privacy in the Information Society. Studies in Applied Philosophy, Epistemology and Rational Ethics, vol. 3, pp. 155–170. Springer, Heidelberg (2013). https://doi.org/10.1007/978-3-642-30487-3_8

23. Kasirzadeh, A., Smart, A.: The use and misuse of counterfactuals in ethical machine learning. In: Proceedings of the 2021 ACM Conference on Fairness, Accountability, and Transparency (FAccT 2021), pp. 228–236 (2021)

24. Lee, M.S.A., Floridi, L., Singh, J.: Formalising trade-offs beyond algorithmic fairness: lessons from ethical philosophy and welfare economics. SSRN Scholarly Paper ID 3679975, Social Science Research Network, July 2020. https://papers.ssrn.com/abstract=3679975

25. Lundberg, S.M., Lee, S.I.: A unified approach to interpreting model predictions. In: Proceedings of the 31st International Conference on Neural Information Processing Systems, NIPS 2017, pp. 4768–4777 (2017)

26. McGrath, R., et al.: Interpretable credit application predictions with counterfactual explanations. CoRR abs/1811.05245 (2018). http://arxiv.org/abs/1811.05245

27. Mehrabi, N., Morstatter, F., Saxena, N., Lerman, K., Galstyan, A.: A survey on bias and fairness in machine learning. CoRR abs/1908.09635 (2019). http://arxiv.org/abs/1908.09635

28. Miller, T.: Explanation in artificial intelligence: insights from the social sciences. Artif. Intell. **267**, 1–38 (2019)

29. Molnar, C., Casalicchio, G., Bischl, B.: Interpretable machine learning - a brief history, state-of-the-art and challenges. arXiv:2010.09337 [cs, stat], October 2020

30. Piano, S.L.: Ethical principles in machine learning and artificial intelligence: cases from the field and possible ways forward. Humanit. Soc. Sci. Commun. **7**(1), 1–7 (2020). https://doi.org/10.1057/s41599-020-0501-9

31. Robert Jr., L.P., Bansal, G., Melville, N., Stafford, T.: Introduction to the special issue on AI fairness, trust, and ethics. AIS Trans. Hum.-Comput. Interact. **12**(4), 172–178 (2020). https://doi.org/10.17705/1thci.00134

32. Rudin, C., Wang, C., Coker, B.: The age of secrecy and unfairness in recidivism prediction. Harv. Data Sci. Rev. **2**(1) (2020). https://doi.org/10.1162/99608f92. 6ed64b30, https://hdsr.mitpress.mit.edu/pub/7z10o269

33. Saxena, N.A., Huang, K., DeFilippis, E., Radanovic, G., Parkes, D.C., Liu, Y.: How do fairness definitions fare? Examining public attitudes towards algorithmic definitions of fairness. In: Proceedings of the 2019 AAAI/ACM Conference on AI, Ethics, and Society, AIES 2019, pp. 99–106 (2019)

34. Schmidt, P., Biessmann, F.: Quantifying interpretability and trust in machine learning systems. In: Proceedings of AAAI Workshop on Network Interpretability for Deep Learning 2019 (2019)

35. Schneeberger, D., Stöger, K., Holzinger, A.: The European legal framework for medical AI. In: Holzinger, A., Kieseberg, P., Tjoa, A.M., Weippl, E. (eds.) CD-MAKE 2020. LNCS, vol. 12279, pp. 209–226. Springer, Cham (2020). https://doi.org/10.1007/978-3-030-57321-8_12

36. Schumann, C., Foster, J.S., Mattei, N., Dickerson, J.P.: We need fairness and explainability in algorithmic hiring. In: Proceedings of the 19th International Conference on Autonomous Agents and MultiAgent Systems, AAMAS 2020, pp. 1716–1720 (2020)

37. Shin, D., Park, Y.J.: Role of fairness, accountability, and transparency in algorithmic affordance. Comput. Hum. Behav. **98**, 277–284 (2019)

38. Starke, C., Baleis, J., Keller, B., Marcinkowski, F.: Fairness perceptions of algorithmic decision-making: a systematic review of the empirical literature (2021)

39. Wang, X., Yin, M.: Are explanations helpful? A comparative study of the effects of explanations in AI-assisted decision-making, pp. 318–328. ACM (2021)

40. Warner, R., Sloan, R.H.: Making artificial intelligence transparent: fairness and the problem of proxy variables. Crim. Just. Ethics **40**(1), 23–39 (2021)

41. Zhao, J., Wang, T., Yatskar, M., Ordonez, V., Chang, K.W.: Men also like shopping: reducing gender bias amplification using corpus-level constraints. In: Proceedings of the 2017 Conference on Empirical Methods in Natural Language Processing, pp. 2979–2989. Copenhagen, Denmark, September 2017

42. Zhou, J., Chen, F.: 2D transparency space—bring domain users and machine learning experts together. In: Zhou, J., Chen, F. (eds.) Human and Machine Learning. HIS, pp. 3–19. Springer, Cham (2018). https://doi.org/10.1007/978-3-319-90403-0_1

43. Zhou, J., Chen, F. (eds.): Human and Machine Learning: Visible, Explainable, Trustworthy and Transparent. Human-Computer Interaction Series, Springer, Cham (2018). https://doi.org/10.1007/978-3-319-90403-0

44. Zhou, J., Gandomi, A.H., Chen, F., Holzinger, A.: Evaluating the quality of machine learning explanations: a survey on methods and metrics. Electronics **10**(5), 593 (2021)

45. Zhou, J., Khawaja, M.A., Li, Z., Sun, J., Wang, Y., Chen, F.: Making machine learning useable by revealing internal states update—a transparent approach. Int. J. Comput. Sci. Eng. **13**(4), 378–389 (2016)

Logic and Pragmatics in AI Explanation

Chun-Hua Tsai[1]([✉]) and John M. Carroll[2]

[1] University of Nebraska at Omaha, Omaha, NE, USA
chunhuatsai@unomaha.edu
[2] Pennsylvania State University, University Park, PA, USA
jmc56@psu.edu

Abstract. This paper reviews logical approaches and challenges raised for explaining AI. We discuss the issues of presenting explanations as accurate computational models that users cannot understand or use. Then, we introduce pragmatic approaches that consider explanation a sort of speech act that commits to felicity conditions, including intelligibility, trustworthiness, and usefulness to the users. We argue Explainable AI (XAI) is more than a matter of accurate and complete computational explanation, that it requires pragmatics to address the issues it seeks to address. At the end of this paper, we draw a historical analogy to usability. This term was understood logically and pragmatically, but that has evolved empirically through time to become more prosperous and more functional.

Keywords: Explainable AI · Pragmatics · Conversation · Causability

1 Introduction

Artificial intelligence (AI) technology has advanced many human-facing applications in our daily lives. As one of the most widely used AI-driven intelligent systems, recommendation systems have been an essential part of today's digital ecosystems. For example, recommendation systems have been widely adopted for suggesting relevant items or people to the users on social media [8]. Billion people have adopted or interacted with these AI systems every day. Effective recommender systems typically exploit multiple data sources and ensemble intelligent inference methods, e.g., machine learning or data science approaches. However, it is usually difficult to comprehend the internal processes of how the recommendation was made for the end-users. The *reasons* of receiving specific recommendations usually stay in a *black box*, which frequently makes the resulting recommendations less *trustworthy* to the users [1]. The users generally have little understanding of the mechanism behind these systems, so these recommendations are not yet transparent to the users. The opaque designs are known to negatively affect users' satisfaction and impair their trust in the recommendation systems [25]. Moreover, in this situation, processing this output could produce user behavior that can be confusing, frustrating, or even dangerous in life-changing scenarios [1].

© The Author(s) 2022
A. Holzinger et al. (Eds.): xxAI 2020, LNAI 13200, pp. 387–396, 2022.
https://doi.org/10.1007/978-3-031-04083-2_19

We argue providing explainable recommendation models and interfaces may not assure the users will *understand* the underlying rationale, data, and logic [26]. The *scientific explanations*, which are based on accurate AI models, might not *comprehensible* to the users who are lack competent AI literacy. For instance, a software engineer would appreciate inspecting the approximated probability in a recommendation model. However, this information could be less meaningful or even overloaded to lay users with varied computational knowledge, beliefs, and even biases [2]. We believe the nature of an explanation is to help the users to understand and to build a working mental model of using AI applications in everyday lives [5]. We urgently need more work on empowering lay users by providing comprehensible explanations in AI applications to benefit from the daily collaboration with AI.

In this paper, we aim to review *logical* approaches to Explainable AI (XAI). We would review the logic of explanation and challenges raised for explaining AI using generic algorithms. Specifically, we are interested in presenting such explanations to users, for instance, explaining accurate system models that users cannot understand or use. Then, we would discuss pragmatic approaches that consider explanation a sort of speech act that commits to felicity conditions, including intelligibility, trustworthiness, and usefulness to the listener. We argue XAI is more than a matter of accurate and complete explanation, that it requires pragmatics of explanation to address the issues it seeks to address. We then draw a historical analogy to usability. This term was understood logically and pragmatically, but that has evolved empirically through time to become more prosperous and functional.

2   The Logic of Explanations

Explainable AI (XAI) has drawn more and more attention in the broader field of human-computer interaction (HCI) due to the extensive social impact. With the popularity of AI-powered systems, it is imperative to provide users with effective and practical transparency. For instance, the newly initiated European Union's General Data Protection Regulation (GDPR) requires the owner of any data-driven application to maintain a "right to the explanation" of algorithmic decisions [7]. Enhancing transparency in AI systems has been studied in the XAI research to improve AI systems' explainability, interpretability, or controllability [14,16]. Researchers have explored a range of user interfaces and explainable models to support exploring, understanding, explaining, and controlling recommendations [10,25,26]. In many user-centered evaluations, these explanations positively contribute to the user experience, i.e., trust, understandability, and satisfaction [25]. Self-explainable recommender systems have been proved to increase user perception of system transparency and acceptance of the system suggestions [14]. These explanations were usually post-hoc and one-shot with an obvious challenge of when, why, and how to explain the system to the users based on their information needs and beliefs.

Another stream of research has identified the effects of making the recommendation process more transparent. It could improve the user's conceptual model by enhancing the recommendation system's controllability [22,26]. In these attempts, users were allowed to *influence* the presented recommendations by interacting with different visual interfaces. The interactive recommender systems demonstrated that users appreciate *controllability* in their interactions with the recommender systems [14]. The similar effects applied to visualization that users can understand how their actions can impact the system, which contributes to the overall *inspectability* [14] and *causability* [13] of the recommendation process. The transparent recommendation process could accelerate the information-seeking process but does not guarantee the comprehension of the target system's inner logic. These solutions empowered the user to control the system for accessing the desired recommendations. However, these controllable interfaces may not fulfill the explanation needs and help the users build a mental model to tell how the system works.

The user's mental model represents the knowledge of information systems generated and evolved through the interaction with the system [18]. The idea was founded in cognitive science and HCI discipline in the 1980s. For instance, Norman [21] argued the user could invent a mental model to simulate system behavior and make assumptions or predictions about the interaction outcome based on a target system. Follow Norman's definition, the user's mental models are constructed, incomplete, limited, unstable and sometime "superstitions" [21]. The user's mental model interacts with the *conceptual model* that the system designer used to develop the system. HCI researchers have considered the user's mental model in designing the usable system or interfaces in the past two decades. However, only a few studies have examined the user's mental model while interacting with the context of AI-powered recommender systems and algorithmic decisions [20].

We argue that these controllable and explainable user interfaces may not always ensure that users understand the underlying rationale of each contributing data or method [26]. The users could perceive the system's usefulness but still lack the *predictability* or *causability* [13] that to approximate the behavior of the target system [21]. In our observation, the users could build different mental models while interacting with an explainable system. For instance, users with more robust domain knowledge, such as trained computer sciences students, would be more judgmental in using the explainable system through their computational knowledge. However, the *naive* users would be more willing to accept and trust the recommendations [26]. We also observe controllable interfaces would lead the user to *compare* the recommendations in their decision-making process. Still, it does not mean the users could understand or predict the system's underlying logic. These findings demonstrate that personal factors and mental models (such as education, domain experience, and familiarity with technology) could significantly affect the system's user perception and cognitive process of machine-generated explanations.

3   The Pragmatics of Explanations

Miller [18] and Mittelstadt et al. [19] suggest that the AI and HCI researchers need to differentiate *scientific* and *everyday* explanations. To provide the *everyday* explanations, researchers need to consider cross-discipline knowledge (e.g., HCI, social science, cognitive science, psychology, etc.) and the user's mental model. Instead of the scientific intuition to provide prediction approximations (e.g., the global or local surrogate XAI models). For example, as HCI researchers, we already know the success explanation should be iterative, sound, complete, and not overwhelm the user. Social science researchers defined the everyday explanation through three principles [24]. 1) *human explanations are contrastive*: perceiving abnormality played an important role in seeking an explanation, i.e., the users would be more like to figure out an unexpected recommendation [18]. 2) *human explanations are selective*: the users may not seek a "complete cause" of an event; instead, the users tend to seek useful information in the given context. The selective could reduce long causal chains' effort and the cognitive load of processing countless modern AI models' parameters. 3) *human explanations are social*: the process of seeking an explanation should be interactive, such as a conversation. The explainer and explained can engage in information transfer through dialogue or other means [12].

Specifically, we propose to explore the *pragmatics of Explanations in AI*, i.e., the known mechanism of how the user requests an explanation from AI applications. The HCI community has long been interested in the interaction benefits of conversational interfaces. The design space could be situated within a rich body of studies on conversational agents or chatbot applications, e.g., AI-driven personal assistant [17]. The design of conversational agents offers several advantages over traditional WIMP (Windows, Icons, Menus, and Pointers) interfaces. The interface could provide a natural and familiar way for users to tell the system about themselves, which improves the system's usability and updates the user's mental model to the system. Moreover, the design is flexible (like a dialogue) and can accommodate diverse user requests without requiring users to follow a fixed path (e.g., the controllable interfaces [26]). The interaction could augment by a personified persona, in which the anthropomorphic features could help attract user attention and gain user trust.

In this section, we present two case studies to introduce our early investigation on pragmatics of AI explanations.

3.1   Case 1: Conversational Explanations

Online symptom checkers (OSCs) are intelligent systems using machine learning approaches (e.g., clinical decision tree) to help patients with self-diagnosis or self-triage [27]. These systems have been widely used in various health contexts, e.g., patients could use OSCs to check their early symptoms. The patient could learn their symptoms before a doctor visit, and to identify the appropriate care level and services and whether they need medical attention from healthcare providers [23]. The AI-powered symptom checkers promise various benefits,

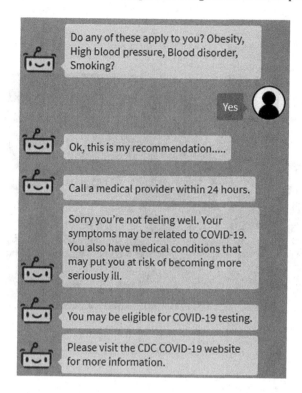

Fig. 1. Example of the conversational AI explanations [27]

such as providing quality diagnosis and reducing unnecessary visits and tests. However, unlike real healthcare professionals, most OSCs do not *explain why* the OSCs provide such diagnosis or *why* a patient falls into a disease classification. OSCs' data and clinical decision models are usually neither transparent nor comprehensible to lay users.

We argue *explanations* could be used to promote diagnostic transparency of online symptom checkers in a conversational manner. First, we conducted an interview study to explore *what explanation needs exist in the existing use of OSCs*. Second, informed by the first study's results, we used a design study to investigate *how explanations affect the user perception and user experience with OSCs*. We designed an COVID-19 OSC (shown in Fig. 1) and tested it with three styles of explanations in a lab-controlled study with 20 subjects. We found that conversational explanations can significantly improve overall user experiences of trust, transparency perception, and learning. Besides, we showed that by interweaving explanations into conversation flow, OSC could facilitate users' comprehension of the diagnostics in a dynamic and timely fashion.

The findings contributed empirical insights into user experiences with explanations in healthcare AI applications. Second, we derived conceptual insights into OSC transparency. Third, we proposed design implications for improving transparency in healthcare technologies, and especially explanation design in conversational agents.

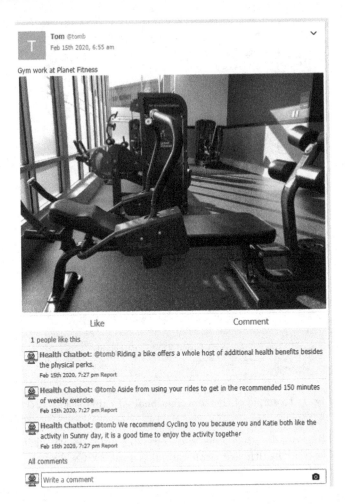

Fig. 2. Example of the Explainable AI-Mediated communication.

3.2 Case 2: Explainable AI-Mediated Communication (XAI-MC)

The integral part of modern health promotion initiatives for non-collocated members is computer-mediated communications [11]. The concept has been extensively adopted as an interpersonal communication medium in public health research such as telemedicine and mental health supports. Today, the Artificial Intelligence-Mediated Communication (AI-MC) between people could be augmented by computational agents to achieve different communication goals [9]. For instance, the interpersonal text-based communications (e.g., email) could be augmented by auto-correct, auto-completion, or auto-response. AI-MC has received more and more attention in recent socially efficacious research. For example, an AI agent could undermine the writers' message by altering the negative keywords (e.g., "sorry") to encourage the user to normalize language as the right way of speaking. AI agent could mitigate interpersonal biases by triggering alert messages when the agents detected the users intend to post negative messages on social media [15]. The introduction of AI brings new opportunities to adopt computational agents in family health collaboration and communications. AI-MC could be used to engage family members' health conversation better, and the communication may translate into healthy behavioral changes [6]. We can introduce a designate agent to mediate the communication by *recommending and explaining the health information* to the family members. Little attention has been paid to the question of how computational agents ought to disclose to users in AI-MC and the effects on family health promotion.

We explored the effects of promoting non-collocated family members' healthy lifestyle through *Explainable AI-meditated Communication (XAI-MC)*. We examined how XAI-MC would help non-collocated family members to engage in conversations about health, to learn more about each other's healthy practices, and as a result to encourage family collaboration via an online platform. We are particularly interested in exploring the effect of bringing transparent AI agents to the family communication. Specifically, we proposed to design a transparent AI agent to mediate the non-collocated family members' communication on healthy lifestyles. In our design, the users could share healthy activities information for enhancing family health awareness and engagement in a social media application. In the platform (shown in Fig. 2), a designate AI-powered health chat bot was used to mediate family members' communication on social media by explaining the health recommendations to them. We adopted the explainable health recommendations to address existing challenges related to remote family collaboration on health through XAI-MC. The findings could help to generate insights into designing transparent AI agents to support collaborating and sharing health and well-being information with online conversation.

We conducted a week-long field study with 26 participants who have at least one non-collocated family member or friend willing to join the study together. Based on a within-subject design, participants were assigned to two study phases: *1) AI-MC with non-explainable health recommendation* and *2) XAI-MC with explainable health recommendation*. We adopted a mixed-method to evaluate our design by collecting quantitative and qualitative feedback. We found evidence to support that providing transparent AI agents helped individuals gain health awareness, engage in conversations about healthy living practices, and promote collaboration among family members. Our findings provide insights into developing effective family-centered health interventions that aid non-collocated families in cultivating health together. The experiment results help to explain how transparent AI agents could mediate the health conversation and collaboration within non-collocated families.

4   Usability, Explaniability and Causability

The two case studies present our preliminary findings to support our arguments on the XAI is more than a matter of accurate explainable or interpretable models. Here we would like to draw a historical analogy to usability. One tension in contemporary AI is the perception that core system qualities like speed, efficiency, accuracy and reliability might be compromised by pursuing objectives like transparency and accountability for some form of diffuse *explanatory value* [9]. But though our understanding of qualities like transparency and accountability is limited, this can be directly addressed to enhance the causability.

The trend of Explainable AI can be seen as analogous to usability: merely simplifying a user interface (in a logical/formal sense) may or may not make it more usable, instead the key to usability is a set of pragmatic conditions. It must be satisfying, challenging, informative, intuitive, etc. We could conclude that XAI is more than a matter of accurate and complete explanation, that it requires pragmatics of explanation in order to address the issues it seeks to address. One specific issue in XAI is that AI should be able to explain how it is fair. Such an explanation will necessarily intersect with an accurate system model but would be much more focused on interaction scenarios and user experiences.

On the history in age of 1980 simple noting of usability. Only saying keep simple as stupid? Directly pursue the simple solution is not the same as usability. User's ability to trust of understand AI is not sufficient. Usability we don't really have a theory in these aspects. Usability is not equal to empirical evidence, to do experiment with kind of explanations and exploratory interaction and explore the consequence. The consequence could be part of the usability. Something goes wrong, and the users need an explanation, i.e., we want to know what is happening. Explanations could be an engagement. Active thinking and active learning, user interaction and usability, and wrong and addition situations. Try to understand the system model, but why do uses want to get this explanation?

Carroll and Aaronson [3] investigated a Wizard of Oz simulation of intelligent help. They studied interactions with a popular database application and identified 13 critical user errors, including the application state people were in when they made these errors. In this way, the help simulation recognized and guided recovery from a set of serious mistakes. Carroll and Aaronson designed two kinds of helpful information: "how-it-works," explaining how the system model worked to allow the error and leaving it to the user to discover what to do, and "how to do it," describing procedures the user should follow to recover from the error and continue their task. They found that people often preferred help messages explaining how the database application worked, for example, when it noted the distinction between forms and data when users entered both field labels and numeric values. When puzzled by the system, such interactions were satisfying to users, but "how-it-works" messages particularly pleased users in answering questions just as they were being formulated. Simplifying a user interface in a logical/formal sense may or may not make it more usable. Key usability also considers pragmatic conditions - systems must be satisfying, challenging, informative, intuitive, etc.

The field of human-computer interaction (HCI) coalesced around the concept of usability in the early 1980s, but not because the idea was already defined clearly, or could be predictably achieved in system design. It was instead because the diffuse and emerging concept of usability evoked a considerable amount of productive inquiry into the nature and consequences of usability, fundamentally changing how technology developers, users, and everyone thought about what using a computer could and should be [4]. Suppose that AI technologies were correctly reconceptualized, including the capability to effectively explain what they are doing, how they are doing it, and what courses of action they are considering. Adequate, in this context, would mean codifying and reporting on plans and activities in a way that is intelligible to humans. The standard would not be a superficial Turing-style simulacrum but a depth-oriented investigation of human-computer interaction to fundamentally advance our understanding of accountability and transparency. We have already seen how such a program of inquiry can transform computing.

References

1. Amershi, S., et al.: Guidelines for human-ai interaction. In: Proceedings of the 2019 CHI Conference on Human Factors in Computing Systems, p. 3. ACM (2019)
2. Anderson, A., et al.: Mental models of mere mortals with explanations of reinforcement learning. ACM Trans. Interact. Intell. Syst. (TiiS) **10**(2), 1–37 (2020)
3. Carroll, J., Aaronson, A.: Learning by doing with simulated intelligent help. Commun. ACM **31**(9), 1064–1079 (1988)
4. Carroll, J.M.: Beyond fun. Interactions **11**(5), 38–40 (2004)
5. Craik, K.J.W.: The Nature of Explanation, vol. 445. CUP Archive, Cambridge (1952)
6. Dragoni, M., Donadello, I., Eccher, C.: Explainable AI meets persuasiveness: translating reasoning results into behavioral change advice. Artif. Intell. Med. **105**, 101840 (2020)
7. Eiband, M., Schneider, H., Bilandzic, M., Fazekas-Con, J., Haug, M., Hussmann, H.: Bringing transparency design into practice. In: 23rd International Conference on Intelligent User Interfaces, pp. 211–223. ACM (2018)
8. Guy, I.: Social recommender systems. In: Ricci, F., Rokach, L., Shapira, B. (eds.) Recommender Systems Handbook, pp. 511–543. Springer, Boston, MA (2015). https://doi.org/10.1007/978-1-4899-7637-6_15
9. Hancock, J.T., Naaman, M., Levy, K.: Ai-mediated communication: definition, research agenda, and ethical considerations. J. Comput. Mediat. Commun. **25**(1), 89–100 (2020)
10. Herlocker, J.L., Konstan, J.A., Riedl, J.: Explaining collaborative filtering recommendations. In: Proceedings of the 2000 ACM Conference on Computer Supported Cooperative Work, pp. 241–250. ACM (2000)
11. Herring, S.C.: Computer-mediated communication on the internet. Ann. Rev. Inf. Sci. Technol. **36**(1), 109–168 (2002)
12. Hilton, D.J.: Conversational processes and causal explanation. Psychol. Bull. **107**(1), 65 (1990)
13. Holzinger, A., Carrington, A., Müller, H.: Measuring the quality of explanations: the system causability scale (scs). KI-Künstliche Intelligenz **34**(2), 193–198 (2020)
14. Knijnenburg, B.P., Bostandjiev, S., O'Donovan, J., Kobsa, A.: Inspectability and control in social recommenders. In: Proceedings of the Sixth ACM Conference on Recommender Systems, pp. 43–50. ACM (2012)
15. Levy, K., Barocas, S.: Designing against discrimination in online markets. Berkeley Technol. Law J. **32**(3), 1183–1238 (2017)

16. Liao, Q.V., Gruen, D., Miller, S.: Questioning the AI: informing design practices for explainable AI user experiences. In: Proceedings of the 2020 CHI Conference on Human Factors in Computing Systems, pp. 1–15 (2020)
17. Liao, Q.V., et al.: All work and no play? In: Proceedings of the 2018 CHI Conference on Human Factors in Computing Systems, pp. 1–13 (2018)
18. Miller, T.: Explanation in artificial intelligence: insights from the social sciences. Artif. Intell. **267**, 1–38 (2019)
19. Mittelstadt, B., Russell, C., Wachter, S.: Explaining explanations in AI. In: Proceedings of the Conference on Fairness, Accountability, and Transparency, pp. 279–288 (2019)
20. Ngo, T., Kunkel, J., Ziegler, J.: Exploring mental models for transparent and controllable recommender systems: A qualitative study. In: Proceedings of the 28th ACM Conference on User Modeling, Adaptation and Personalization, pp. 183–191 (2020)
21. Norman, D.A.: Some observations on mental models. Ment. Models **7**(112), 7–14 (1983)
22. O'Donovan, J., Smyth, B., Gretarsson, B., Bostandjiev, S., Höllerer, T.: Peerchooser: visual interactive recommendation. In: Proceedings of the SIGCHI Conference on Human Factors in Computing Systems, pp. 1085–1088. ACM (2008)
23. Powley, L., McIlroy, G., Simons, G., Raza, K.: Are online symptoms checkers useful for patients with inflammatory arthritis? BMC Musculoskelet. Disord. **17**(1), 362 (2016)
24. Ruben, D.H.: Explaining Explanation. Routledge, London (2015)
25. Tintarev, N., Masthoff, J.: Explaining recommendations: design and evaluation. In: Ricci, F., Rokach, L., Shapira, B. (eds.) Recommender Systems Handbook, pp. 353–382. Springer, Boston, MA (2015). https://doi.org/10.1007/978-1-4899-7637-6_10
26. Tsai, C.-H., Brusilovsky, P.: The effects of controllability and explainability in a social recommender system. User Model. User-Adapt. Interact. **31**(3), 591–627 (2020)
27. Tsai, C.H., You, Y., Gui, X., Kou, Y., Carroll, J.M.: Exploring and promoting diagnostic transparency and explainability in online symptom checkers. In: Proceedings of the 2021 CHI Conference on Human Factors in Computing Systems, pp. 1–17 (2021)

Author Index

Printed in the United States
by Baker & Taylor Publisher Services